Seven Years of Membranes: Feature Paper 2017

Seven Years of Membranes: Feature Paper 2017

Special Issue Editor

Spas D. Kolev

MDPI • Basel • Beijing • Wuhan • Barcelona • Belgrade

MDPI

Special Issue Editor
Spas D. Kolev
University of Melbourne
Australia

Editorial Office
MDPI
St. Alban-Anlage 66
Basel, Switzerland

This is a reprint of articles from the Special Issue published online in the open access journal *Membranes* (ISSN 2077-0375) from 2017 to 2018 (available at: http://www.mdpi.com/journal/membranes/special_issues/membr_feature_paper_2017)

For citation purposes, cite each article independently as indicated on the article page online and as indicated below:

LastName, A.A.; LastName, B.B.; LastName, C.C. Article Title. *Journal Name* **Year**, *Article Number*, Page Range.

ISBN 978-3-03842-991-3 (Pbk)
ISBN 978-3-03842-992-0 (PDF)

Contents

About the Special Issue Editor vii

Preface to "Seven Years of Membranes: Feature Paper 2017" ix

Yanfei Liu, Tonghu Xiao, Chenghuan Bao, Jifei Zhang and Xing Yang
Performance and Fouling Study of Asymmetric PVDF Membrane Applied in the Concentration
of Organic Fertilizer by Direct Contact Membrane Distillation (DCMD)
Reprinted from: *Membranes*, **2018**, *8*, 9, doi: 10.3390/membranes8010009 1

Tiziana Marino, Francesca Russo, Lina Rezzouk, Abderrazak Bouzid and Alberto Figoli
PES-Kaolin Mixed Matrix Membranes for Arsenic Removal from Water
Reprinted from: *Membranes*, **2017**, *57*, , doi: 10.3390/membranes7040057 14

Juliana I. Clodt, Volkan Filiz and Sergey Shishatskiy
Perfluorinated Compounds as Test Media for Porous Membranes
Reprinted from: *Membranes*, **2017**, *7*, 51, doi: 10.3390/membranes7030051 29

**Francois-Marie Allioux, Oana David, Miren Etxeberria Benavides, Lingxue Kong,
David Alfredo Pacheco Tanaka and Ludovic F. Dumée**
Preparation of Porous Stainless Steel Hollow-Fibers through Multi-Modal Particle Size Sintering
towards Pore Engineering
Reprinted from: *Membranes*, **2017**, *7*, 40, doi: 10.3390/membranes7030040 41

Joerg Winter, Benoit Barbeau and Pierre Bérubé
Nanofiltration and Tight Ultrafiltration Membranes for Natural Organic Matter
Removal—Contribution of Fouling and Concentration Polarization to Filtration Resistance
Reprinted from: *Membranes*, **2017**, *7*, 34, doi: 10.3390/membranes7030034 56

Jochen A. Kerres and Henning M. Krieg
Poly(vinylbenzylchloride) Based Anion-Exchange Blend Membranes (AEBMs): Influence of
PEG Additive on Conductivity and Stability
Reprinted from: *Membranes*, **2017**, *7*, 32, doi: 10.3390/membranes7020032 70

Mohammed Kadhom, Weiming Hu and Baolin Deng
Thin Film Nanocomposite Membrane Filled with Metal-Organic Frameworks UiO-66 and
MIL-125 Nanoparticles for Water Desalination
Reprinted from: *Membranes*, **2017**, *7*, 31, doi: 10.3390/membranes7020031 94

Gisela Bengtson, Silvio Neumann and Volkan Filiz
Membranes of Polymers of Intrinsic Microporosity (PIM-1) Modified by Poly(ethylene glycol)
Reprinted from: *Membranes*, **2017**, *7*, 28, doi: 10.3390/membranes7020028 110

Xuezhong He
Fabrication of Defect-Free Cellulose Acetate Hollow Fibers by Optimization of
Spinning Parameters
Reprinted from: *Membranes*, **2017**, *7*, 27, doi: 10.3390/membranes7020027 131

Alejandro Ruiz-García, Noemi Melián-Martel and Ignacio de la Nuez
Short Review on Predicting Fouling in RO Desalination
Reprinted from: *Membranes*, **2017**, *7*, 62, doi: 10.3390/membranes7040062 140

Armineh Hassanvand, Kajia Wei, Sahar Talebi, George Q. Chen and Sandra E. Kentish
The Role of Ion Exchange Membranes in Membrane Capacitive Deionisation
Reprinted from: *Membranes*, **2017**, 7, 54, doi: 10.3390/membranes7030054 . . . . **157**

Patrizia Formoso, Elvira Pantuso, Giovanni De Filpo and Fiore Pasquale Nicoletta
Electro-Conductive Membranes for Permeation Enhancement and Fouling Mitigation:
A Short Review
Reprinted from: *Membranes*, **2017**, 7, 39, doi: 10.3390/membranes7030039 . . . . **180**

About the Special Issue Editor

Spas D. Kolev, PhD, FRACI, is a Professor in the School of Chemistry at the University of Melbourne. His research is focused mainly in the areas of membrane-based separation and flow analysis techniques. He has co-authored 195 refereed articles, four book chapters, and he has co-edited a book. He has received a number of prestigious awards: the Ronald Belcher Memorial Award from the journal Talanta (1988), the Lloyd Smythe Medal (2009) and the Environmental Chemistry Medal (2017) from the Analytical and Environmental Chemistry Division of the Royal Australian Chemical Institute, the Medal of the Japanese Association for Flow Injection Analysis (2010), the Grimwade Prize in Industrial Chemistry from the University of Melbourne (2012), and the Max O'Connor Prize for Chemistry from La Trobe University (2017). He is the Founding Editor-in-Chief of the journal Membranes, Co-Editor-in-Chief of the Journal of Chemical Sciences, and a member of the Editorial/Advisory Boards of Analytica Chimica Acta, Talanta, Environmental Modeling and Assessment, Sensors, Challenges, and the International Journal of Analytical Chemistry.

Preface to "Seven Years of Membranes: Feature Paper 2017"

For the last seven years *Membranes* has provided an outstanding platform for the publication of articles at the forefront of research in the areas of membrane fabrication, characterization and application. This Special Issue, entitled "Seven Years of Membranes: Feature Paper 2017," celebrates this achievement. The articles included in it, by prominent researchers in the field, provide an authoritative and up-to-date account of advances in membrane science and technology. They describe new methods for the fabrication of organic, inorganic and mixed matrix membranes and their utilization in improving the efficiency of membrane-based separation processes, such as membrane distillation, nanofiltration, ultrafiltration, reverse osmosis, and gas permeation. A number of articles are focused on water treatment, which, because of its significance to sustainable development, is one of the main areas of membrane research and application. These articles report novel techniques for the clean-up of contaminated waters, and the desalination of industrial effluents, brackish water and seawater.

Spas D. Kolev
Special Issue Editor

![membranes logo] *membranes*

MDPI

Article

Performance and Fouling Study of Asymmetric PVDF Membrane Applied in the Concentration of Organic Fertilizer by Direct Contact Membrane Distillation (DCMD)

Yanfei Liu [1], Tonghu Xiao [1,*], Chenghuan Bao [1], Jifei Zhang [1] and Xing Yang [2,*]

[1] Faculty of Materials Science and Chemical Engineering, Ningbo University, Ningbo 315211, China; liuyanfei2716@163.com (Y.L.); 17855828741@163.com (C.B.); zhangjifei_nbu@163.com (J.Z.)

[2] College of Engineering and Science, Victoria University, P.O. Box 14428, Melbourne, VIC 8001, Australia

* Correspondence: xiaotonghu@nbu.edu.cn (T.X.); xing.yang@vu.edu.au (X.Y.); Tel.: +86-136-8589-2736 (T.X.); +61-9919-7690 (X.Y.)

Received: 11 January 2018; Accepted: 13 February 2018; Published: 16 February 2018

Abstract: This study proposes using membrane distillation (MD) as an alternative to the conventional multi-stage flushing (MSF) process to concentrate a semi-product of organic fertilizer. By applying a unique asymmetric polyvinylidene fluoride (PVDF) membrane, which was specifically designed for MD applications using a nonsolvent thermally induced phase separation (NTIPS) method, the direct contact membrane distillation (DCMD) performance was investigated in terms of its sustainability in permeation flux, fouling resistance, and anti-wetting properties. It was found that the permeation flux increased with increasing flow rate, while the top-surface facing feed mode was the preferred orientation to achieve 25% higher flux than the bottom-surface facing feed mode. Compared to the commercial polytetrafluoroethylene (PTFE) membrane, the asymmetric PVDF membrane exhibited excellent anti-fouling and sustainable flux, with less than 8% flux decline in a 15 h continuous operation, i.e., flux decreased slightly and was maintained as high as 74 kg·m^{-2}·h^{-1} at 70 °C. Meanwhile, the lost flux was easily recovered by clean water rinsing. Overall 2.6 times concentration factor was achieved in 15 h MD operation, with 63.4% water being removed from the fertilizer sample. Further concentration could be achieved to reach the desired industrial standard of 5x concentration factor.

Keywords: direct contact membrane distillation; asymmetric PVDF membrane; concentration of organic fertilizer; anti-fouling

---

1. Introduction

Livestock manure and crop straw have been used as fertilizer as they are rich in nitrogen, phosphorus, and organic matter that can improve the physical and chemical properties of soil and provide nutrients essential to crops [1]. However, such fertilizers need to be concentrated to a certain level, i.e., at least 3–5 times concentration from an initial organic content of 3%, to achieve the desired nutrient strength. Currently, several technologies have been applied in industry to concentrate liquid fertilizers, such as multistage flash distillation (MSF), multiple-effect distillation (MED), or reverse osmosis (RO). Both MSF and MED plants are known to be inefficient, energy-intensive, and land-consuming [2]. The major limitation of RO in such applications is its relatively low water removal (~35%) due to the high osmotic pressure limited by the concentration effect and thus the low overall concentration factor (<1.0) [3,4]. Also, RO is highly susceptible to membrane fouling [5].

Membrane distillation (MD) is an alternative emerging technology that combines the comparative advantages of thermal distillation and membrane processes and involves the transport of water vapor across a microporous hydrophobic membrane [6,7]. The driving force of MD is supplied by the vapor pressure difference generated by the temperature gradient imposed between the liquid/vapor interfaces [8]. Compared to other separation processes, MD has many advantages [9]. It exhibits a complete rejection of dissolved, non-volatile species, and lower (ambient) operating pressure than the pressure-driven membrane processes. Highly saturated solutions can be treated in MD [10]. Meanwhile, MD has the potential to achieve a high concentration factor while operating at low temperature differences that are achievable using waste-grade waste heat [11] or a renewable energy source, such as solar and geothermal energy [12,13]. MD has the potential to concentrate and recover valuable resources. For example, MD has been widely investigated for desalination [14], concentration of juices [15], crystallization of minerals [16], recovery of volatiles such as nitrogen [17], and waste water purification [18] and treatment [19] in recent years. However, the concentration of organic fertilizer using MD has not been studied thus far, where abundant waste heat (70–80 °C) will be available from the fertilizer production process [20].

Although MD has great potential for treating highly concentrated solutions, membrane fouling in MD is inevitable in the treatment of real industrial effluents [21]. Fouling results in a decrease in membrane permeability due to a deposition of suspended or dissolved substances, including organic and inorganic components, on the membrane surface and within its pores, reducing the effective vapor transport area and causing potential pore wetting problems that are detrimental to MD performance [22]. In the dewatering process of aqueous solutions such as juice [23] and RO brines [24], the occurrence of fouling on the MD membrane surface is highly possible but this aspect has not been thoroughly investigated [21]. Fouling control in MD lies in the process operating strategies (i.e., hydrodynamics) [25] and membrane properties [26] (i.e., surface roughness and hydrophobicity, etc.). In particular, the development of suitable MD membranes for sustaining the concentration processes is desirable. The long-term stability of the membranes in terms of deterioration of hydrophobicity and pore wetting needs to be resolved. To date, no commercial membranes with superior anti-fouling have been specifically developed for MD applications. Overall, the implementation of MD on an industrial scale is limited by the availability of robust membranes.

Recent studies showed that most of the MD membranes currently used are fabricated for other processes, such as microfiltration (MF), due to the similar hydrophobic nature and the microporous structure [27]. The desired MD performance with high permeability, long-term stability, and high energy efficiency is typically associated with the following membrane characteristics: a relatively small maximum pore size, the highest possible porosity, a narrow pore size distribution with a high degree of pore interconnectivity, and good anti-wetting properties with high liquid entry pressure of water (LEPw) [28,29]. The membrane properties directly affect the membrane performance and, therefore, an optimized membrane specifically designed for MD is vital for implementing industrial applications [27]. Common membrane materials include poly (vinylidene fluoride) (PVDF), which is widely used for fabricating MD membranes via various fabrication methods, such as conventional nonsolvent induced phase separation (NIPS) [30] and thermally induced phase separation (TIPS) [31], as well as the recently proposed nonsolvent thermally induced phase separation (NTIPS, also referred to as combined NIPS and TIPS) [32]. Our recent work [32] showed that a unique asymmetric PVDF membrane could be fabricated via the NTIPS method to achieve an ultra-thin separation skin layer with a highly porous and interconnected pore structure. Such a membrane exhibited extraordinary permeability as high as 85.6 kg·m^{-2}·h^{-1} at 80 °C.

In this MD study a previously developed polyvinylidene fluoride (PVDF) membrane was applied in the concentration of liquid organic fertilizer. The membrane was prepared by the nonsolvent thermally induced phase separation (NTIPS) method and exhibited superior permeability and anti-wetting properties with a unique asymmetric structure. Firstly, the effects of operating parameters in direct contact membrane distillation (DCMD) were investigated with the as-prepared PVDF membrane, such as flow rate, membrane orientation, and solution salinity. Secondly, the application of the membrane in the dewatering of real organic fertilizer stream to the desired concentration was examined in terms of the process stability and membrane fouling behavior, which was then compared with the commercial polytetrafluoroethylene (PTFE) membrane after previous systematic research into industrial applications [33–35].

2. Experimental

2.1. Membranes

The asymmetric poly(vinylidene fluoride) membranes used in this study were prepared by the nonsolvent thermally induced phase separation (NTIPS) method, with 15 wt % PVDF polymer (Model: 1015, Solvay Co, Brussels, Belgium) dissolved into water-soluble diluent ε-Caprolactam (CPL, Sinopharm Reagent Inc, Shanghai, China) at 150 °C. The nascent membrane was obtained at 20 °C in a coagulation bath with deionized water. Details on the membrane preparation and characterization can be found elsewhere [32]. The commercial polytetrafluoroethylene (PTFE) membrane provided by Ningbo Changqi Porous Membrane Technology Co., Ltd. (Ningbo, China) was also used in this work.

2.2. Feed Solutions

Three synthetic solutions were prepared as feed in MD with various salt concentrations C_f: (1) bitter salt solution: 1.7 wt % sodium chloride (NaCl, 99.5%, Sinopharm Reagent Inc, Shanghai, China); (2) synthetic seawater: 3.5 wt % NaCl; (3) 6.0 wt % NaCl solution.

The organic fertilizer sample was obtained from the Environmental Technology Development Co., Ltd. (Ningbo, China). It is a semi-product in the fertilizer production process, made of mixed solution of manure and milled crop straw after purification, pressurized hydrolysis, and pH adjustment. This stream coming from the pressurized hydrolysis process carries certain thermal energy (70 °C) that could be readily used in MD for dewatering. Based on the "Organic Fertilizer Content Standard" (DB33/699-2008) formulated by the National Center for Fertilizer Inspection and Supervision (Beijing, China), the company expected to concentrate the organic matter of the fertilizer product to 15%. The adjusted pH of the sample is within the range of 4.0–8.0 with minimal volatile ammonia nitrogen present. However, the semi-product of the organic fertilizer has only low organic matter around 3%, which needs to be concentrated about 5x to achieve useful strength for industrial applications. Thus, dewatering or concentration of the semi-product will be conducted in MD.

2.3. Membrane Characterization

The top/bottom surface and cross-sections of PVDF flat sheet membrane were observed using a scanning electron microscope (SEM, NOVA NANOSEM 450, FEI, Hillsboro, OR, USA). Prior to the scan, membrane samples were immersed in liquid nitrogen, fractured, and then coated with platinum using a coater (VACUUM DEVICE MSP-1S, FEI, Hillsboro, OR, USA).

The overall membrane porosity (ε) was calculated from the ratio of the pore volume to the total volume of the membrane. The membrane pore volume was determined by measuring the dry and wet weights of membrane using isopropyl alcohol (IPA) as a wetting agent [36].

The measurement of liquid entry pressure of water (LEP_w) of the membranes was conducted using a customized setup with synthetic seawater (i.e., 3.5 wt % NaCl solution, conductivity ~60 ms·cm^{-1}) as the testing liquid on the feed side and DI water (conductivity < 10 µs·cm^{-1}) as the reference at the permeate side to detect the occurrence of pore wetting. During testing, the pressure of the NaCl solution side was increased steadily using compressed N$_2$ gas, by 0.01 MPa increments every 15 min. The pressure at which there was a drastic initial increase in the conductivity of the permeate side and a continuous conductivity increase was taken as the LEP. The conductivity of the solution was monitored by a conductivity meter (DDSJ-308A, INESA Instrument, Shanghai, China).

The mean pore size of the PVDF membrane was determined by the liquid–liquid displacement method based on an isobutanol–DI water system. The detailed experimental procedure can be found elsewhere [37].

The contact angle (CA) of prepared PVDF membranes is measured by a goniometer (Kruss DSA100, Hamburg, Germany). Five points on each membrane are tested and the average of the measured values is reported.

2.4. DCMD Experiments

The DCMD experiments were conducted with the laboratory setup shown in Figure 1. In all DCMD experiments, the membrane was installed into a flat sheet membrane cell, giving an effective membrane area of 10×10^{-4} m^2. The feed and permeate were flowing counter-currently, with the feed pumped through a magnetic drive pump at a flow rate range of Q_f = 50–110 L/h and the permeate recirculated through another centrifugal pump at Q_p = 50–110 L/h. A magnetic stirrer was used in the feed tank to improve the mixing of solutions. The feed temperature T_f is in the range of 50–80 °C and permeate temperature T_p was kept constant at 16 °C. Both synthetic solutions and real industrial samples were tested under the identified operating conditions through this study. The continuous weight gain of the distillate was measured using a digital balance (EK-2000i, A&D Co. Ltd., Tokyo, Japan) for membrane flux calculation. The total dissolved solids (TDS) of the permeate stream was monitored by the conductivity meter to calculate rejection of non-volatiles. For each membrane, DCMD experiments were repeated three times to ensure reproducibility.

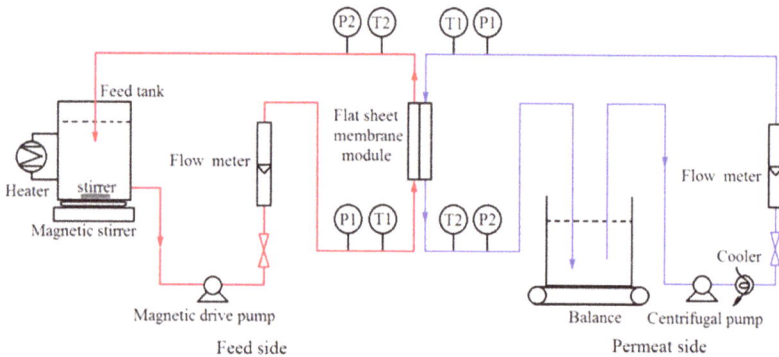

Figure 1. DCMD experimental setup.

2.5. Evaluation of DCMD Performance

The permeation flux (J, Kg·m^{-2}·h^{-1}) in MD was calculated by Equation (1):

$$J = \frac{\Delta W}{A \bullet \Delta t}, \tag{1}$$

where ΔW (Kg) is the weight of permeation, A (m^2) is the total effective membrane area, and Δt (h) is the operation time.

The normalized/relative flux (%) before and after fouling was calculated by Equation (2):

$$J_N = \frac{J_i}{J_0} \times 100\%, \tag{2}$$

where J_0 (Kg·m^{-2}·h^{-1}) is the initial flux, and J_i (Kg·m^{-2}·h^{-1}) is the instantaneous flux during the filtration of real industrial sample, which could cause flux decline due to fouling.

The rejection (R) of solute was calculated by Equation (3):

$$R = \frac{C_{f0} - C_{pt}}{C_{f0}}, \tag{3}$$

where C_{f0} (mg/L) is the total dissolved solids (TDS) in the original feed, and C_{pt} (mg/L) is the TDS concentration in the permeate water collected at time t.

3. Results and Discussion

3.1. Membrane Characterization

The membranes used in this study are asymmetric PVDF and commercial PTFE membranes. The characteristics of the PVDF membrane are given in Table 1; the characteristics of the commercial PTFE membrane were provided by the manufacturer and the relevant literature [33,34]. The PVDF membrane has a much smaller mean pore size (r_m) of 34 nm than that of PTFE (450 nm) [33,34], leading to a much higher liquid entry pressure of water (LEP_w) of 3.5 bar, indicating excellent anti-wetting properties. Aside from the high porosity (ε) of 86%, which is similar to that of the PTFE membrane [33,34], the total membrane thickness (δ) is as thin as 95 μm, which is indicative of high permeability. SEM images of both membranes are shown in Figure 2. In Figure 2a,d, it was found that the asymmetric PVDF membrane exhibits a dense and smooth top surface, which is significantly different from the rough fibrous structure of the commercial PTFE membrane. The asymmetric structure of the PVDF membrane with an ultra-thin skin top surface, finger-like pores, and a bicontinuous network beneath the skin are observed in the cross section in Figure 2c.

Table 1. Characterization of asymmetric PVDF membrane and PTFE membrane.

Membrane Type	Porosity (ε, %)	LEP_w (Bar)	Mean Pore Size (r_m, nm)	Total Thickness (δ, μm)	Contact Angle (θ, °)
Asymmetric PVDF	86 ± 1	3.5 ± 0.1	34 ± 3	95 ± 5	85 ± 3
Commercial PTFE	92.5 ± 0.5	0.8 ± 0.05	450 ± 50	36 ± 1 (PTFE layer)	140 ± 2.5

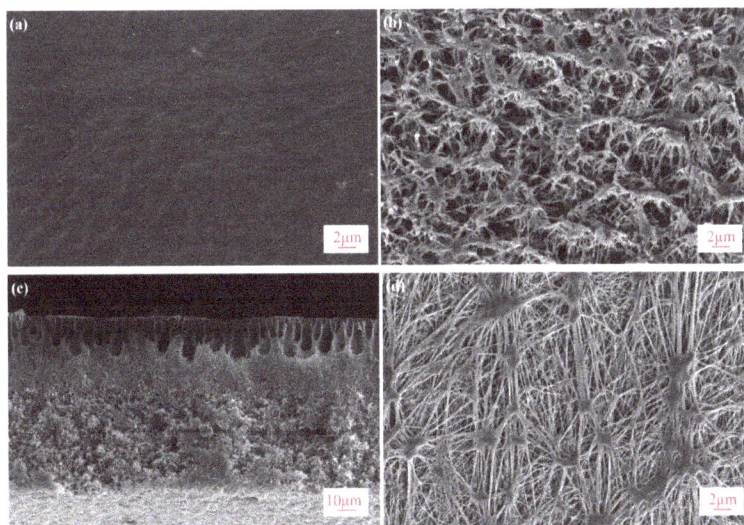

Figure 2. SEM images of membranes (**a**) top surface of virgin PVDF membrane (5000×); (**b**) bottom surface of virgin PVDF membrane (5000×); (**c**) cross section of virgin PVDF membrane (1000×); (**d**) top surface of virgin PTFE membrane (5000×).

3.2. Effects of Operating Parameters in DCMD

3.2.1. Effect of Flow Rate

Figure 3 shows the relationship between permeation flux and the flow rates of feed and permeate, where the flow rate of both sides were kept the same. The membrane flux increases as the flow rate increases from 50 to 110 L/h (linear velocity: 0.28–0.61 m/s). This is because the increase of flow rate helps reduce the thickness of the liquid boundary layer adjacent to the membrane surface, which alleviates the effect of concentration and temperature polarization, resulting in enhanced mass and heat transfer coefficients [38]. Thus, it improves the process driving force and subsequently permeation flux. Similar investigations on flow rate have been reported [12,39]. Therefore, the highest flow rate of 110 L/h within the testing range was selected for the following tests. It is also noted that during the above DCMD experiments the salt rejection was stable at 99.99% to ensure membrane integrity.

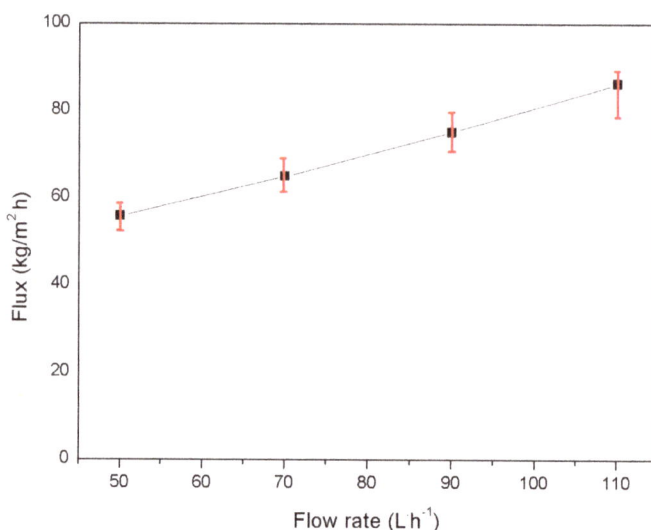

Figure 3. Effect of flow rate on permeation flux of asymmetric PVDF membrane in DCMD (C_f = 3.5 wt % NaCl, T_f = 70 °C, T_p = 16 °C, $Q_f = Q_p$).

3.2.2. Effect of Membrane Orientation

The effect of membrane orientation was investigated at varying feed temperatures from 50 to 80 °C, i.e., with the feed solution facing the top or bottom surface of the membrane. The results are presented in Figure 4. Compared to the bottom-surface-facing feed mode, the top-surface facing the feed solution produces at least 25% higher flux. For example, the flux of the top-surface facing the feed mode showed up to 123 kg·m^{-2}·h^{-1} at 80 °C. This is due to the different pore structure of the two surfaces of the asymmetric PVDF membrane fabricated by the NTIPS method, producing an ultra-thin, dense and smooth top surface exhibiting no macropores that is potentially smaller than the mean free path (<0.11 μm) of the water molecules and thus will likely follow the Knudsen diffusion mechanism in the classic MD mass transfer model [40,41]; the highly porous and rough bottom surface of the membrane exhibits much larger pores and hence may fall into the regime of combined Knudsen/molecular diffusion [39,42]. Thus, MD flux involving the Knudsen mechanism is considered higher than that of the combined Knudsen/molecular diffusion mechanism, as reported in the literature [32,43]. Hence, the orientation of top surface facing the feed was used in subsequent investigations. It is noted that the salt rejection was stable at 99.99% in the above DCMD tests for both orientations.

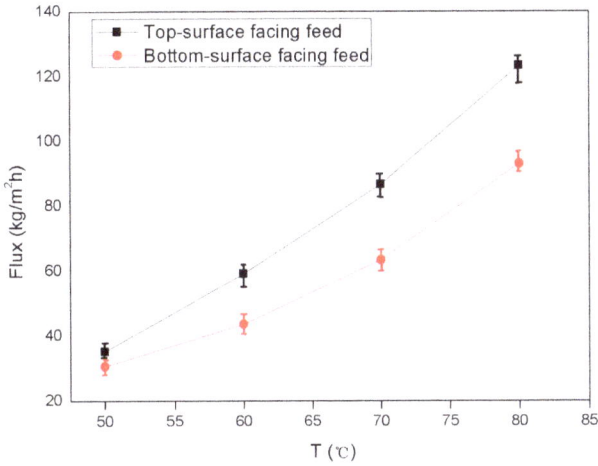

Figure 4. Effect of membrane orientation on permeation flux of asymmetric PVDF in DCMD at varying feed temperature (C_f = 3.5 wt % NaCl, T_p = 16 °C, Q_f = Q_p = 110 L·h^{-1}).

3.2.3. Effect of Feed Salinity

Four synthetic solutions with varying salinity from 0 to 60 g/L were tested in DCMD with the asymmetric PVDF membranes. The influence of salinity on the permeation flux is presented in Figure 5. It was found that the permeate flux decreased slightly by 12%, i.e., from 88.6 to 77.6 kg·m^{-2}·h^{-1}, as the salt concentration increased from 0 to 6 wt %. This can be explained by the reduction of vapor pressure and water activity coefficient of the feed when increasing the solute concentration [9], which leads to decreased driving force for vapor transport in MD. However, within a given salinity range, the concentration polarization effect was not known to significantly affect the flux. Overall, the membrane performance was only slightly influenced by the salt concentration of the feed, up to the salinity level of the real industrial sample to be investigated in this work.

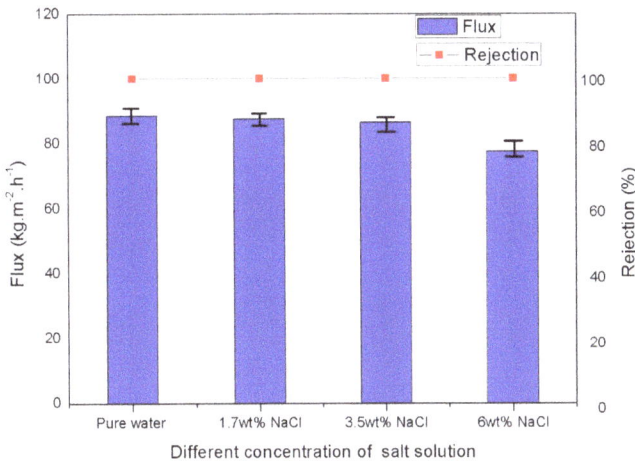

Figure 5. Effects of different salt concentrations of feed (T_f = 70 °C, T_p = 16 °C, δ = 95 μm, Q_f = Q_p = 110 L·h^{-1}).

3.3. Concentration of Real Organic Fertilizer by DCMD

In this section, with the synthetic seawater testing as a benchmark, the concentration of real organic fertilizer was measured using both the as-prepared PVDF and commercial PTFE membranes.

3.3.1. Permeation Flux of Organic Fertilizer as Feed

The PVDF membrane performance was evaluated by testing both the NaCl solution (3.5 wt % at the beginning) and an organic fertilizer in 15-h continuous DCMD runs in batch mode to concentrate the fertilizer. The concentration results are presented in terms of permeation flux, as illustrated in Figure 6. The initial flux of the organic fertilizer feed was around 80 kg·m^{-2}·h^{-1}, which is similar to that of the synthetic seawater, i.e., 86 kg·m^{-2}·h^{-1}. Although a minor decrease in the flux was observed for the industrial sample after 15 h of operation, it still remained around 74 kg·m^{-2}·h^{-1}, where 2.6x concentration has been achieved to obtain a fertilizer of 7.8% organic matter, i.e., 63.4% water was removed from the fertilizer sample containing approximately 3% organic nutrients. It is noted that the TDS rejection of the membrane was stable at 99.99% for both feed solutions. Figure 7 shows a comparison of the original organic fertilizer sample (A) and the permeate (B). The feed solution is turbid and dark brown, which is in contrast to the transparent permeate solution.

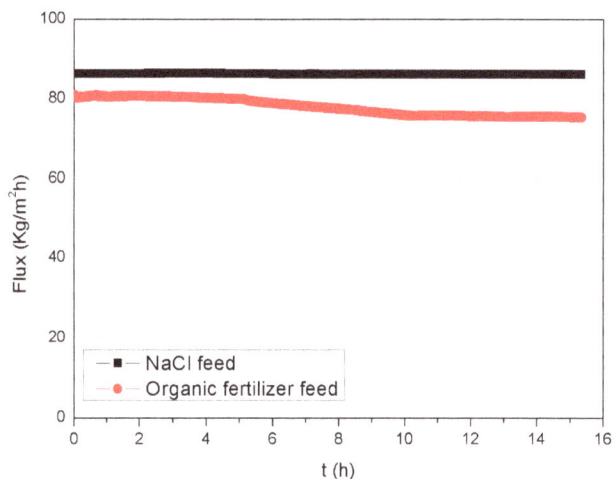

Figure 6. Continuous DCMD runs of organic fertilizer and NaCl feed using asymmetric PVDF membrane (T$_f$ = 70 °C, T$_p$ = 16 °C, Q$_f$ = Q$_p$ = 110 L·h^{-1}).

Figure 7. Images of original organic fertilizer sample (**A**) and MD permeate sample (**B**).

3.3.2. Membrane Fouling and Surface Inspection

As a result of the 15-h continuous operation of the organic fertilizer, flux decline was observed. This may be due to the reduction in vapor pressure of the feed solution and hence the transmembrane driving force as concentration increased. Also, the build-up of the fouling layer on the surface of the PVDF membrane could cause further flux decline, as evidenced by the surface inspection by SEM in Figures 8 and 9. To reveal the fouling behavior of the PVDF membrane, surface inspection was carried out. Figure 8 shows the SEM images of the fouled membrane (a) and the cleaned membrane rinsed with pure water (b). Correspondingly, Figure 9 shows pictures of the fouled (A) and cleaned membrane (B). As shown in Figures 8a and 9a, the entire surface of the PVDF membrane was covered by a layer of amorphous deposition, which has the same dark brown color as the fertilizer feed; the fouling layer was almost completely removed through clean water rinsing, as shown in Figures 8b and 9b. The easy cleaning of the membrane after 15 h of running may be attributed to the unique dense and smooth top surface structure of the asymmetric PVDF membrane.

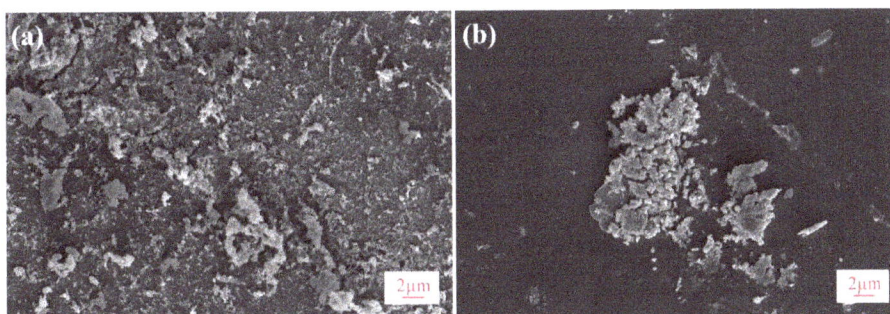

Figure 8. SEM images of PVDF membrane after 15-h continuous DCMD experiments (**a**) fouled membrane after 15 h operation (5000×); (**b**) cleaned membrane rinsed with pure water (5000×).

Figure 9. Pictures of as-prepared PVDF membrane (**a**) fouled membrane after fertilizer testing and (**b**) cleaned membrane rinsed with pure water.

3.3.3. Comparison of the Anti-Fouling Performance of Different Membranes

The anti-fouling property of the asymmetric PVDF membrane has been further investigated in comparison to the commercial PTFE membrane. The normalized fluxes (Equation (2)) were used to evaluate the fouling tendency associated with performance deterioration of both membranes. Figure 10 shows the normalized flux of both membranes, where both a synthetic 3.5 wt % salt solution and organic fertilizer feed were tested in the initial 1-h experiments. Compared to 14% flux decrease of the PTFE membrane, the PVDF membrane showed only a minor flux decline of 1.8%, indicating a more sustainable performance in treating challenging feed solutions.

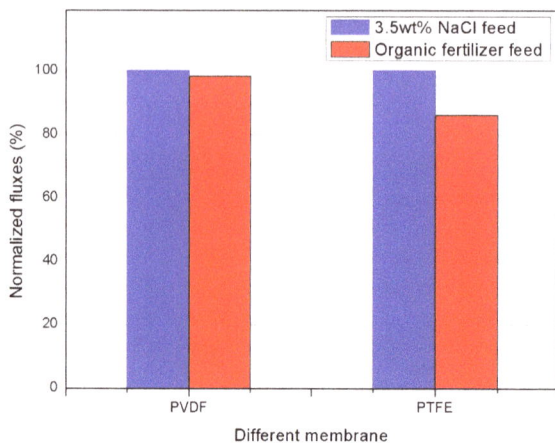

Figure 10. Comparison of initial normalized fluxes of as-prepared asymmetric PVDF and commercial PTFE membrane with synthetic and organic fertilizer (T_f = 70 °C, T_p = 16 °C, Q_f = Q_p = 110 L·h^{-1}).

The comparison of normalized fluxes of the asymmetric PVDF and commercial PTFE membrane was further investigated with an organic fertilizer feed in a 15-h continuous operation. The results are shown in Figure 11, in which the normalized flux of the PVDF membrane exhibits a very slow and minor decrease of 8% and remains relatively constant after a 15-h continuous operation. In contrast, a rapid decrease was observed with the PTFE membrane, resulting in 56% flux decline after 15 h. This could be due to the quick build-up of the fouling layer on the surface of the PTFE membrane, which exhibits a rough and fibrous surface structure, as indicated in Figure 2. Pictures of the fouled membranes after the 15-h operation are given in Figure 11: the deposit on the smooth surface of the PVDF membrane (top picture) was minor and relatively loose, while the cake layer on the PTFE membrane was dense.

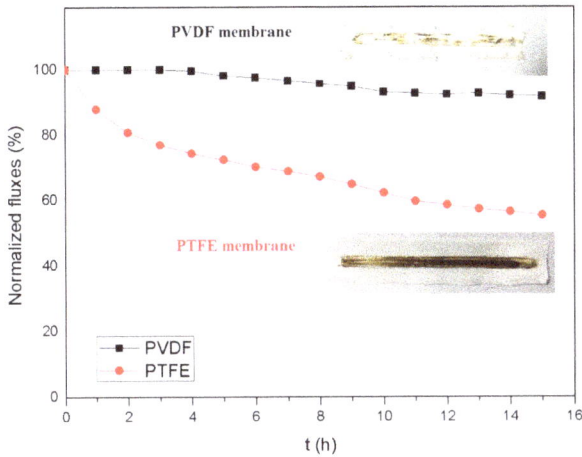

Figure 11. Comparison of normalized flux of asymmetric PVDF and commercial PTFE membranes with organic fertilizer tested in 15-h continuous MD runs (C_f = organic fertilizer feed, T_f = 70 °C, T_p = 16 °C, $Q_f = Q_p$ = 110 L·h^{-1}).

4. Conclusions

Membrane distillation is an emerging technology for solute concentration and value recovery from aqueous streams. As an alternative to the conventional MSF process, the potential of MD to be applied in the concentration of a semi-product of organic fertilizer was evaluated with a unique asymmetric PVDF membrane prepared by the unconventional NTIPS method. The MD process's stability was examined in terms of the membrane integrity and fouling tendency associated with flux loss and membrane cleaning. Investigations revealed that the asymmetric PVDF membrane exhibited superior permeability up to 86 kg·m^{-2}·h^{-1} at 70 °C. Consistent with the literature data, the membrane flux increased with increasing flow rate and decreasing solution salinity. Interestingly, the selection of membrane orientation, i.e., top-surface- or bottom-surface-facing feed mode, was proven to be important in determining the membrane permeability. As a result, the top-surface-facing feed mode was chosen due to the smaller pore size contributed by the Knudsen diffusion mechanism of mass transport. Furthermore, compared to the commercial PTFE membrane, the asymmetric PVDF membrane showed superior sustainability in permeability and fouling propensity, maintaining more than 92% membrane flux after a 15-h continuous operation. The flux was easily recovered by simple water rinsing. As a result, a 2.6x concentration factor was achieved in one MD run. Thus, the potential to achieve a much higher concentration factor is feasible with MD due to the ease of flux recovery and the excellent anti-fouling and anti-wetting properties of the as-developed PVDF membrane.

Acknowledgments: This study was financially supported by the Social Development Research Projects of the Science and Technology Department of Zhejiang Province (No. 2016C33023), the Natural Science Foundation of Ningbo (No. 2017A610042), and the Ningbo International Science and Technology Cooperation Project (No. 2014D10017). This work was also sponsored by the K.C. Wong Magna Fund from Ningbo University. Xing Yang gratefully acknowledges the Australian Research Council Discovery Project (No. DP170102391).

Author Contributions: Tonghu Xiao and Yanfei Liu conceived and defined the problem. Yanfei Liu, Tonghu Xiao, and Chenghuan Bao developed the methodology and case studies. Yanfei Liu wrote the paper. Jifei Zhang and Xing Yang contributed to the discussion. Xing Yang supervised Yanfei Liu through the experiments and technical writing, and reviewed and finalized the paper structure.

Conflicts of Interest: The authors declare no conflict of interest.

References

1. Qureshi, A.; Lo, K.V.; Liao, P.H.; Mavinic, D.S. Real-time treatment of dairy manure: Implications of oxidation reduction potential regimes to nutrient management strategies. *Bioresour. Technol.* **2008**, *99*, 1169–1176. [CrossRef] [PubMed]
2. Al-Karaghouli, A.; Kazmerski, L.L. Energy consumption and water production cost of conventional and renewable-energy-powered desalination processes. *Renew. Sustain. Energy Rev.* **2013**, *24*, 343–356. [CrossRef]
3. Fritzmann, C.; Löwenberg, J.; Wintgens, T.; Melin, T. State-of-the-art of reverse osmosis desalination. *Desalination* **2007**, *216*, 1–76. [CrossRef]
4. Jiao, B.; Cassano, A.; Drioli, E. Recent advances on membrane processes for the concentration of fruit juices: A review. *J. Food Eng.* **2004**, *63*, 303–324. [CrossRef]
5. Malaeb, L.; Ayoub, G.M. Reverse osmosis technology for water treatment: State of the art review. *Desalination* **2011**, *267*, 1–8. [CrossRef]
6. Khayet, M. Membranes and theoretical modeling of membrane distillation: A review. *Adv. Colloid Interface Sci.* **2011**, *164*, 56–88. [CrossRef] [PubMed]
7. Drioli, E.; Ali, A.; Macedonio, F. Membrane distillation: Recent developments and perspectives. *Desalination* **2015**, *356*, 56–84. [CrossRef]
8. Al-Obaidani, S.; Curcio, E.; Macedonio, F.; Di Profio, G.; Al-Hinai, H.; Drioli, E. Potential of membrane distillation in seawater desalination: Thermal efficiency, sensitivity study and cost estimation. *J. Membr. Sci.* **2008**, *323*, 85–98. [CrossRef]
9. El-Bourawi, M.S.; Ding, Z.; Ma, R.; Khayet, M. A framework for better understanding membrane distillation separation process. *J. Membr. Sci.* **2006**, *285*, 4–29. [CrossRef]
10. Martinetti, C.R.; Childress, A.E.; Cath, T.Y. High recovery of concentrated RO brines using forward osmosis and membrane distillation. *J. Membr. Sci.* **2009**, *331*, 31–39. [CrossRef]
11. Khraisheh, M.; Benyahia, F.; Adham, S. Industrial case studies in the petrochemical and gas industry in qatar for the utilization of industrial waste heat for the production of fresh water by membrane desalination. *Desalin. Water Treat.* **2013**, *51*, 1769–1775. [CrossRef]
12. Cath, T.Y.; Adams, V.D.; Childress, A.E. Experimental study of desalination using direct contact membrane distillation: A new approach to flux enhancement. *J. Membr. Sci.* **2004**, *228*, 5–16. [CrossRef]
13. Suárez, F.; Tyler, S.W.; Childress, A.E. A theoretical study of a direct contact membrane distillation system coupled to a salt-gradient solar pond for terminal lakes reclamation. *Water Res.* **2010**, *44*, 4601–4615. [CrossRef] [PubMed]
14. Hsu, S.T.; Cheng, K.T.; Chiou, J.S. Seawater desalination by direct contact membrane distillation. *Desalination* **2002**, *143*, 279–287. [CrossRef]
15. Quist-Jensen, C.A.; Macedonio, F.; Conidi, C.; Cassano, A.; Aljlil, S.; Alharbi, O.A.; Drioli, E. Direct contact membrane distillation for the concentration of clarified orange juice. *J. Food Eng.* **2016**, *187*, 37–43. [CrossRef]
16. Christensen, K.; Andresen, R.; Tandskov, I.; Norddahl, B.; du Preez, J.H. Using direct contact membrane distillation for whey protein concentration. *Desalination* **2006**, *200*, 523–525. [CrossRef]
17. Yang, X.; Pang, H.; Zhang, J.; Liubinas, A.; Duke, M. Sustainable waste water deammonification by vacuum membrane distillation without ph adjustment: Role of water chemistry. *Chem. Eng. J.* **2017**, *328*, 884–893. [CrossRef]
18. Macedonio, F.; Drioli, E. Pressure-driven membrane operations and membrane distillation technology integration for water purification. *Desalination* **2008**, *223*, 396–409. [CrossRef]
19. Khayet, M. Treatment of radioactive wastewater solutions by direct contact membrane distillation using surface modified membranes. *Desalination* **2013**, *321*, 60–66. [CrossRef]
20. Antal, M.J., Jr.; Allen, S.G.; Deborah Schulman, A.; Xu, X.; Divilio, R.J. Biomass gasification in supercritical water. *Ind. Eng. Chem. Res.* **2000**, *39*, 4040–4053. [CrossRef]
21. Tijing, L.D.; Woo, Y.C.; Choi, J.S.; Lee, S.; Kim, S.H.; Shon, H.K. Fouling and its control in membrane distillation—A review. *J. Membr. Sci.* **2015**, *475*, 215–244. [CrossRef]
22. Gryta, M. Fouling in direct contact membrane distillation process. *J. Membr. Sci.* **2008**, *325*, 383–394. [CrossRef]
23. Bhattacharjee, C.; Saxena, V.K.; Dutta, S. Fruit juice processing using membrane technology: A review. *Innov. Food Sci. Emerg. Technol.* **2017**, *43*, 136–153. [CrossRef]

24. Ge, J.; Peng, Y.; Li, Z.; Chen, P.; Wang, S. Membrane fouling and wetting in a DCMD process for RO brine concentration. *Desalination* **2014**, *344*, 97–107. [CrossRef]

25. Ding, Z.; Liu, L.; Liu, Z.; Ma, R. Fouling resistance in concentrating TCM extract by direct contact membrane distillation. *J. Membr. Sci.* **2010**, *362*, 317–325. [CrossRef]

26. Guillen-Burrieza, E.; Thomas, R.; Mansoor, B.; Johnson, D.; Hilal, N.; Arafat, H. Effect of dry-out on the fouling of PVDF and PTFE membranes under conditions simulating intermittent seawater membrane distillation (SWMD). *J. Membr. Sci.* **2013**, *438*, 126–139. [CrossRef]

27. Eykens, L.; De Sitter, K.; Dotremont, C.; Pinoy, L.; Van der Bruggen, B. Membrane synthesis for membrane distillation: A review. *Sep. Purif. Technol.* **2017**, *182*, 36–51. [CrossRef]

28. Yang, X.; Fane, A.G.; Wang, R. *Membrane Distillation: Now and Future*; John Wiley & Sons: Hoboken, NJ, USA, 2014; pp. 373–424.

29. Wang, P.; Teoh, M.M.; Chung, T.S. Morphological architecture of dual-layer hollow fiber for membrane distillation with higher desalination performance. *Water Res.* **2011**, *45*, 5489–5500. [CrossRef] [PubMed]

30. Deshmukh, S.P.; Li, K. Effect of ethanol composition in water coagulation bath on morphology of PVDF hollow fibre membranes. *J. Membr. Sci.* **1998**, *150*, 75–85. [CrossRef]

31. Gu, M.; Zhang, J.; Wang, X.; Tao, H.; Ge, L. Formation of poly(vinylidene fluoride) (PVDF) membranes via thermally induced phase separation. *Desalination* **2006**, *192*, 160–167. [CrossRef]

32. Xiao, T.; Wang, P.; Yang, X.; Cai, X.; Lu, J. Fabrication and characterization of novel asymmetric polyvinylidene fluoride (PVDF) membranes by the nonsolvent thermally induced phase separation (NTIPS) method for membrane distillation applications. *J. Membr. Sci.* **2015**, *489*, 160–174. [CrossRef]

33. Zhang, J.; Li, J.D.; Gray, S. Effect of applied pressure on performance of PTFE membrane in DCMD. *J. Membr. Sci.* **2011**, *369*, 514–525. [CrossRef]

34. Zhang, J.; Gray, S.; Li, J.D. Predicting the influence of operating conditions on DCMD flux and thermal efficiency for incompressible and compressible membrane systems. *Desalination* **2013**, *323*, 142–149. [CrossRef]

35. Hausmann, A.; Sanciolo, P.; Vasiljevic, T.; Weeks, M.; Schroën, K.; Gray, S.; Duke, M. Fouling mechanisms of dairy streams during membrane distillation. *J. Membr. Sci.* **2013**, *441*, 102–111. [CrossRef]

36. Smolders, K.; Franken, A.C.M. Terminology for membrane distillation. *Desalination* **1989**, *72*, 249–262. [CrossRef]

37. Hu, N.; Xiao, T.; Cai, X.; Ding, L.; Fu, Y.; Yang, X. Preparation and characterization of hydrophilically modified PVDF membranes by a novel nonsolvent thermally induced phase separation method. *Membranes* **2016**, *6*, 47. [CrossRef] [PubMed]

38. Bouguecha, S.; Chouikh, R.; Dhahbi, M. Numerical study of the coupled heat and mass transfer in membrane distillation. *Desalination* **2003**, *152*, 245–252. [CrossRef]

39. Srisurichan, S.; Jiraratananon, R.; Fane, A. Mass transfer mechanisms and transport resistances in direct contact membrane distillation process. *J. Membr. Sci.* **2006**, *277*, 186–194. [CrossRef]

40. Phattaranawik, J.; Jiraratananon, R.; Fane, A.G. Effect of pore size distribution and air flux on mass transport in direct contact membrane distillation. *J. Membr. Sci.* **2003**, *215*, 75–85. [CrossRef]

41. Alklaibi, A.M.; Lior, N. Heat and mass transfer resistance analysis of membrane distillation. *J. Membr. Sci.* **2006**, *282*, 362–369. [CrossRef]

42. Qtaishat, M.; Matsuura, T.; Kruczek, B.; Khayet, M. Heat and mass transfer analysis in direct contact membrane distillation. *Desalination* **2008**, *219*, 272–292. [CrossRef]

43. Khayet, M.; Matsuura, T.; Mengual, J.I.; Qtaishat, M. Design of novel direct contact membrane distillation membranes. *Desalination* **2006**, *192*, 105–111. [CrossRef]

membranes

MDPI

Article

PES-Kaolin Mixed Matrix Membranes for Arsenic Removal from Water

Tiziana Marino [1], Francesca Russo [1], Lina Rezzouk [2], Abderrazak Bouzid [2] and Alberto Figoli [1,*]

[1] Institute on Membrane Technology, ITM-CNR, Via P. Bucci 17C, 87036 Cosenza, Italy; t.marino@itm.cnr.it (T.M.); ru.francy88@hotmail.it (F.R.)

[2] Materials and Electronic Systems Laboratory (LMSE), University of Bordj Bou Arreridj, El-Anasser 34030, Bordj Bou Arreridj, Algeria; lina.r@live.fr (L.R.); a_bouzid34@hotmail.com (A.B.)

* Correspondence: a.figoli@itm.cnr.it; Tel.: +39-0984-49-2027

Received: 11 August 2017; Accepted: 26 September 2017; Published: 30 September 2017

Abstract: The aim of this work was the fabrication and the characterization of mixed matrix membranes (MMMs) for arsenic (As) removal from water. Membrane separation was combined with an adsorption process by incorporating the kaolin (KT2) Algerian natural clay in polymeric membranes. The effects of casting solution composition was explored using different amounts of polyethersufone (PES) as a polymer, polyvinyl-pyrrolidone (PVP K17) and polyethylene glycol (PEG 200) as pore former agents, N-methyl pyrrolidone (NMP) as a solvent, and kaolin. Membranes were prepared by coupling Non-solvent Induced Phase Separation and Vapour Induced Phase Separation (NIPS and VIPS, respectively). The influence of the exposure time to controlled humid air and temperature was also investigated. The MMMs obtained were characterized in terms of morphology, pore size, porosity, thickness, contact angle and pure water permeability. Adsorption membrane-based tests were carried out in order to assess the applicability of the membranes produced for As removal from contaminated water. Among the investigated kaolin concentrations (ranging from 0 wt % to 5 wt %), a content of 1.25 wt % led to the MMM with the most promising performance.

Keywords: arsenic; kaolin KT2; phase inversion; polyethersulfone; mixed matrix membranes; water treatment

1. Introduction

Arsenic (As) is a natural element, which behaves like a dangerous agent to human health and the environment. Several studies have shown a relationship between As exposure and teratogenicity, mutagenicity carcinogenic effects [1,2]. The harmful effects concern skin, bladder, lung liver, prostate cancer, as well as cardiovascular, pulmonary, immunological, neurological and endocrine diseases [3]. Drinking water represents the most As polluted source, especially in countries such as Argentina, Banghadesh, China, Chile, India and Mexico [4,5]. The World Health Organization [6] has set the maximum concentration limit to 10 µg/L or 10 ppb [6]. Arsenic exists in two primary forms: inorganic and organic. Inorganic compounds occur in two valence states (arsenite As(III) and arseniate As(V)), while organic compounds mainly are monomethyl arsenic acid and dimethyl arsenic acid [7]. Arsenic removal from water has been carried out in different ways. Classic options include coagulation, ion exchange, adsorption and membrane-based operations. Although these techniques are suitable for As removal, some of them still present disadvantages. In fact, coagulation requires a pre-oxidation step [8]. The efficiency of ion exchange strongly depends on the pH solution and concentration of anions [9,10]. Adsorption may represent one of the most promising removal process, offering the possibility to operate with high-removal efficiency, lower cost, simplicity, easy operation [11,12]. Molecules are removed from the aqueous solution by adsorption into solid surfaces. However, as reported in literature [8] this process is highly pH sensitive and requires periodical adsorbents regeneration. Moreover, this technique needs the recovery of adsorbents particles, contributing to the increase of cost maintenance [13]. Membrane technology should offer a promising solution for water treatment, especially when coupled with other removal techniques [5,14–16]. In this context, the mixed matrix membranes (MMMs) are particularly interesting; they pose a valid alternative, combining the positive properties of polymeric membrane and adsorbents. A MMM is composed of inorganic filler particles embedded in the polymeric matrix. Gohari et al. [17] and Zheng et al. [18] highlighted that the presence of adsorbents inside the membrane bulk should improve the separation selectivity, decrease fouling and increase membrane hydrophilicity. Mohan and Pittman [19] identified different adsorbents to remove As, such alumina, iron oxide, manganesia, titania and ferric phosphate. Metal oxides have large adsorption capacities due to their extremely high surface areas [20]. Fe(II) is present in different natural materials, such as magnetite, siderite and hematite, which have been investigated for removing As from water. The removal mechanism is mainly based on electrostatic attraction and adsorption on the surface of Fe(II)-bearing minerals [21,22]. Mohan and Pittman [19] studied the possibility of using activated carbon, showing that As(V) can be removed more efficiently than As(III). Among the naturally occurring materials, kaolinite, montmorillionite and illite [23,24] represent particularly attracting clay minerals with good adsorption capabilities. Kaolinite is a group of alumina silicates clay minerals which comprise the principal ingredients of kaolin and it is characterized by fine particle size, brightness and whiteness, chemical inertness, structure [24].

"KT2" (Table 1), is an Algerian kaolin coming from the original EL Milia deposit "TAMAZERT" in the region of Jijel (Algeria). This kaolin is abundant in soils and sediments [25], it is odorless white to yellowish or grayish powder and it has density 2.65 g/cm^3 [26]. It has been selected for water treatment studies for its high capacity adsorbents on the surface, very low cost material and alumino silicate-based compositions $(Al_2O_3(SiO_2)_2(H_2O)_2)$ [24,25,27–31].

Table 1. Chemical composition of KT2 kaolin [28].

Components (wt %)								
SiO_2	Al_2O_3	Fe_2O_3	MgO	CaO	Na_2O	K_2O	TiO_2	LOI
49.30	33.50	1.59	0.40	0.08	0.09	2.75	0.24	10.50

Zen et al. [28] tested two Algerian kaolin clays, DjebelDebagh "DD3" and Tamazert "KT2" for the adsorption of Derma Blue R67 acid dye, commonly used in the tanning industry. The results obtained evidenced the efficiency and the feasibility of dye adsorption at ambient temperature, thus demonstrating that natural clays work as promising adsorbent candidates for waste water treatments.

Sarbatly [30] described the preparation of kaolin–polyethersufone (PES) membranes via NIPS and sintering techniques, investigating the effect of the clay/polymer ratio and sintering temperature on pore size and porosity. The registered data showed that the pore size changed from 20 to 8 μm and the porosity decreased from 26% to 11% when additive/polymer and sintering temperature increased from 1 to 3.5 and from 1100 to 1500 °C, respectively.

Han et al. [31] worked on the production of Al_2O_3, Al_2O_3–SiO_2 and Al_2O_3–kaolin hollow fibers using a wet-spinning method using PES as polymeric material. Hollow fiber membranes were prepared by combining preheating and sintering processes. The addition of SiO_2 was beneficial for Al_2O_3 hollow fiber membranes, even if the amount of SiO_2 should be rigorously controlled. When kaolin was used instead of SiO_2, the introduction of both the reactions of Al_2O_3–SiO_2 were achieved. In fact, kaolin, acting as an inorganic binder, contributed to the sintering of the membranes, thus the structure was maintained and the particles fixed up by the reaction of alumina and silica in kaolin.

In this work, kaolin was chosen as a natural adsorbent for the preparation of a MMM flat sheet membrane using PES as polymeric material.

The adsorption capability of the natural clay was tested for As removal from water. Considering the outstanding property of kaolin as an inorganic binder in the membrane preparation process, its natural abundance, fine texture, grain size of 0.2–1 μm, high chemical and thermal stability and low cost, kaolin should be adopted in membrane preparation as substitute of Al_2O_3, TiO_2 and ZrO_2 [28–31].

PES is among the most used amorphous thermoplastics which finds applications in many industrial sectors, such as electronic, medical, aerospace, food service [32]. It presents excellent thermal and chemical stability, dimensional stability and resistance in a wide range of pH [33].

The aim of this work was the fabrication and the characterization of organic-inorganic PES–kaolin membranes for arsenic removal from water. Membrane separation was coupled with the adsorption process by including Algerian natural clay (kaolin KT2) into the polymeric solution.

MMMs were prepared by coupling NIPS (Non-solvent Induced Phase Separation) and VIPS (Vapour Induced Phase Separation) procedures in order to study the influence of several parameters which affect the final membrane structure and properties, such as kaolin and pore former agent content, as well as the exposure time to relative humidity.

2. Materials and Methods

2.1. Materials

Polyethersulfone (PES) polymer (Veradel®3000 P; Mw = 60 kg/mol) was kindly supplied by Solvay Specialty Polymers (Bollate, Italy) and N-methyl pyrrolidone (NMP) synthetic grade was purchased by Sigma Aldrich (Milan, Italy).

Polyvinylpyrrolidone (PVP, Luviskol K17; Mw = 9 kg/mol) was purchased by BASF (Ludwigshafen, Germany) and as well as for the polymer, was desiccated under vacuum at 50 °C for 24 h before its use. Polyethylene glycol (PEG; Mw = 0.2 kg/mol) was purchased by Sigma Aldrich. The precipitation medium was *i*-distilled water (at 15 °C). Kaolin (KT2), was obtained from Guelma region (Algeria).

2.2. Membrane Preparation

Membranes were prepared via phase inversion, based on the separation of an initially stable solution, in a polymer-rich phase and a polymer poor-phase [34].

Thermodynamic instability during phase inversion can be induced by:

- Adding a non-solvent (non-solvent induced phase separation, NIPS).
- Exposing the polymeric solution to a non-solvent vapour (vapour induced phase separation, VIPS) up to the complete polymer precipitation or prior to immersion in a non-solvent coagulation bath.
- Evaporating the solvent (evaporation induced phase separation, EIPS).
- Cooling down a polymer solution obtained at elevated temperature promoting polymer precipitation and the formation of the final membrane (thermally induced phase separation, TIPS).

In this work, membranes were produced via NIPS or by coupling NIPS and VIPS.

In particular, MMMs with different morphology and properties were prepared by changing in the dope solution the kaolin and the PEG content, and by varying the exposure time of the nascent film to humid air (Table 2).

Table 2. Compositions of the casting solutions and exposure time to controlled temperature (25 °C) and RH %.

Membrane Code	PES (wt %)	KT2 Kaolin (wt %)	PVP K17 (wt %)	PEG 200 (wt %)	NMP (wt %)	Exposure Time to RH % (min)
M1	12	0	5	35	48	0
M2						5
M3	12	1.25	5	35	46.75	0
M4						5
M5	12	2.5	5	35	45.5	0
M6						5
M7	12	5	5	35	43	0
M8						5

The optimized procedure for preparing casting solution was the following:

1. The liquid phase (NMP and PEG) was magnetically stirred.
2. Kaolin was added and the suspension sonicated in an ultrasonic bath (for 90 min at 25 °C) in order to assure a homogeneous nanoparticles dispersion.
3. The solid components (i.e., PES and PVP) were added and kept under stirring until a homogeneous casting suspension was observed.

The dope solution were cast on a glass plate using a manual casting knife (Elcometer 3700/1 Doctor Blade, Aalen, Germany; adjustable gap size: 30–4000 µm) with a 300 µm gap, at 25 °C and 55% of relative humidity (RH %) in a climatic chamber (DeltaE srl, Rende, Italy).

Membranes were washed three consecutive times with hot water (60 °C), dried at room temperature for 4 h and then dried in an oven at 40 °C for 24 h before characterize them.

2.3. Experimental Set up

The experimental set-up is composed of Amicon Model 8200 (EMD Millipore, Billerica, MA, USA), equipped with a filtration unit (180 mL). 1 ppm of Sodium arsenate dibasic heptahydrate was placed into 200 mL of water. As aqueous solution was forced to pass through the membrane by means of a peristaltic pump (Masterflex®7518-10, Cole-Parmer, Vernon Hills, IL, USA) (Figure 1).

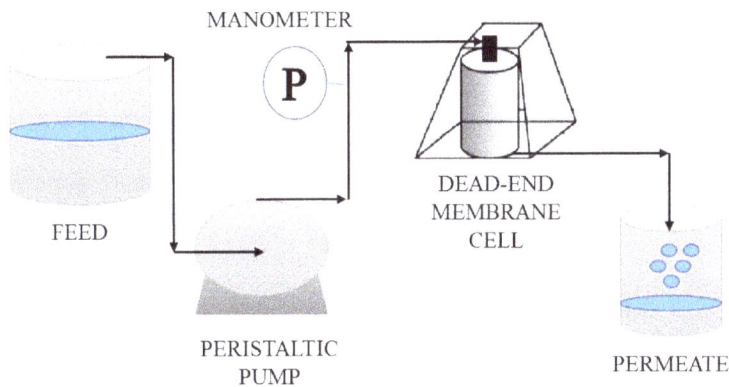

Figure 1. Experimental set up used for Arsenic (As) removal from aqueous solutions.

Both the feed and the permeate composition were evaluated by Inductively Coupled Plasma (ICP) analysis (performed by GEOLAB S.r.l. laboratory, Rende, Italy) after specific contact time. Before ICP analysis, all the samples were stabilized with ultrapure nitric acid (1.0% HNO_3).

The percentage of Arsenic removal %R was determined as follows:

$$\%R = ((C_0 - C_t) \times 100)/C_0 \qquad (1)$$

where C_0 is the initial As concentration and C_t is the As content at time t.

2.4. Pure Water Permeability (PWP)

The membrane performance was tested in a cross flow system in recycle mode. In this set up permeate is pumped back to the feed water solution. The permeate flux was determined by monitoring the permeate weight during time. PWP was measured at 25 °C by means of a cross-flow cell (DeltaE srl, Italy). Pure water at 25 °C was driven across the membrane with 8 cm^2 area, by using Tuthill Pump Co. (Alsip, IL, USA) gear pump. *PWP* was calculated using the following equation:

$$PWP = \frac{Q}{A\, t\, \Delta P} \tag{2}$$

where Q is the permeate volume (L), A is the membrane area (m^2), t is the time (h), and ΔP is the pressure difference across the membrane sides.

Three different transmembrane pressures (i.e., 1.0/0.8/0.6 bar separated by a stabilization period of one to another of 20 min), were used to calculate PWP.

2.5. Thickness

The thickness was determined by the multiple point method using a digital micrometer (Carl Mahr, Göttingen, Germany; precision of ±0.001 mm). Thickness was verified for seven regions of membrane, the average value and the standard deviation were taken into account.

2.6. Porosity

Membrane porosity was determined by cutting three different samples of the same membrane, drying them in a vacuum oven at 50 °C for 12 h and measuring their weight by means of an analytical balance. Then, the samples were soaked in Kerosene for 24 h before verifying the wet membrane weight. The average and the standard deviation values were then calculated. Membrane porosity (ε) is defined as the volume of the pores divided by the total volume of the membrane. The porosity was calculated by the following equation [35]:

$$\varepsilon = \frac{\frac{wt_w - wt_d}{\rho_k}}{\frac{wt_w - wt_d}{\rho_k} + \frac{wt_d}{\rho_p}} \times 100 \tag{3}$$

where ε is the porosity (%), wt_w and wt_d are the wet and dry weight of the membrane, respectively; ρ_k and ρ_p are the density of kerosen and PES, respectively.

2.7. Pore Size and Pore Size Distribution

The analysis was performed three times for each membrane, and by extrapolating the average and the standard deviation. A capillary flow porometer (CFP-1500 AEXL, Porous Materials Inc., Ithaca, NY, USA) was used. Dried membranes were kept in a wetting liquid (Fluorinert®FC-40, Sigma-Aldrich) having low surface tension (16 dynes/cm) for 24 h prior analyzing the pore size.

The microflow porometer measures pressure close to the membrane. Differential pressures and gas flow rates through dry and wet samples were recorded. The pore structure characteristics include bubble point, mean flow pore size, pore fraction distribution and air permeability.

2.8. Mechanical Properties of Membranes

The mechanical properties which include the Young's modulus and elongation at break of the prepared membranes were measured using a Zwick Roell Z2.5 test unit (Zwick Roell Group, Ulm, Germany) with pneumatic clamps and 50 N maximum load cell. The stress (σ) calculated as the ratio between the applied force (F) per unit area (A).

While the strain is defined as:

$$\varepsilon = \ln(L)/L_0 \tag{4}$$

where L is the sample length at break and L_0 is the initial sample length, allow to calculate Young's modulus (E) corresponding to: $E = \sigma/\varepsilon$ from the initial part of the slope from the stress/strain curve [36]. Membranes were cut into strips of 1 cm width and 5 cm length. Tests were carried out by starting with an initial 50 mm gauge length and by stretching the membranes unidirectional at a constant rate of 5 mm/min. 5 specimens were used for each membrane and the average and the standard deviation were extrapolated.

Three samples (1 cm × 5 cm) of each type of membrane were tested at ambient temperature with a cross head speed of 0.5 mm/min. The tension strength and elongation at break average values as well as their standard deviations were determined.

2.9. Scanning Electron Microscopy (SEM) Analysis

Cross section and surface of the membranes produced were observed with a Zeiss-EVO MA10 instrument (Zeiss, Milano, Italy). The membranes were first freeze fractured using liquid nitrogen for cross section preparation. Then for both surface and cross-section analyses, the samples were sputtered for making a thin gold layer (by using a Quorum Q150 RS sputter (Quorum Technologies, Laughton, East Sussex, UK) in order to enhance membrane conductivity and prevent electrical charging.

2.10. Contact Angle

The static contact angle of water can be defined as the angle comprised between the membrane surface and the tangent line at the water droplet contact point on the membrane surface. An optical contact angle meter (CAM100, KSV Instruments, Helsinki, Finland) was used to measure the contact angle of the membranes prepared. A water droplet of 5 μL was positioned on the membrane top surface. For each membrane, five measurements were carried out, at the end of which the average and the standard deviation were calculated.

3. Results and Discussions

3.1. Membrane Morphology

The influence of the concentration of kaolin, as well as the adopted membrane preparation procedure, was investigated using scanning electron microscopy (SEM) apparatus. The SEM images of the cross-sections of pristine PES and MMMs membranes prepared with different concentrations of kaolin via NIPS are represented in Figure 2. PES membrane (M1) exhibited a typical asymmetric structure, composed of a thin skin layer and a porous bulk with a finger-like structure. Similarly to what was observed for the pristine PES membrane, by increasing the kaolin content in the dope solution, the membrane exhibited porous finger-like morphology (Figure 3). The bounding of PES chains and kaolin nanoparticles could be observed with further SEM magnification (1000×), which showed that the kaolin particles were entrapped in the membrane structure by means of interactions with polymer chains.

By coupling the NIPS and VIPS procedures, M2, M4, M6 and M8 membranes were obtained by changing the kaolin concentration (Figure 3).

During VIPS process, the cast film is first exposed to humidity prior to immersion in the coagulation bath. The non-solvent in liquid and vapour form leads to the phase separation. Typically, by adjusting some parameters, such as temperature, relative humidity and exposure time, membranes with large pores on the top surface can be obtained [37].

Although the presence of non-solvent in the vapour phase may reduce the exchange rate between solvent and non-solvent, leading to sponge-like symmetric structures, the presence of hydrophilic kaolin nanoparticles may contribute to the faster water–NMP demixing, promoting the formation of fingers.

Kaolin nanoparticle dispersion was visibly homogeneous without evident defects across the membrane morphology. However, as reported in the literature [31], a small number of aggregations formed with nanoparticles with particularly high surface would contribute to reduce the surface energy, thus enhancing the system stability.

All the membranes prepared displayed a very compact polymeric layer on the top and the bottom surface, with few bare kaolin particles. This might be related to the intrinsic binder properties of kaolin [38]. In fact, it is used in ceramic preparation thanks to its tremendous molding performance. At lower kaolin content, inorganic binding of the membranes could be established by inorganic nanoparticles, promoting membrane formation and particle dispersion. However, at higher concentrations, kaolin would negatively affect the particle dispersion, giving rise to the formation of some defects in the membrane structure [31].

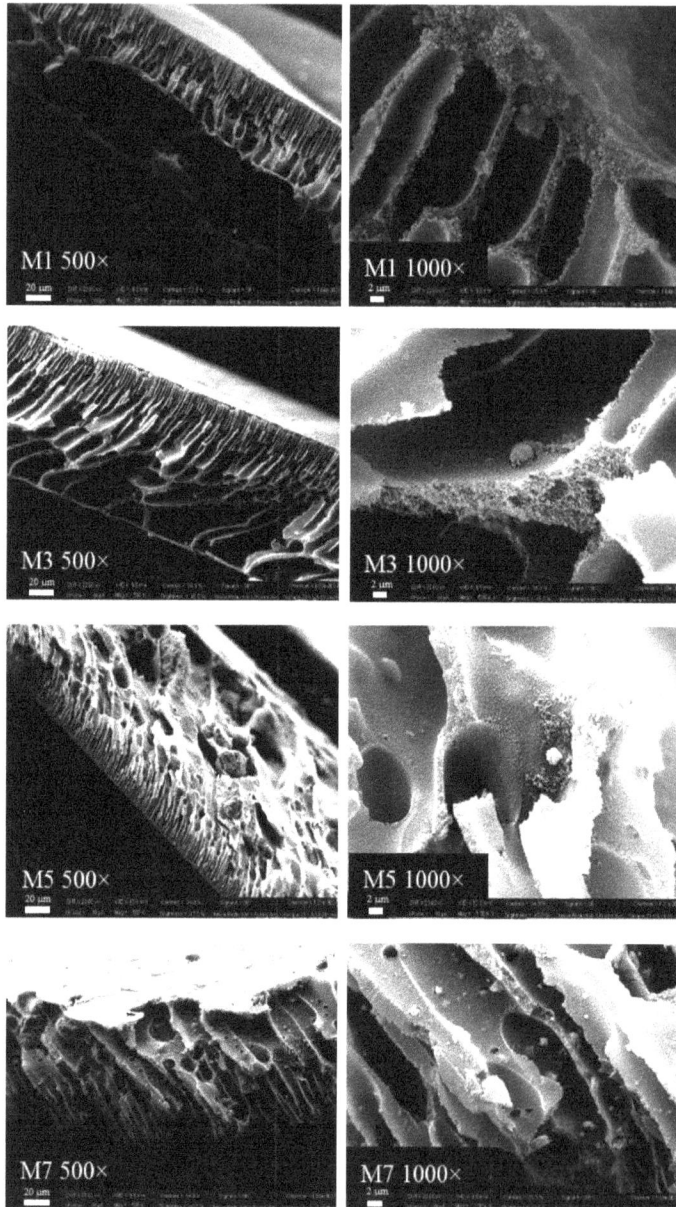

Figure 2. Scanning electron microscopy (SEM) images of the PES and PES–kaolin membranes prepared via the Non-solvent Induced Phase Separation (NIPS) procedure.

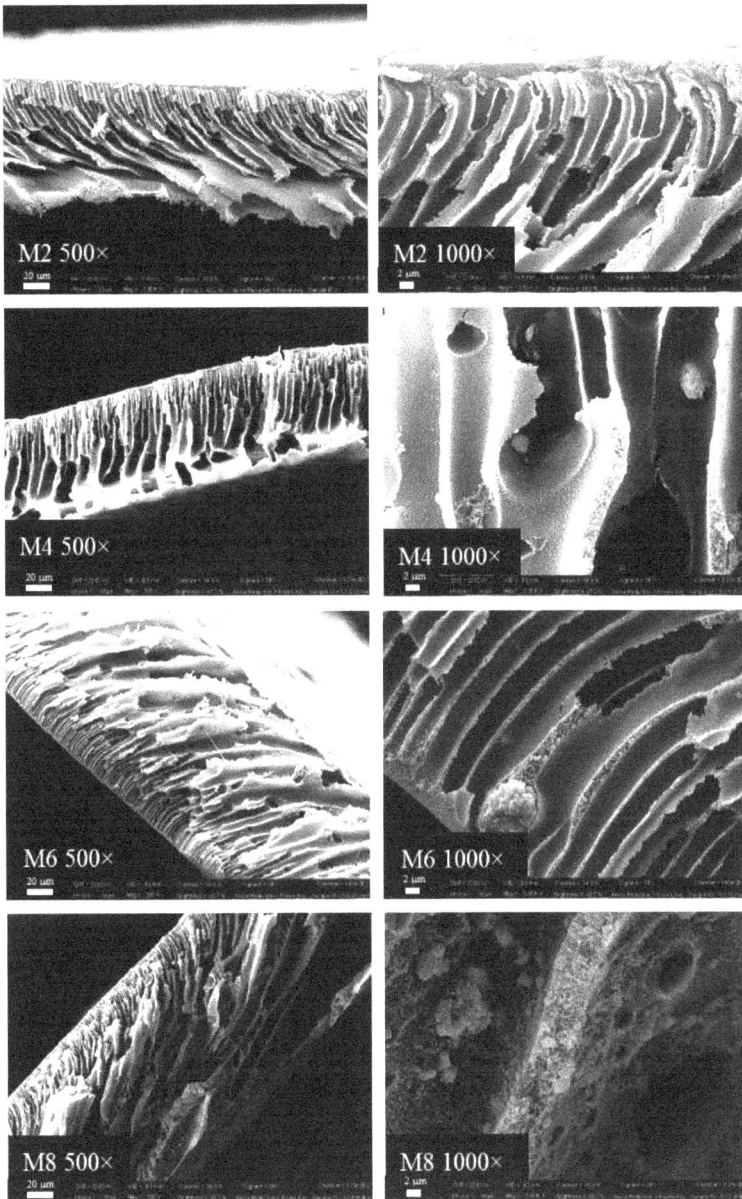

Figure 3. SEM images of the PES and PES–kaolin membranes prepared via NIPS–Vapour Induced Phase Separation (VIPS) procedure.

3.2. Membrane Pore Size

In order to achieve separation, membrane systems rely on membranes which are able to remove undesirable compounds or allow the passage only of selected kinds of solutes. It is evident that pore size represents an important parameter to achieve this separation. The pore size of the produced PES–kaolin membranes is reported in Table 3.

Table 3. Smallest, mean flow and largest pore diameter of the prepared membranes.

Membrane Code	Smallest Pore Size (µm)		Mean Flow Pore Diameter (µm)		Largest Pore Size (µm)	
	Average	Std. Dev.	Average	Std. Dev.	Average	Std. Dev.
M1	0.03	0.01	0.05	0.01	0.06	0.02
M2	0.13	0.02	0.14	0.01	0.15	0.02
M3	0.05	0.01	0.10	0.01	0.18	0.01
M4	0.06	0.02	0.17	0.01	0.23	0.03
M5	0.05	0.02	0.12	0.02	0.23	0.01
M6	0.08	0.02	0.23	0.01	0.31	0.01
M7	0.05	0.03	0.21	0.02	0.33	0.02
M8	0.02	0.01	0.26	0.01	0.32	0.02

Membranes prepared without kaolin (M1 and M2) present noticeable differences in the pore size depending on the preparation method adopted. In particular, the mean flow pore diameter was lower (0.05 ± 0.01 µm) when the membrane was produced via NIPS and increased up to 0.14 ± 0.01 µm by exposing the nascent film to controlled humid air and temperature for 5 min before immersion in the water bath (NIPS–VIPS procedure). In accordance with what has been reported in the literature [39], the non-solvent in vapour form promoted a reduction in the polymer precipitation rate, due to a delayed solvent/non-solvent exchange, thus allowing the growth of membrane pores.

The addition of kaolin in the casting solution further promotes the increase in membrane pore size, due to the increase in dope solution viscosity, which changed in the following order: 1120 cP (0 wt % kaolin) > 1640 cP (1.25 wt % kaolin) > 2130 cP (2.50 wt % kaolin) > 3550 cP (5 wt % kaolin). The increased viscosity hindered the exchange process between water in the coagulation bath and NMP in polymer dope, enhancing, at the same time, water inflow rate in the membrane matrix. Similar results were reported by Ma et al. [40].

Pore formation is further supported by the presence of the non-solvent in vapour form, which caused the increase in the pore size when NIPS was coupled with VIPS. In fact, during VIPS, the presence of water in vapour form was found to strongly affect membrane formation, especially in cases where water represents a strong non-solvent for the polymer and the affinity with the solvent is high (such in this case for the water–NMP system) [41]. By increasing the exposure time to humid air, membranes with higher pore size could be produced [39].

In this case the pore size changed from 0.14 ± 0.01 µm (M2) to 0.17 ± 0.01 µm (M4) and 0.23 ± 0.01 µm (M6) up to 0.26 ± 0.01 µm (M8) by increasing the kaolin content in the dope solution.

3.3. Thickness, Porosity, Contact Angle and Mechanical Properties

Thickness, porosity, contact angle and mechanical features are key parameters, which strongly influence the efficacy of separation operations and are correlated to the membrane morphology and PWP. Membrane thickness is a significant factor affecting the separation rate across the membrane. As presented in Table 4, the thickness of the membranes prepared with and without kaolin was almost unchanged, even when the nascent films were exposed to controlled humidity and temperature before polymer precipitation in the coagulation bath.

Table 4. Thickness, porosity, contact angle and mechanical properties of the prepared PES and PES–kaolin membranes.

Membrane Code	Thickness (mm)	Porosity (%)	Contact Angle (°) Top Surface	Mechanical Properties Young's Modulus (n/mm²)	Elongation at Break (%)
M1	0.099 ± 0.000	85.40 ± 0.18	52.30 ± 0.14	98.85 ± 1.69	2.48 ± 0.70
M2	0.094 ± 0.001	86.45 ± 0.99	66.11 ± 1.17	96.77 ± 1.04	2.62 ± 0.49
M3	0.103 ± 0.013	89.36 ± 0.45	66.48 ± 0.24	77.35 ± 1.72	2.81 ± 0.40
M4	0.099 ± 0.002	89.74 ± 0.69	62.04 ± 0.89	76.52 ± 1.88	2.29 ± 0.54
M5	0.109 ± 0.004	89.67 ± 0.00	71.97 ± 2.38	74.85 ± 2.54	2.81 ± 0.95
M6	0.110 ± 0.001	89.97 ± 0.43	75.32 ± 1.09	72.41 ± 2.09	2.94 ± 0.55
M7	0.111 ± 0.008	90.54 ± 0.78	73.80 ± 3.66	72.15 ± 0.75	1.17 ± 0.76
M8	0.100 ± 0.013	90.57 ± 2.30	67.26 ± 2.45	70.94 ± 2.55	1.15 ± 0.36

The contact angle is an important parameter, which proves the hydrophilic or hydrophobic membrane characteristic [42,43]. The contact angles of the MMMs obtained are reported in Table 4. By preparing membrane via NIPS, the addition of kaolin nanoparticles into the polymeric solution, resulted in an increase in contact angle from about 60° for the control membrane (M1) to 70° (M7) for the highest investigated kaolin content (5.00 wt %). Similarly, also by producing membranes via NIPS–VIPS, membranes exhibited increasing contact angle values by increasing kaolin concentration in the dope solution. In this case, the contact angle passed from 65° (M2) to 70° (M8). These results may be related to increases in membrane roughness, and, as reported by Mierzwa et al. [33], the increase in contact angle may also depend on nanosize effects. However, a more in-depth investigation on factors affecting membrane hydrophilicity would be necessary to completely understand the effect of kaolin on the final membrane properties. Mierzwa et al. [33] also reported the preparation of PES–clay ultrafiltration membranes via NIPS, varying the clay content in the range between 1 wt % and 5 wt %. In agreement with what was observed in this work, the contact angles of the membranes prepared increased with the increase in clay content. However, in other studies clay-doped membranes were reported to have lower or similar contact angle to the undoped polymeric membranes. Anadao et al. [44] described the preparation and the characterization of nanocomposite polysulfone (PSf) and sodium montmorillonite (MMT) membranes, for which the water contact angle slightly decreased with the increase in MMT content. Also Ghaemi et al. [45] reported the production of significantly more hydrophilic PES–organically modified MMT membranes when the clay mineral content increased, with corresponding increases in water flux. In contrast, Monticelli et al. [46] registered minor changes in contact angle for the membranes fabricated with pristine PSf and PSf–cloisite 93A mixtures.

The presence of kaolin had a positive effect on membrane porosity (Table 4). In fact, the porosity gradually increased with the increasing clay content in the polymeric solution, passing from 85.40 ± 0.18 for M1 to 90.57 ± 2.30 for M8. Similar observations were reported by Ma et al. [40], who attributed the porosity increase to the lower solvent-non solvent exchange rate which in turn was caused by the incorporation of clay nanoparticles in the polymeric matrix. The increased water inflow rate in the nascent film was responsible for the formation of more porous membranes. The membranes prepared via NIPS–VIPS procedure presented higher porosity than those obtained by NIPS. Also in this case the reasons may lie with the decreased demixing rate between NMP and water due to the delayed water vapour adsorption into the nascent membrane, which promote the formation of membranes with more pores and larger pore size.

Elastic modulus and elongation at break are two important factors to define the mechanical resistance of membranes. The effects of kaolin concentration on mechanical strengths of PES-kaolin membranes is shown in Table 4. By increasing the clay content, Young's modulus decreased from 98.85 ± 1.69 (M1) to 70.94 ± 2.55 (M8) MPa, while the elongation at break passed from 2.48% ± 0.70% (M1) to 1.15% ± 0.36% (M8). The increased pore size and porosity of MMMs may induce worsening of the mechanical properties. Moreover, by increasing the kaolin dosage in the dope solution, the inhomogeneity of filler distribution inside the membrane matrix may lead to particle agglomeration, which further deteriorates membrane mechanical resistance. A similar effect was also reported by Ma et al. [40].

3.4. Pure Water Permeability (PWP)

Pure water permeability (PWP), or pure water flux, represents a critical parameter for porous membranes, and it is directly correlated with membrane pore size and porosity [31,40,47]. Figure 4 shows the effects of kaolin content on PWP (average and standard deviation) for the membranes produced. PES membranes prepared via NIPS and NIPS–VIPS methods (M1 and M2, respectively) displayed similar PWP. In contrast, the preparation methods had a critical influence on PWP for the PES–kaolin membranes. In fact, MMMs exposed to controlled humid air and temperature before the immersion in the coagulation bath exhibited considerably higher PWP than those obtained via NIPS, and showed higher values than the pristine PES membranes. PWP changed from 843 ± 29 to 8781 ± 32 L/m^2·h·bar for M3 and M8, respectively. The results obtained indicated that PWP reflected pore size and porosity data, but may be ascribed also to the presence of kaolin in the membrane matrix. Ma et al. [40] reported that the addition of clay in the polymeric solution led to an increase in the large pore ratio in the skin layer, which explained the registered increased pure water flux with the increase of clay dosage.

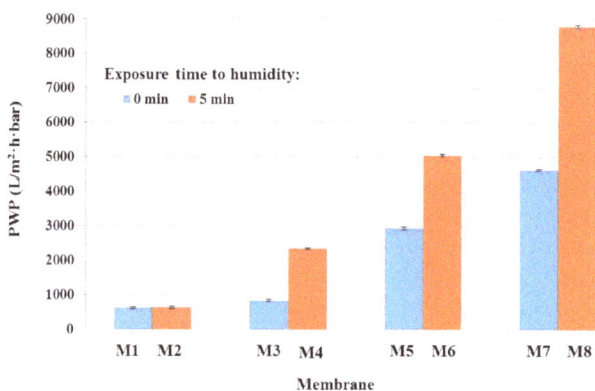

Figure 4. Pure water permeability (PWP) of the prepared PES and PES–kaolin membranes.

3.5. Membrane Performance Evaluation: As Removal from Water

In order to evaluate the applicability of PES–kaolin membranes on water treatment, As adsorption tests were carried out using 4 h as contact time. As reported by Zen et al. [28], As removal capacity is a parameter ruled by physical adsorption over the membrane, and the adsorption rate for the contact time is influenced by several parameters, such as the clay particle size entrapped into the membrane, the ionic strength and the size of As compounds. Figure 5a illustrates As adsorption capacity after 4 h for the unmodified PES membranes as well as for PES–kaolin membranes. It can be noted that As removal by adsorption through the PES membranes was very low (5% for M1 and 1.5% for M2). When kaolin was used, As removal significantly increased, reaching a value of up to 30% when 2.5 wt % kaolin was incorporated into the polymeric solution and the membrane was prepared via NIPS method (M5).

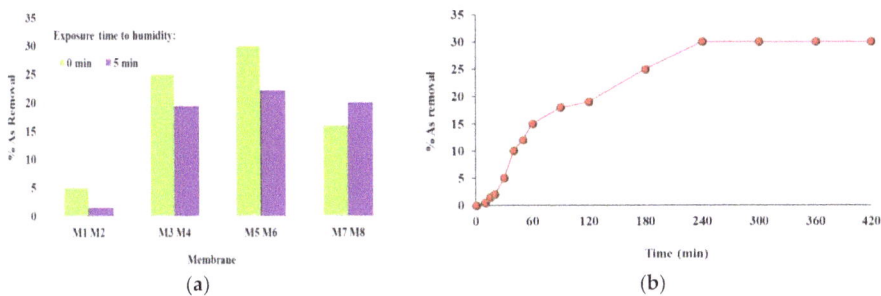

Figure 5. (a) As adsorption after 240 min test by using PES and PES–kaolin membranes prepared via NIPS and NIPS–VIPS; (b) Kinetic of As adsorption on the M5 PES–kaolin membrane.

By using MMMs membranes, As removal firstly increased by increasing the clay dosage from 1.25 wt % to 2.5 wt %, then decreased when kaolin concentration was further increased up to 5 wt %. These results may be related to the active adsorbent sites of clay particles distributed in the membrane bulk. In fact, an initial rapid increase in the adsorption upon a higher kaolin dosage may be related to a higher number of As-adsorption available sites. In presence of 2.5 wt % kaolin, the removal of As reached the maximum value of 30%. However, by increasing the adsorbent load, there was a decrease in the As removal, which means that, among the kaolin dosages investigated, 2.5 wt % was the optimal amount of clay necessary to remove As from water. When a 5 wt % clay was used, the active sites were probably occluded due to the kaolin particles' aggregation. A similar behavior was reported for Derma Blue R67 acid dye [48], cationic starch derivatives [49], methyl orange, Ponceau 6R, and Congo Red adsorbed on pullulan microspheres [50]. All the membranes prepared via NIPS presented a better performance in comparison to that of the analogue membranes obtained via NIPS–VIPS. This may be related to the combination of some membrane characteristics, such as morphology, pore size and porosity, but also PES–kaolin–As interactions that favored the arsenic adsorption in the membrane matrix. However, further investigations are needed to better understand how the operational conditions can affect the adsorption process.

On the basis of the adsorption results discussed above, the adsorption dynamic kinetics of As on M5 was analyzed (Figure 5b).

The adsorption capacity increased rapidly in the first hour (15% As removal after 60 min) and reached a value of 30% after 250 min. After this time, an adsorption equilibrium was observed, and the variation of As removal was maintained unchanged during the extended contact time (420 min). These observations may arise from the accessibility of kaolin bonding sites, which resulted completely hindered by the adsorbed As.

4. Conclusions

Flat sheet PES–kaolin membranes with different clay dosages were prepared by NIPS and NIPS–VIPS procedure. The membranes prepared exhibited asymmetric structure, in which the incorporation of well-dispersed kaolin nanoparticles can be clearly observed. The combination of the operational conditions adopted to produce the membranes allowed us to tailor their properties. The presence of kaolin promoted pore formation and the increase in its content led to an increase in the membrane pore size and PWP. In contrast, membrane hydrophilicity and mechanical resistance worsened when the clay was included in the polymeric solution. Preliminary tests on As removal from water were carried out, showing that inorganic-organic membranes had a better performance than the analogue PES membranes. Even though further investigations are needed to improve the membrane-based As adsorption, this work demonstrated that kaolin can be efficiently used as a low cost natural material for the preparation of MMMs to be used for water treatment.

Author Contributions: Alberto Figoli, Abderrazak Bouzid and Tiziana Marino conceived and designed the experiments; Lina Rezzouk, Tiziana Marino and Francesca Russo performed the experiments; Tiziana Marino, Alberto Figoli and Francesca Russo analyzed the data; Alberto Figoli and Abderrazak Bouzid contributed reagents/materials/analysis tools; Tiziana Marino, Francesca Russo and Alberto Figoli wrote the paper.

Conflicts of Interest: The authors declare no conflict of interest.

References

1. Mandal, B.K.; Suzuki, T.K. Arsenic round the world—A review. *Talanta* **2002**, *58*, 201–235. [CrossRef]
2. Jain, C.K.; Ali, I. Arsenic: Occurence, toxicity and speciation techniques. *Water Res.* **2000**, *34*, 4304–4312. [CrossRef]
3. United States Environmental Protection Agency (EPA). *Technical Fact Sheet: Final Rule for Arsenic in Drinking Water*; 815-F-00-016; United States Environmental Protection Agency: Washington, DC, USA, January 2001.
4. Mukherjee, A.; Bhattacharya, P.; Savage, K.; Foster, A.; Bundschuh, J. Distribution of geogenic arsenic in hydrologic systems: Controls and challenges. *J. Contam. Hydrol.* **2008**, *99*, 1–7. [CrossRef] [PubMed]
5. Marino, T.; Figoli, A. Arsenic removal by liquid membranes. *Membranes* **2014**, *4*, 1–21. [CrossRef] [PubMed]
6. World Health Organization. *Guidelines for Drinking Water Quality*; World Health Organization: Geneva, Switzerland, 2006.
7. Criscuoli, A.; Carnevale, M. *Membrane Distillation for the Treatment of Waters Contaminated by Arsenic, Fluoride and Uranium in Membrane Technologies for Water Treatement: Removal of Toxic Trace Elements whit Emphasis on Arsenic, Fluoride and Uranium*; Figoli, A., Hoinkis, H., Bundschuh, J., Eds.; CRC Press: Boca Raton, FL, USA, 2016; Volume 13, pp. 238–252.
8. Norton, M.V.; Chang, Y.J.; Galeziewski, T.; Kommineni, S.; Chowdhury, Z. Throwawayiron and aluminium sorbents versus conventional activate dalumina for arsenic removal—Pilot testing results. In *Proceedings—Annual Conference*; Water Works Association: Washington, DC, USA, 2001; pp. 2073–2087.
9. Oehmen, A.; Viegas, R.; Velizarov, S.; Reis, M.A.M.; Crespo, J.G. Removal of heavy metals from drinking water supplie sthrough the ionex change membrane bioreactor. *Desalination* **2006**, *199*, 405–407. [CrossRef]
10. Mahmoudi, H.; Ghaffour, N.; Goosen, M. *Fluoride Arsenic and Uranium Removal from Water Using Adsorbent Materials and Inegrated Membrane Systems*; Figoli, A., Hoinkis, H., Bundschuh, J., Eds.; CRC Press: Boca Raton, FL, USA, 2016; Volume 13, pp. 91–113.
11. Mar, K.K.; Karnawati, D.; Putra Sarto, D.P.E.; Igarashi, T.; Tabelina, C.B. Comparison of arsenic adsorption on lignite, bentonite, shale, and ironsand from Indonesia. *Procedia Earth Planet. Sci.* **2013**, *6*, 242–250. [CrossRef]
12. Hua, J. Synthesis and characterization of bentonite base dinorgano–organo-composites and their performances for removing arsenic from water. *Appl. Clay Sci.* **2015**, *114*, 239–246. [CrossRef]
13. Basso, M.C.; Cerrella, E.G.; Cukierman, A.L. Empleo de algas marinas para la biosorcion de metales pesados de aguas contaminadas. *Avances en Energias Renovables y. Medio Ambiente* **2002**, *6*, 669–674.
14. Criscuoli, A.; Figoli, A.; Leopold, A.; Simone, S.; Benamor, M.; Drioli, E. Removal of As(V) by PVDF hollow fibers membrane contactors using Aliquat-336 as extractant. *Desalination* **2010**, *264*, 193–200.
15. Criscuoli, A.; Bafaro, P.; Drioli, E. Vacuum membrane distillation for purifying waters containing arsenic. *Desalination* **2013**, *323*, 17–21. [CrossRef]
16. Mondal, P.; Bhowmick, S.; Chatterjee, D.; Figoli, A.; Van der Bruggen, B. Remediation of inorganic arsenic in groundwater for safe water supply: A critical assessment of technological solutions. *Chemosphere* **2013**, *92*, 157–170. [CrossRef] [PubMed]
17. Gohari, R.J.; Lau, W.J.; Matsuura, T.; Ismail, A.F. Fabrication and characterization of novel PES/Fe–Mn binary oxide UF mixed matrix membrane for adsorptive removal of As(III) from contaminated water solution. *Sep. Purif. Technol.* **2013**, *118*, 64–72. [CrossRef]
18. Zheng, Y.-M.; Zhou, S.-W.; Nanayakkara, K.G.N.; Matsuura, T.; Paul Chen, J. Adsorptive removal of arsenic from aqueous solution by a PVDF/zirconia blend flat sheet membrane. *J. Membr. Sci.* **2011**, *374*, 1–11. [CrossRef]
19. Mohan, D.; Pittman, C.U. Arsenic removal from water/waste water using adsorbents—A critical review. *J. Hazard. Mater.* **2007**, *142*, 1–53. [CrossRef] [PubMed]
20. Dixit, S.; Hering, J.G. Comparison of arsenic(V) and arsenic(III) sorption onto iron oxide minerals: Implications for arsenic mobility. *Environ. Sci. Technol.* **2003**, *37*, 4182–4189. [CrossRef] [PubMed]

21. Guo, H.; Stuben, D.; Berner, Z. Removal of arsenic from aqueous solution by natural siderite and hematite. *Appl. Geochem.* **2007**, *22*, 1039–1051. [CrossRef]
22. Jönsson, J.; Sherman, D.M. Sorption of As(III) and As(V) to siderite, green rust (fougerite) and magnetite: Implications for arsenic release in anoxic ground waters. *Chem. Geol.* **2008**, *255*, 173–181. [CrossRef]
23. Lin, Z.; Pulse, R.W. Adsorption, desorption and oxidation of arsenic affected by clay minerals and aging process. *Environ. Geol.* **2000**, *39*, 753–759. [CrossRef]
24. Chandrasekhar, S.; Ramaswamy, S. Influence of mineral impurities on the properties of kaolin and its thermally treated products. *Appl. Clay Sci.* **2002**, *21*, 133–142. [CrossRef]
25. Hosseini, M.R.; Ahmadi, A. Biological beneficiation of Kaolin: A review on ironremoval. *Appl. Clay Sci.* **2015**, *107*, 238–245. [CrossRef]
26. PubChem Home Page. Available online: https://pubchem.ncbi.nlm.nih.gov/compound/kaolin (accessed on 11 August 2017).
27. De Mesquita, L.M.S.; Rodrigues, T.; Gomes, S.S. Bleaching of Brazilian kaolin using organic acids and fermented medium. *Miner. Eng.* **1996**, *9*, 965–971. [CrossRef]
28. Zen, S.; El Berrichi, F.Z. Adsorption of tannery anionic dyes by modified kaolin from aqueous solution. *Desalin. Water Treat.* **2014**, *52*, 1–9. [CrossRef]
29. Gürses, A.; Karaca, S.; Dogar, C.; Bayrak, R.; Acikyildiz, M.; Yalcin, M. Determination of adsorptive properties of clay/water system: Methylene blue sorption. *J. Colloid Int. Sci.* **2004**, *269*, 310–314. [CrossRef]
30. Sarbatly, R. Effect of kaolin/PESf ration and sintering temperature on pore size and porosity of the kaolin membrane support. *J. Appl. Sci.* **2011**, *11*, 2306–2312.
31. Han, L.F.; Xu, Z.L.; Cao, Y.; Wei, Y.M.; Xu, H.T. Preparation, characterization and permeation property of Al$_2$O$_3$, Al$_2$O$_3$–SiO$_2$ and Al$_2$O$_3$–kaolin hollow fiber membranes. *J. Membr. Sci.* **2011**, *372*, 154–164. [CrossRef]
32. Dizman, C.; Tasdelen, M.A.; Yagci, Y. Recent advances in the preparation of functionalized polysulfones. *Polym. Int.* **2013**, *62*, 991–1007. [CrossRef]
33. Mierzwa, J.C.; Arieta, V.; Verlage, M.; Carvalho, J.; Vecitis, C.D. Effect of clay nanoparticles on the structure and performance of polyethersulfone ultra filtration membranes. *Desalination* **2013**, *314*, 147–158. [CrossRef]
34. Yip, Y.; McHugh, A.J. Modeling and simulation of nonsolvent vapor-induced phase separation. *J. Membr. Sci.* **2006**, *271*, 163–176. [CrossRef]
35. Marino, T.; Blefari, S.; Di Nicolò, E.; Figoli, A. A more sustainable membrane preparation using triethyl phosphate as solvent. *Green Process. Synth.* **2017**, *6*, 295–300. [CrossRef]
36. Goetz, L.A.; Jalvo, B.; Rosal, R.; Mathew, A. Superhydrophilic anti-fouling electrospun cellulose acetate membranes coated with chitin nanocrystals for water filtration. *J. Membr. Sci.* **2016**, *510*, 238–248. [CrossRef]
37. Khare, V.P.; Greenberg, A.R.; Krantz, W.B. Vapor-induced phase separation—Effect of the humid air exposure step on membrane morphology: Part I. Insights from mathematical modeling. *J. Membr. Sci.* **2005**, *258*, 140–156. [CrossRef]
38. Chen, C.Y.; Lan, G.S.; Tuan, W.H. Preparation of mullite by the reaction sintering of kaolinite and alumina. *J. Eur. Ceram. Soc.* **2000**, *20*, 2519–2525. [CrossRef]
39. Susanto, H.; Stahra, N.; Ulbricht, M. High performance polyethersulfone microfiltration membranes having high flux and stable hydrophilic property. *J. Membr. Sci.* **2009**, *342*, 153–164. [CrossRef]
40. Ma, Y.; Shi, F.; Wang, Z.; Wu, M.; Ma, J.; Gao, C. Preparation and characterization of PSf/clay nanocomposite membranes with PEG 400 as a pore forming additive. *Desalination* **2012**, *286*, 131–137. [CrossRef]
41. Tan, P.C.; Low, S.C. Role of hygroscopic triethylene glycol and relative humidity in controlling morphology of polyethersulfone ultrafiltration membrane. *Desalin. Water Treat.* **2015**, *57*, 1–11. [CrossRef]
42. Keurentjes, J.T.F.; Harbrecht, J.G.; Brinkman, D.; Hanemaaijer, J.H.; Cohen Stuart, M.A.; van't Riet, H. Hydrophobicity measurements of MF and UF membranes. *J. Membr. Sci.* **1989**, *47*, 333–337. [CrossRef]
43. Palacio, L.; Calvo, J.I.; Prádanos, P.; Hernández, A.; Väisänen, P.; Nyström, M. Contact angles and external protein adsorption onto ultrafiltration membranes. *J. Membr. Sci.* **1999**, *152*, 189–201. [CrossRef]
44. Anadão, P.; Sato, L.F.; Wiebeck, H.; Valenzuela-Díaz, F.R. Montmorillonite as a component of polysulfone nanocomposite membranes. *Appl. Clay Sci.* **2010**, *48*, 127–132. [CrossRef]
45. Ghaemi, N.; Madaeni, S.S.; Alizadeh, A.; Rajabi, H.; Daraei, P. Preparation, characterization and performance of polyethersulfone/organically modified montmorillonite nanocomposite membranes in removal of pesticides. *J. Membr. Sci.* **2011**, *382*, 135–147. [CrossRef]

46. Monticelli, O.; Bottino, A.; Scandale, I.; Capanelli, G.; Russo, S. Preparation and propertiesof polysulfone-clay composite membranes. *J. Appl. Polym. Sci.* **2007**, *103*, 3637–3644. [CrossRef]

47. Han, M.-J.; Nam, S.-T. Thermodynamic and rheological variation in polysulfone solution by PVP and its effect in the preparation of phase inversionmembrane. *J. Membr. Sci.* **2002**, *202*, 55–61. [CrossRef]

48. Zen, S.; El. Berrichi, F.Z.; Abidi, N.; Duplay, J.; Jada, A.; Gasmi, B. Activated Algerian kaolin's as low-coast potential adsorbents for the removal from industrial effluents of Derma Blue R67 acid dye. Kinetic and thermodynamic studies. In Proceedings of the 5th International Conference on Sustainable Solid Waste Management, Athens, Greece, 21–24 June 2017.

49. Khalil, M.I.; Aly, A.A. Use of cationic starch derivatives for the removal of anionic dyes from textile effluents. *J. Appl. Polym. Sci.* **2004**, *93*, 227–234. [CrossRef]

50. Constantina, M.; Asmarandeia, I.; Harabagiua, V.; Ghimicia, L.; Ascenzib, P. Removal of anionic dyes from aqueous solutions by an ion-exchange based on pullulan microspheres. *Carbohyd. Polym.* **2013**, *91*, 74–84. [CrossRef] [PubMed]

membranes

MDPI

Article

Perfluorinated Compounds as Test Media for Porous Membranes

Juliana I. Clodt, Volkan Filiz and Sergey Shishatskiy *

Helmholtz-Zentrum Geesthacht, Institute of Polymer Research, Max-Planck-Str.1, 21502 Geesthacht, Germany;
Juliana.Clodt@hzg.de (J.I.C.); volkan.filiz@hzg.de (V.F.)
* Correspondence: sergey.shishatskiy@hzg.de; Tel.: +49-4152-87-2467

Received: 19 July 2017; Accepted: 29 August 2017; Published: 5 September 2017

Abstract: We suggest a failure-free method of porous membranes characterization that gives the researcher the opportunity to compare and characterize properties of any porous membrane. This proposal is supported by an investigation of eight membranes made of different organic and inorganic materials, with nine different perfluorinated compounds. It was found that aromatic compounds, perfluorobenzene, and perfluorotoluene, used in the current study show properties different from other perfluorinated aliphatics. They demonstrate extreme deviation from the general sequence indicating the existence of π-π-interaction on the pore wall. The divergence of the flow for cyclic compounds from ideal e.g., linear compounds can be an indication of the pore dimension.

Keywords: ultrafiltration; membrane characterization; perfluorinated compounds; porous membranes; permeance

1. Introduction

Ultrafiltration (UF) membranes have attracted the increasing interests of scientists during the last decades. In the meantime, they are common in several applications as waste-water treatment [1–3], biomolecule separation [1,4], and controlled drug release [5]. UF membranes can be made from organic and inorganic materials by different techniques suitable for the material nature including casting, phase inversion, track-etching, anodizing, sintering, and film-stretching as conventional examples. Since each membrane is limited in their separation performance [6–8], new methods for the formation of UF membranes, especially of isoporous membranes [9], will be still present on the agenda of membrane scientists.

For many years the only type of commercially available isoporous membranes was track-etched membranes. One success story to implement other methods to fabricate isoporous membranes starting ten years ago is the combination of self-assembly of amphiphilic block copolymers (S), and the non-solvent induced phase separation process (NIPS) called SNIPS [10]. With this method, UF membranes from various block copolymers can be prepared in a fast one-step process, leading to a new type of integral-asymmetric membrane with highly ordered, hexagonally arranged pores on the surface [11–14]. During the last few years, significant advances towards mass production of isoporous structures based on amphiphilic polymers has been achieved [15–17]. Nevertheless, the characterization of such membranes will cause problems if swellable polymer blocks are implemented and a standard solvent, e.g., water, will be used. In consequence, flux decline can occur as shown in Figure 1 [11,18–21].

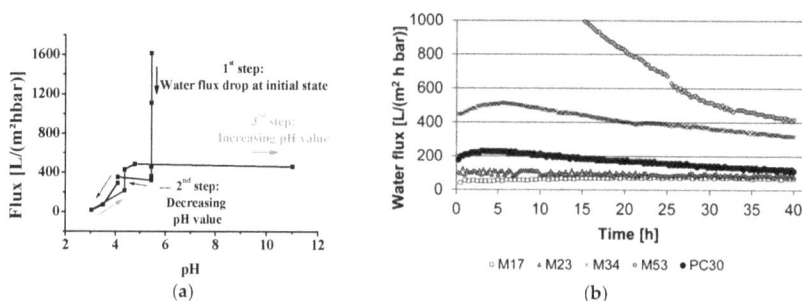

Figure 1. (a) pH-dependent water flux measurement of an isoporous block copolymer membrane with flux decline at initial state (water flux drop). (Reproduced from [11] with permission; Copyright 2012 Wiley-VCH Verlag GmbH & Co., KGaA). (b) Water flux measurements for membranes made from polystyrene-*block*-poly(4-vinylpyridine) with different pore sizes and with different flux decline. (Reproduced from [20] with permission; Copyright 2014 Royal Society of Chemistry).

Other problems for the characterization of UF membranes using water as liquid include fouling [22–24] and the unstable flow at different pressures [20]. The quality of water also plays an important role since deionized water made with different techniques, general tap water, or UV-treated water are used in different laboratories, dependent on the application and available facilities. The water flux values for UF membranes given in the literature are measured so far with several techniques, involving different: Pressures, time to stabilize the system until the value is measured, temperature, pre-treatment of the membranes in case of hydrophobic material, and so on.

In order to solve the aforementioned problems, one important question arises: How can we characterize UF membranes and get comparable data independent of the membranes' chemical composition? The probing liquid which is inert for as many membrane materials as possible is needed, as well as an easy to implement method of the membranes permeance measurement. From the variety of liquid compounds available on the market, perfluorinated (PF) compounds were found to be the most promising. PF compounds are known for their very limited interaction/affinity with/to substances not containing fluorine, offering low surface tension, high density, and the wide variation in dynamic viscosity. In this work, PF compounds were chosen as a test media for the characterization of UF membranes of various nature. The PF compounds were chosen to cover as big a range of viscosity as possible, and included linear, cyclic and aromatic compounds. As examples of commercial membranes, ceramic anodic alumina, polymeric track-etched, and standard ultrafiltration membranes with isotropic and anisotropic morphology were chosen.

The results of our experiments confirm that porous membranes of almost any nature can be tested using PF liquids, and membrane properties can be compared. The membranes do not swell during the experiments and are not altered by PF compounds.

2. Results and Discussion

In this work, we want to justify the method suitable for standard membrane characterization with examples of membranes of different nature, cross-sectional morphology, and membrane preparation method: Polymeric membranes prepared from homopolymers by the phase inversion process and resulting in a broad pore size distribution, two mostly isoporous membranes: Polymeric track-etched and inorganic anodic alumina and highly isoporous membrane prepared from amphiphilic diblock copolymer manufactured by SNIPS. The PF compounds used for the membrane characterization have a carbon backbone of different arrangements, which allows for investigating the dependence of the fluid flow through the pores on the shape of the fluid molecule.

The first section will be dedicated to the verification of the method of permeance measurements with a water sensitive membrane and PF hexane. Later, we apply the method for the characterization of different membranes with various PF compounds.

2.1. Verification of the Experimental Method for Sensitive Membranes

Since block copolymer membranes (BCPM) are the most sensitive of all membranes under investigation in this work, BCPM will be used to verify the method by using PF hexane for flux measurements. We chose one of the most common in-house made isoporous block copolymer membrane system, membranes made from polystyrene-*b*-poly(4-vinylpyridine) (PS-*b*-P4VP) with 4-vinylpyridine as a water swellable block.

Figure 2 depicts the permeance of PF hexane for a BCPM made from $PS_{82.8}$-*b*-$P4VP_{17.2}^{190k}$ diblock copolymer. One membrane stamp was measured three times in a row using the same PF hexane in order to evaluate the stability of the permeance which was found to be similar for all three measurements. Only a small deviation of less than 2% was found coming from the accuracy of the measurement collecting the data each second. Compared to water flux measurements, where big flux declines can be found at the beginning of the measurements (compare to Figure 1), this is an improvement for the characterization of performances of BCPM made from copolymers containing water swellable blocks. Deviation of the permeances at the beginning of the measurement may occur due to the wetting of the membrane and the instability of the pressure when the device is set on.

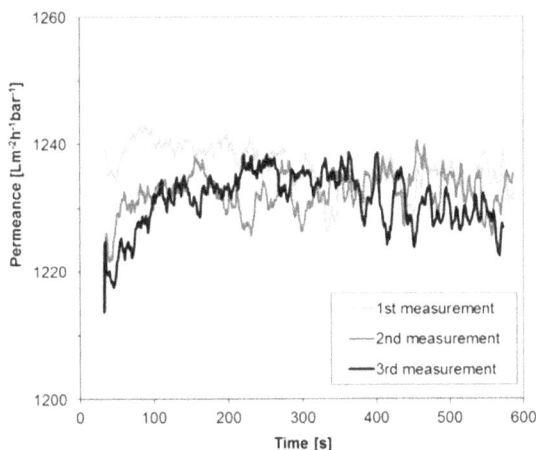

Figure 2. Permeance of PF hexane through the block copolymer membranes (BCPM) of $PS_{82.8}$-*b*-$P4VP_{17.2}^{190k}$.

For the purpose of examining the stability of permeances depending on the transmembrane pressure, measurements were carried out at different pressures between 0.5 and 2 bar, i.e., within the design pressure range of the flux measurement facility. Figure 3 depicts the flux of three different PF compounds through the BCPM examined for the permeance stability (Figure 2) as a function of a trans-membrane pressure. For three PF compounds chosen to cover the full range of dynamic viscosity the permeance increases linearly, with the pressure indicating that in the selected pressure range any pressure suitable for the measurement can be chosen in dependence on the liquid viscosity.

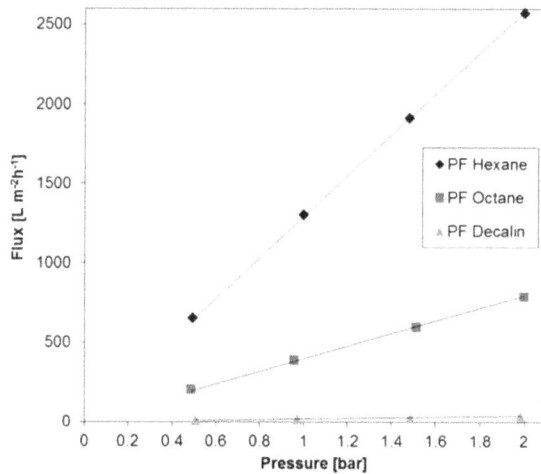

Figure 3. Flux through the BCPM of three perfluorinated (PF) compounds of various viscosity plotted against the trans-membrane pressure.

2.2. Permeance of Ultrafiltration Membranes

In general, the theoretical volume flow rates of membranes may be calculated by the Equation (1) of Hagen-Poiseuille for a laminar flow in a simple straight cylinder:

$$\text{Volume flow rate} \; = \; \frac{\pi r^4 \Delta p}{8 \eta L} \tag{1}$$

where r is the radius of the open pore, Δp is the transmembrane pressure, η is the dynamic viscosity of a liquid, and L the length of the cylindrical pore. Assuming that the r and L will be constant if one membrane sample is used for experiments with different liquids, and Δp is kept constant through the experiment, the volume flow rate normalized to the transmembrane pressure, here labeled as permeance, is proportional to the reciprocal dynamic viscosity (See Equation (2)):

$$\text{Permeance} \; \sim \; \frac{1}{\eta} \tag{2}$$

The permeances of different linear aliphatic, cyclic aliphatic, and aromatic PF compounds were studied for an anodic alumina membrane, Anodisc® (GE Healthcare, Chalfont St Giles, UK) with 20 nm pore diameter, and a UPZP05205 polyethylene membrane (Millipore, Billerica, MA, USA) in order to prove applicability of the Equation (2) for PF compounds in general. The results are depicted in Figure 4 and at first sight a linear trend can be seen for both types of membranes, as expected. On the other hand, a small difference in the flow of molecules with different shapes is apparent. The permeance of aromatic PF compounds is lower than those of the linear aliphatic compounds, and additionally, a discontinuity in the trend of the linear and cyclic compounds appears.

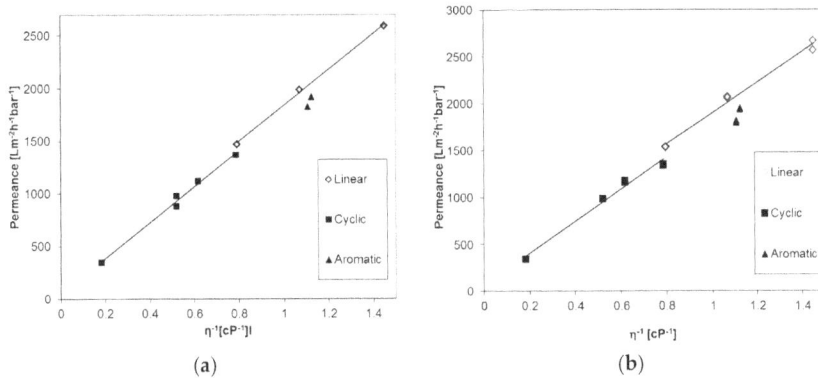

Figure 4. Permeances of Anodisc® membrane with 20 nm pore diameter (**a**) and PE membrane with 50 nm pore diameter (**b**) against reciprocal dynamic viscosity of linear, cyclic aliphatic and aromatic PF compounds.

In order to explore further the break in the Hagen-Poiseuille trend for permeances found for the compounds with different molecule shapes, three PC track-etched membranes with different pore sizes were investigated. Figure 5 summarizes the relative permeances of different PF compounds for the case of three PC track-etched membranes (pore diameters 30, 50 and 80 nm) in dependence of the reciprocal viscosity. The relative permeance is shown in the graph where the highest permeance of PF hexane for each membrane is set to unity, and the others are set in proportion to it. The following tendencies were observed: Linear aliphatic as well as cyclic aliphatic PF compounds indicate each a proportional relative permeance to the reciprocal viscosity, as expected from Equation (2). However, track-etched PC membranes show considerable differences of relative permeances of PF compounds depending on the shape of the PF molecule. The permeances of cyclic aliphatic PF compounds, and especially aromatic PF compounds, are significantly smaller than the permeances of linear PF aliphatics, indicating, most probably, that the formation of immobilized molecular layers on the pore walls. Linear compounds are more flexible, and additionally, can be more easily oriented in the direction of the liquid flow, while in the case of cyclic compounds, steric hindrance may play an additional role leading to more distinctive immobilization on the pore wall. Immobilization of the molecules on the pore wall can cause a significant change of the permeance since the pore diameter of the studied membranes is comparable to the characteristic size of the molecule [25]. This trend increases with the decreasing pore diameter of the PC track-etched membranes and could further be used to determine the pore size without microscopy studies. Very small or neglectable differences of relative permeances between cyclic and linear PF compounds one indicate that the pore sizes are below 30 nm for PC track-etched membranes. This behavior could be implemented for other membrane materials as well, especially membranes with isoporosity, probably dependent on the materials nature and substructure. Aromatic compounds undergo immobilization more readily than aliphatic compounds since they have a tendency of π-π-interaction, which can be induced by the shear force near (in molecular size terms) the pore wall where the flow velocity is minimal [25]. Since PC track-etched membranes are, in general, made from polycarbonates with an aromatic backbone, π-π-interaction with the pore wall itself will also occur especially for track-etched membranes containing straight cylindrical pores.

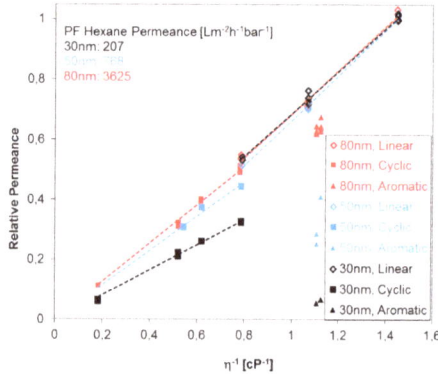

Figure 5. Relative permeance (dimensionless) of PC track-etched membranes with 30, 50 and 80 nm pore diameters against reciprocal dynamic viscosity of linear, cyclic, and aromatic PF compounds.

In order to investigate liquid permeances for phase inversion membranes with broader pore size distribution as compared to the membranes discussed before, a PAN (polyacrylonitrile) ultrafiltration anisotropic membrane was selected exemplarily, and the results are depicted in Figure 6. The permeance for aromatic PF compounds has the same trend as we discussed for PC track-etched and the Anodisc® membranes before. Interestingly, the gap between linear and cyclic PF compounds is lower than for the PC30, even though the PAN membrane has a smaller average pore diameter of 22 nm. For this type of membranes, the aforementioned influence of substructure may play an additional role for the immobilization of PF molecules on pore or substructural walls. The porosity of PAN membranes is much higher than for track-etched membranes, 12.4% as compared with 0.4% for PC30, and the PAN membranes have a wide pore size distribution that plays an important role [26].

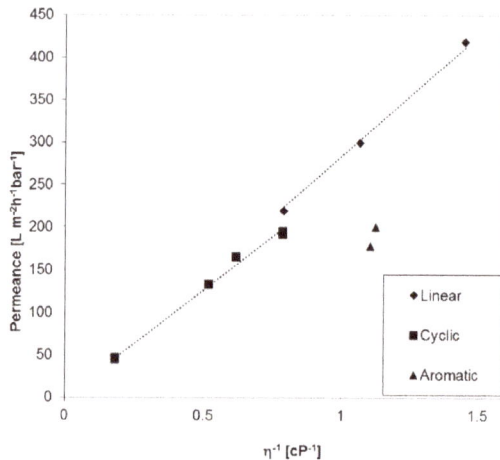

Figure 6. Permeance of a PAN membrane against reciprocal viscosity of linear, cyclic, and aromatic PF compounds.

The question of whether the observed behavior of aromatic, cyclic, and linear PF compounds can be found for the aforementioned BCPM arises. Therefore, the permeances were measured for the set of different PF compounds and water, as depicted in Figure 7. Each compound was measured at least two times in a row to ensure the reproducibility of the obtained results. Similar to the other polymeric membranes used in this work, the permeance of aromatic PF compounds is lower as compared to aliphatic compounds. Interestingly, the permeances of cyclic aliphatic are again lower when compared to linear aliphatic compounds, but in this case the corresponding slopes overlap around 0.8 cP^{-1}, and a steeper slope is observed for linear compounds as compared to the slope of cyclic compound without a gap of the trend line between linear and cyclic compounds. Furthermore, the trend for linear PF compounds unexpectedly does not approach zero at infinite dynamic viscosity, a fact to be examined further. On the one hand the average pore diameter of this BCPM is 42 nm, which should lead only to small differences in the pemeances of linear and cyclic compounds as observed for PC track-etched membranes discussed above. On the other hand, the substructure of BCPM plays an important role in their performances, as studied before [26]. Nevertheless, permeance measurements of PF compound could help to understand the behavior of BCPM, and should be studied in detail in future works depending on the pore sizes and substructure. In case of water, the flux decreases to about 30% already during the second measurement (compare first and second measurement in Figure 7), leading to further problems when the performance of such membranes is under investigation.

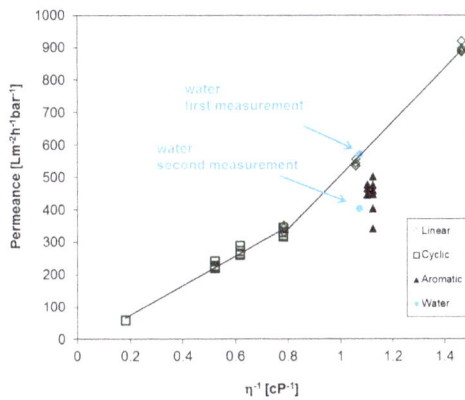

Figure 7. Permeances of BCPM made from PS$_{82.8}$-*b*-P4VP$_{17.2}$^{190k} against reciprocal viscosity of linear, cyclic, and aromatic PF compounds, respectively, and water. Each point represents the average value for one experiment with a liquid.

In order to compare the experimental and theoretical fluxes, the theoretical fluxes were calculated according to Hagen-Poiseulle's law as follows, taking the porosities of the membranes into account (See Equation (2)):

$$\text{Theoretical water fluxes} = \frac{\pi r^4 \Delta p \text{Porosity}}{8\eta LA\Delta p} = \frac{r^2 \text{Porosity}}{8\eta L} \tag{3}$$

where A is the area of the pore corresponding $A = \pi r^2$, L the length of the cylindrical pore, and the average surface porosity values were used as listed in Table 1. The tortuosity of the membrane is not taken into account since it is impossible to obtain the reliable value from any source of information either flow experiment or microscopy. Theoretical fluxes were calculated according to Equation (3) for the track-etched membranes, PC30, PC50, and PC80, which should have straight cylindrical pores through the whole membrane thickness. Surprisingly, theoretical permeances for PF hexane are around ten times lower than the experimental fluxes, namely theoretical fluxes 350, 80, and 10 L/m^{-2}·h^{-1}·bar^{-1} as compared to the experimental fluxes 3685, 775, and 211 L/m^{-2}·h^{-1}·bar^{-1} for PC80, PC50, and PC30. The calculation was based on the pore diameter of the entrance of the pore given by the supplier, and with the assumption that track-etched membranes have straight cylindrical pores perpendicular to the membrane surface through the whole thickness of the membrane. For track-etched membranes, the surface pore size is smaller than in the bulk of a membrane [27].

On the other hand, double and multiple pores having an origin in the overlapping of tracks of particles used for the polymer bombardment are often found on the membrane surface of the track-etched membranes [28]. Both of the facts are leading to higher experimental fluxes than the calculated one. For example, for the PC50 membrane, the theoretical, and experimental permeances would match each other when the average pore diameter is 88 nm.

Table 1. Membranes and geometrical features used in this work.

Membrane	Abbr. Used in This Work	Average Pore Diameter (nm)	Porosity (%)
Polycarbonate track-etched PCN8CP04700	PC80	80 [a]	2.0 [a]
Polycarbonate track-etched PCN5CP04700	PC50	50 [a]	1.2 [a]
Polycarbonate track-etched PCN3CP04700	PC30	30 [a]	0.4 [a]
Polyethylene UPZP05205	PE50	50 [a]	– [c]
Anodized alumina Anodisc®	Anodisc	20 [a]	40 [b]
Polyacrylonitrile	PAN	22 [b]	12.3 [b]
Block copolymer membranes	BCPM	42 [b]	29.5 [b]

[a] as stated by supplier; [b] calculated from SEM images as shown in previous work [12,29–31]; [c] accurate surface porosity is not possible to acquire from the SEM image due to very complex membrane morphology.

3. Materials and Methods

3.1. Membranes Used in This Work

Polyacrylonitrile membranes were prepared by the phase inversion process [32]. Commercial polycarbonate track-etched membranes were purchased from Pieper (Germany): Supplier number PCN8CP04700, pore size 80 nm, thickness 6 µm, typical water flow rate 2 mL/(cm^2·min) at 0.67 bar (in this work labelled as PC80); supplier number PCN5CP04700, pore size 50 nm, thickness 6 µm, typical water flow rate 1 mL/(cm^2·min) at 0.67 bar (in this work labelled as PC50); supplier number PCN3CP04700 pore size 30 nm, thickness 6 µm, typical water flow rate 0.2 mL/(cm^2·min) at 0.67 bar (in this work labelled as PC30); and, data as stated by the supplier. Polyethylene (PE) membranes, supplier number UPZP05205, pore size 50 nm were purchased from Millipore. An anodized alumina membrane Anodisc® circle with support ring, 47 mm, 0.02 µm pore size was purchased from GE Healthcare GmbH. Isoporous diblock copolymer membranes (BCPM) used in this work were prepared according to a procedure published by Rangou et al. [12]. All membranes, their abbreviation used in this work, and their geometrical features are listed in Table 1.

3.2. Perfluorinated Compounds Used in This Work and Their Physical Data

The PF compounds used for the membrane performance measurements and their physical properties are listed in the Table 2. PF compounds were selected in order to analyze three different groups: Linear aliphatic compounds: Perfluorohexane, hexadecafluoroheptane, perfluorooctane; PF cyclic aliphatic compounds: Perfluoro(methylcyclohexane), perfluoro-1,3-dimethylcyclohexane, perfluorocycloether labeled as FC-77; and, aromatic compounds: Perfluorobenzene and perfluorotoluene. In order to compare the measurement data of the PF compounds with water, commonly used for flux measurements, demineralized water with an electrical conductivity of ≈0.055 µS·cm^{-1} was employed.

Dynamic viscosity was calculated from kinematic viscosity of the PF compounds, measured using a Lauda iVisc Viscometer Version 1.01 at 23 ± 1 °C in agreement with the temperature of the permeance measurement.

Table 2. Liquids used in this work and their physical data.

Substance	Abbreviation	$M_w{}^a$ (g/mol)	Density ^{a} (g/cm^{-3})	BP ^{a} (°C)	Dyn. Viscosity (cP) at 23 °C
Perfluorohexane	PF Hexane	339.04	1.686	59	0.690
Hexadecafluoroheptane	PF Heptane	388.05	1.731	83	0.936
Perfluorooctane	PF Octane	438.06	1.757	103	1.27
Perfluoro(methylcyclohexane)	PF MCH	350.05	1.784	76	1.62
Perfluoro-1,3-imethylcyclohexane	PF DMCH	400.06	1.838	101	1.93
Perfluorodecalin	PF Decalin	462.08	1.926	142	5.50
Perfluorocycloether	FC-77	416	1.767	97	1.27
Perfluorobenzene	PF Benzene	186.05	1.613	81	0.889
Perfluorotoluene	PF Toluene	236.06	1.664	104	0.903
Water	Water	18.02	0.995	100	0.932

^{a} The data as available by material safety data sheet or stated by a supplier.

3.3. Liquid Permeance Measurement

Liquid permeance measurements were performed in a simple dead-end mode at room temperature using an in-house designed and manufactured testing facility, as schematically shown in Figure 8. A Millipore inline stainless steel filter with 20 mm sample diameter was used as a membrane sample holder. Nitrogen at relative pressures from 0.5 to 2.5 bar was used to create a driving force. The liquid flow rate through the membrane was measured gravimetrically every second by weight acquisition from the Mettler Toledo NewClassic MF MS1003S precision balance with 0.001 g accuracy. In our case, all data including pressure, temperature, and weight was transmitted to a computer for an easy evaluation.

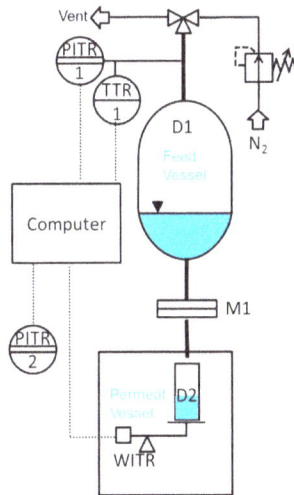

Figure 8. Scheme of the measurement facility for determination of liquid permeance for porous membranes.

The permeance (P) was calculated by normalizing the flux by the trans-membrane pressure (See Equation (4)):

$$P = \frac{\Delta V}{A \, \Delta t \, \Delta p} \qquad (4)$$

where ΔV is the volume of PF compound collected between two mass measurements, A is the membrane surface area, Δt is the time between two mass measurements, and Δp is the trans-membrane pressure.

The flux (*J*) through the membrane was calculated according to the Equation (5):

$$J = \frac{\Delta V}{A\,\Delta t} \tag{5}$$

The trans-membrane pressure was varied in the range 0.5 to 2.5 bar depending on the PF liquid viscosity in order to ensure the collection of the experimental data during similar time intervals for all liquids. The independence of the flux on the trans-membrane pressure was studied and will be described further.

4. Conclusions

In this work we examined the use of PF compounds as possible universal media for the characterization of ultrafiltration membranes in order to avoid e.g., swelling of membranes in water, as discussed before [11,20]. Therefore, PF compounds of different molecular shape and electronic nature were studied. The following results were observed:

1. PF compounds can be used to study the permeances of sensitive membranes made from water or other solvent swellable materials.
2. Pressure dependent measurements show that the permeances of PF compounds are stable in the pressure difference range 0.5 to 2 bar, and the flux does not change significantly with time.
3. All studied PF aliphatic compounds, both linear and cyclic, show permeance gradually changing with the viscosity in full agreement with the Hagen–Poiseuille equation.
4. The Hagen–Poiseuille trends for linear and cyclic aliphatic compounds deviated increasingly with decreasing average pore diameter: The slope for the cyclic compounds is smaller than that for linear compounds and this observation may possibly be used for estimating the pore size of the membrane without microscopic study each time.
5. The permeances of aromatic PF compounds through ultrafiltration membranes made of materials of different nature are lower than the fluxes of aliphatic compounds presumably due to supramolecular interaction between aromatic PF molecules and pore walls.
6. Taking into account the considerations discussed above, one can conclude that most suitable PF compounds for the membrane characterization are perfluorooctane and FC-77 by 3M Company (St. Paul, MN, USA). These two compounds have very similar density and viscosity, while one is linear aliphatic and another cyclic ether.

Acknowledgments: The authors thank Kristian Buhr, Janina Hahn, Adina Jung and Jan Wind for preparing the PAN and BCP membranes; Janina Hahn and Kristian Buhr for their help with the permeance measurements; Jan Pohlmann for data acquisition program; Volker Abetz, Thomas Bucher, Torsten Brinkmann and Sofia Rangou for scientific discussion.

Author Contributions: The authors have contributed equally to the experimental work and to the manuscript. The work load was distributed: Juliana I. Clodt: BCPM membrane preparation, liquid flux measurements, data analysis, manuscript preparation; Volkan Filiz: BCPM membrane preparation, data analysis, manuscript preparation; Sergey Shishatskiy: Commercial membranes choice purchase; liquid flux measurements, experimental data analysis, manuscript preparation.

Conflicts of Interest: The authors declare no conflict of interest.

References

1. Van Reis, R.; Zydney, A. Membrane separations in biotechnology. *Curr. Opin. Biotechnol.* **2001**, *12*, 208–211. [CrossRef]
2. Savage, N.; Diallo, M. Nanomaterials and water purification: Opportunities and challenges. *J. Nanopart. Res.* **2005**, *7*, 331–342. [CrossRef]
3. Van Der Bruggen, B.; Vandecasteele, C.; Van Gestel, T.; Doyen, W.; Leysen, R. A review of pressure-driven membrane processes in wastewater treatment and drinking water production. *Environ. Prog.* **2003**, *22*, 46–56. [CrossRef]

4.   Saxena, A.; Tripathi, B.P.; Kumar, M.; Shahi, V.K. Membrane-based techniques for the separation and purification of proteins: An overview. *Adv. Colloid Interface Sci.* **2009**, *145*, 1–22. [CrossRef] [PubMed]

5.   Jeon, G.; Yang, S.Y.; Kim, J.K. Functional nanoporous membranes for drug delivery. *J. Mater. Chem.* **2012**, *22*, 14814–14834. [CrossRef]

6.   Stroeve, P.; Ileri, N. Biotechnical and other applications of nanoporous membranes. *Trends Biotechnol.* **2011**, *29*, 259–266. [CrossRef] [PubMed]

7.   Clark, W.M.; Bansal, A.; Sontakke, M.; Ma, Y.H. Protein adsorption and fouling in ceramic ultrafiltration membranes. *J. Membr. Sci.* **1991**, *55*, 21–38. [CrossRef]

8.   Robertson, B.C.; Zydney, A.L. Protein adsorption in asymmetric ultrafiltration membranes with highly constricted pores. *J. Colloid Interface Sci.* **1990**, *134*, 563–575. [CrossRef]

9.   Freeman, B. Not all membrane pores are made equal; some are more equal than others. *Nature* **2008**, *454*, 671. [CrossRef] [PubMed]

10. Peinemann, K.V.; Abetz, V.; Simon, P.F. Asymmetric superstructure formed in a block copolymer via phase separation. *Nat. Mater.* **2007**, *6*, 992–996. [CrossRef] [PubMed]

11. Jung, A.; Rangou, S.; Abetz, C.; Filiz, V.; Abetz, V. Structure formation of integral asymmetric composite membranes of polystyrene-*block*-poly(2-vinylpyridine) on a nonwoven. *Macromol. Mater. Eng.* **2012**, *297*, 790–798. [CrossRef]

12. Rangou, S.; Buhr, K.; Filiz, V.; Clodt, J.I.; Lademann, B.; Hahn, J.; Jung, A.; Abetz, V. Self-organized isoporous membranes with tailored pore sizes. *J. Membr. Sci.* **2014**, *451*, 266–275. [CrossRef]

13. Abetz, V. Isoporous block copolymer membranes. *Macromol. Rapid Commun.* **2015**, *36*, 10–22. [CrossRef] [PubMed]

14. Höhme, C.; Hahn, J.; Lademann, B.; Meyer, A.; Bajer, B.; Abetz, C.; Filiz, V.; Abetz, V. Formation of high thermally stable isoporous integral asymmetric block copolymer membranes. *Eur. Polym. J.* **2016**, *85*, 72–81. [CrossRef]

15. Hahn, J.; Clodt, J.I.; Abetz, C.; Filiz, V.; Abetz, V. Thin isoporous block copolymer membranes: It is all about the process. *ACS Appl. Mater. Interfaces* **2015**, *7*, 21130–21137. [CrossRef] [PubMed]

16. Radjabian, M.; Abetz, C.; Fischer, B.; Meyer, A.; Abetz, V. Influence of solvent on the structure of an amphiphilic block copolymer in solution and in formation of an integral asymmetric membrane. *ACS Appl. Mater. Interfaces* **2017**. [CrossRef] [PubMed]

17. Radjabian, M.; Abetz, V. Tailored pore sizes in integral asymmetric membranes formed by blends of block copolymers. *Adv. Mater.* **2015**, *27*, 352–355. [CrossRef] [PubMed]

18. Hahn, J.; Filiz, V.; Rangou, S.; Clodt, J.; Jung, A.; Buhr, K.; Abetz, C.; Abetz, V. Structure formation of integral-asymmetric membranes of polystyrene-*block*-poly(ethylene oxide). *J. Polym. Sci. Polym. Phys.* **2013**, *51*, 281–290. [CrossRef]

19. Hahn, J.; Filiz, V.; Rangou, S.; Lademann, B.; Buhr, K.; Clodt, J.I.; Jung, A.; Abetz, C.; Abetz, V. PtBS-*b*-P4VP and PTMASS-*b*- P4VP isoporous integral-asymmetric membranes with high thermal and chemical stability. *Macromol. Mater. Eng.* **2013**, *298*, 1315–1321. [CrossRef]

20. Hahn, J.; Clodt, J.I.; Filiz, V.; Abetz, V. Protein separation performance of self-assembled block copolymer membranes. *RSC Adv.* **2014**, *4*, 10252–10260. [CrossRef]

21. Shukla, R.; Cheryan, M. Performance of ultrafiltration membranes in ethanol-water solutions: Effect of membrane conditioning. *J. Membr. Sci.* **2002**, *198*, 75–85. [CrossRef]

22. Miller, D.J.; Dreyer, D.R.; Bielawski, C.W.; Paul, D.R.; Freeman, B.D. Surface modification of water purification membranes. *Angew. Chem. Int. Ed. Engl.* **2017**, *56*, 4662–4711. [CrossRef] [PubMed]

23. Sun, W.; Liu, J.; Chu, H.; Dong, B. Pretreatment and membrane hydrophilic modification to reduce membrane fouling. *Membranes* **2013**, *3*, 226–241. [CrossRef] [PubMed]

24. Li, X.; Mo, Y.; Li, J.; Guo, W.; Ngo, H.H. In-situ monitoring techniques for membrane fouling and local filtration characteristics in hollow fiber membrane processes: A critical review. *J. Membr. Sci.* **2017**, *528*, 187–200. [CrossRef]

25. Bitsanis, I.; Somers, S.A.; Davis, H.T.; Tirrell, M. Microscopic dynamics of flow in molecularly narrow pores. *J. Chem. Phys.* **1990**, *93*, 3427–3431. [CrossRef]

26. Clodt, J.I.; Bajer, B.; Buhr, K.; Hahn, J.; Filiz, V.; Abetz, V. Performance study of isoporous membranes with tailored pore sizes. *J. Membr. Sci.* **2015**, *495*, 334–340. [CrossRef]

27. Hernandez, A.; Calvo, J.I.; Pradanos, P.; Tejerina, F. Pore size distributions of track-etched membranes; comparison of surface and bulk porosities. *Colloid Surface A* **1998**, *138*, 391–401. [CrossRef]
28. Calvo, J.I.; Hernández, A.; Caruana, G.; Martínez, L. Pore size distributions in microporous membranes. *J. Colloid Interface Sci.* **1995**, *175*, 138–150. [CrossRef]
29. Grünauer, J.; Filiz, V.; Shishatskiy, S.; Abetz, C.; Abetz, V. Scalable application of thin film coating techniques for supported liquid membranes for gas separation made from ionic liquids. *J. Membr. Sci.* **2016**, *518*, 178–191. [CrossRef]
30. Grünauer, J.; Shishatskiy, S.; Abetz, C.; Abetz, V.; Filiz, V. Ionic liquids supported by isoporous membranes for CO_2/N_2 gas separation applications. *J. Membr. Sci.* **2015**, *494*, 224–233. [CrossRef]
31. Abetz, V.; Brinkmann, T.; Dijkstra, M.; Ebert, K.; Fritsch, D.; Ohlrogge, K.; Paul, D.; Peinemann, K.V.; Pereira-Nunes, S.; Scharnagl, N.; et al. Developments in membrane research: From material via process design to industrial application. *Adv. Eng. Mater.* **2006**, *8*, 328–358. [CrossRef]
32. Scharnagl, N.; Buschatz, H. Polyacrylonitrile (pan) membranes for ultra- and microfiltration. *Desalination* **2001**, *139*, 191–198. [CrossRef]

membranes

MDPI

Article

Preparation of Porous Stainless Steel Hollow-Fibers through Multi-Modal Particle Size Sintering towards Pore Engineering

Francois-Marie Allioux [1,*], Oana David [2], Miren Etxeberria Benavides [2], Lingxue Kong [1], David Alfredo Pacheco Tanaka [2] and Ludovic F. Dumée [1]

[1]   Deakin University, Institute for Frontier Materials, Geelong, VIC 3216, Australia; lingxue.kong@deakin.edu.au (L.K.); ludovic.dumee@deakin.edu.au (L.F.D.)
[2]   Tecnalia, Energy and Environment Division, Mikeletegi Pasealekua 2, 20009 San Sebastian-Donostia, Spain; oana.david@tecnalia.com (O.D.); miren.etxeberria@tecnalia.com (M.E.B.); alfredo.pacheco@tecnalia.com (D.A.P.T.)
*   Correspondence: f.allioux@research.deakin.edu.au; Tel.: +61-479-191-078

Received: 31 May 2017; Accepted: 31 July 2017; Published: 4 August 2017

Abstract: The sintering of metal powders is an efficient and versatile technique to fabricate porous metal elements such as filters, diffusers, and membranes. Neck formation between particles is, however, critical to tune the porosity and optimize mass transfer in order to minimize the densification process. In this work, macro-porous stainless steel (SS) hollow-fibers (HFs) were fabricated by the extrusion and sintering of a dope comprised, for the first time, of a bimodal mixture of SS powders. The SS particles of different sizes and shapes were mixed to increase the neck formation between the particles and control the densification process of the structure during sintering. The sintered HFs from particles of two different sizes were shown to be more mechanically stable at lower sintering temperature due to the increased neck area of the small particles sintered to the large ones. In addition, the sintered HFs made from particles of 10 and 44 μm showed a smaller average pore size (<1 μm) as compared to the micron-size pores of sintered HFs made from particles of 10 μm only and those of 10 and 20 μm. The novel HFs could be used in a range of applications, from filtration modules to electrochemical membrane reactors.

Keywords: porous stainless steel hollow-fiber; metal membrane; multi-modal distributions; coalescence; neck formation

1. Introduction

Membrane separation processes are well-established alternatives to conventional separation processes and physico-chemical treatments, offering lower energy consumption, compact, and scalable module systems [1]. The unique ability of membrane processes to treat chemically and heat sensitive effluents is of particular importance in the food, pharmaceutical, and biofuel industries [2,3]. Porous membrane elements are designed into a variety of shapes and geometries [4]. Among them, hollow-fiber (HF) membranes, which are self-supported structures, offer specific advantages, such as a higher surface to volume ratio and a uniform flow dynamic across the length of the HFs [2,5,6].

The first porous HFs were made of polymer materials, and were developed almost 60 years ago [7]. Since then, and with the use of novel polymeric materials, polymeric HF membranes have found extensive use in industrial gas separation and water treatment, and in the biotechnology field as well as in the medical sector for blood filtration and treatment . Polymeric HF membranes are produced at low cost and in large quantities via a dry–wet spinning process, and can be assembled as one module, which can be composed of thousands of polymeric HFs [3,8]. However, polymeric-based membranes typically exhibit lower tolerance to suspended abrasive particles, chemical cleaning, and steam sterilization procedures as compared to inorganic-based membranes [9]. The development of membranes with a high tolerance to cleaning, disinfection, and sterilization procedures could therefore drastically increase the reliability and life expectancy of such separation systems [3,10].

In a context where processes must achieve high durability and reusability, inorganic-based membranes offer unique advantages, such as high mechanical and thermal resistance, as well as increased stability in harsh chemical and abrasive environments [11,12]. In particular, metal porous HFs exhibit higher mechanical strength and ductility as opposed to the more brittle ceramic and carbon porous HFs [13]. A number of metal and alloy-based materials are also particularly resistant to oxidative chemical and thermal cleaning procedures, providing outstanding reusability characteristics [12,14]. To date, most of the work on the development of metal HFs has been focused on corrosion-resistant metals, such as stainless steel (SS) or titanium alloys, for separation applications [5,15,16]. However, porous HFs made of highly electrically conductive and electro-catalytic materials, such as nickel and copper (outer diameter between 250 and 700 µm), were also prepared and used in electro-membrane reactors for the electrochemical reduction of carbon dioxide or for methane reforming process [8,17,18]. In addition, porous HFs can be used as supports for selective thin films such as palladium–silver (Pd–Ag) materials for water gas shift reactions [19]. Inorganic metal HF membranes (outer diameter ≤ 3 mm) were successfully prepared using different fabrication processes, among which the dry-wet spinning process is commonly preferred [20,21]. The fabrication of inorganic HF membranes is a process with multiple steps whereby a high viscosity dope containing up to 80 wt % of suspended inorganic particles is spun [5,18,20]. Then, the extruded, or so-called green HFs, undergo a series of thermal treatments to remove the polymer binder and simultaneously start the sintering process between metal particles. The key properties of HF membranes and filters are their porosity and mechanical strength, which are directly related to the number, relative density, and volume of necks generated between the particles [22,23]. The stages of the sintering process include: (i) initial neck formation, (ii) growth at the contact point between the particles, and (iii) densification of the structure and ultimately full coalescence of the particles . While the enlargement of the neck area promotes the mechanical strength of the material, the open porosity is also reduced, limiting the materials' performance in separation [24]. One strategy to address this issue is therefore to increase the initial number of contact points per particle in order to form a higher density of small necks during the initial sintering stage. The density of the necks can typically be enhanced by using multi-modal particle distributions in order to increase the packing density of the green material, which is highly dependent on the shape and size distribution of the particles [25]. However, the sintering behaviour and kinetics of the mixed powder matrix will differ from conventional sintering models, where mono-sized spherical particles are normally used. New model developments are therefore required in this area of research [25].

In this study, porous SS HF membranes were prepared by the dry-wet spinning process. The green HFs were composed of a mixture of two powders of different sizes in order to control the densification rate during the sintering process and develop ordered porous structures. The impact of the spinning dope composition and sintering conditions on the final HF morphologies, pore sizes, and mechanical and electrical properties were systematically studied. Pure water permeation experiments were also performed in a dead-end filtration module, and the results were compared to sintered SS HFs obtained from single-sized particles. This work opens the door to the development of new types of inorganic hierarchical porous HF structures with enhanced properties for separation applications, where a better compromise between porosity and mechanical strength is achieved.

2. Results and Discussion

2.1. Metals Powders and Polymer Binder Characterisation

The studies of the shape, roughness, and density of the powders are important steps, since such characteristics will influence the mechanical properties and porosity of the final sintered materials [26].

The 10 and 20 µm 316L stainless steel particles, SS_{10} and SS_{20} respectively, exhibited smooth surfaces with spheroidal shapes and an aspect ratio of 1.09 ± 0.05 and 1.13 ± 0.1, respectively. On the other hand, the shape distribution of the 44 µm 316L stainless steel particles (SS_{44}) was irregular (aspect ratio of 2.17 ± 0.3) and showed greater surface roughness. The particle size distribution of the SS powders and blends was measured in distilled (DI) water and isopropanol (IP) to evaluate potential aggregation mechanisms in the different dispersants. The distributions in both solvents were similar; the SS_{10} powders showed a narrower size distribution as compared to the SS_{20} and SS_{44} particles, which were 40% and 20% wider, respectively. The use of water resulted in little or no clumping of the particles as seen in Figure 1d–f. The information on particle size distributions is shown in Table S1 for the calculated equivalent spherical diameters. The particle size distributions of the SS_{44} particles were found to be almost twice larger than the specifications of the manufacturer, with 90% of the distribution in volume of the particles (d0.9) lying below 72 ± 1.1 µm in water. The surface area of the powder samples was measured using Brunauer-Emmett-Teller (BET) analysis, the data are reported in Table S2. The BET surface areas of the SS_{10} and SS_{20} particles were similar; 0.034 and 0.036 $m^2 \cdot g^{-1}$, respectively. However, the BET surface area of the SS_{44} particles was found to be greater, 0.052 $m^2 \cdot g^{-1}$, which might be due to the textured surface of the SS_{44} particles. The weight ratio of the SS particle blends can therefore be expressed in terms of surface area ratio, corresponding to 72:28 (%). and 92:8 (%) for the SS_{10}:SS_{20} and SS_{10}:SS_{44} blends, respectively. Furthermore, the particle size distributions of the SS powder blends (Figure 2c,d) indicated that the particles did not aggregate in the solutions. The SS_{10} and SS_{44} powders exhibited similar average absolute and relative densities. However, the SS_{20} powders showed a higher relative density due to the wider particle size distribution (Table S2).

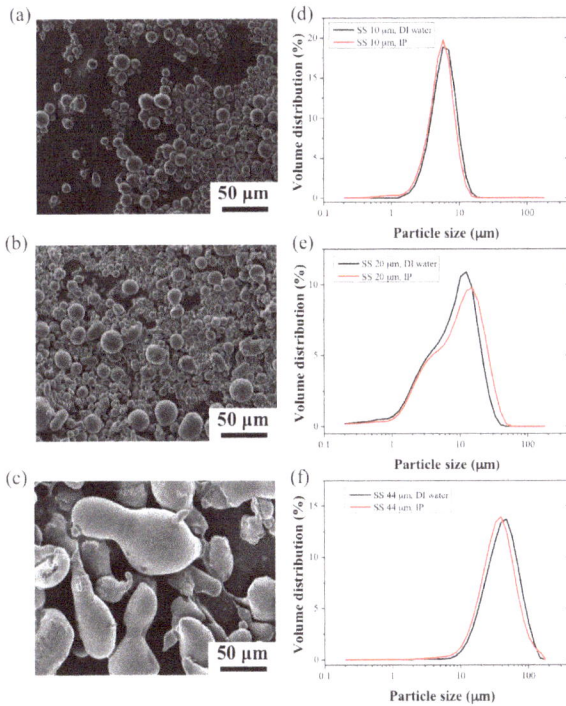

Figure 1. SEM images of (**a**): the 10 µm stainless steel (SS) powder, (**b**): the 20 µm SS powder and (**c**): the 44 µm SS powder. (**d–f**) are their respective size measurements in distilled (DI) water and isopropanol (IP).

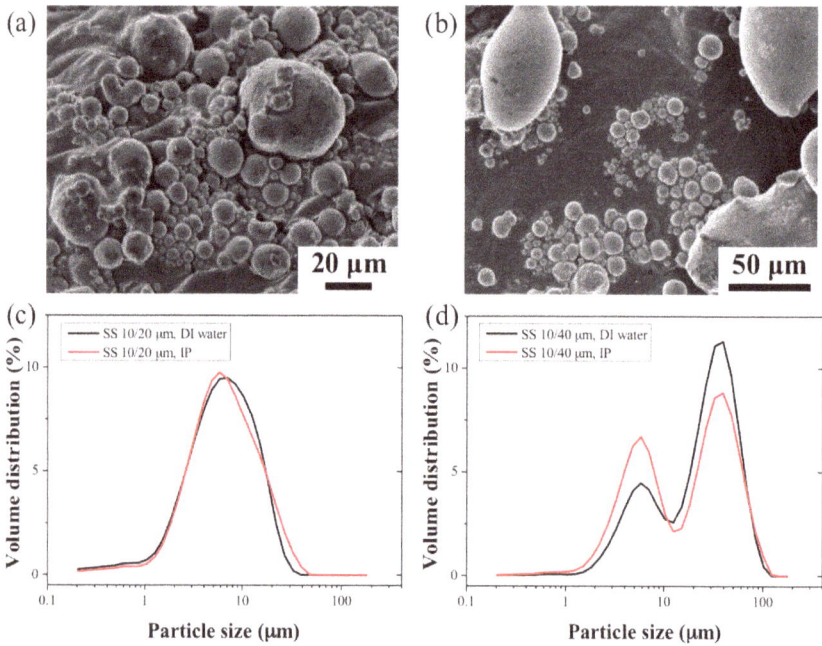

Figure 2. SEM images of (**a**): the 10/20 μm SS powder blend and (**b**): the 10/44 μm SS powder blend. (**c**,**d**) are the respective size measurements of the powder blends in DI water and isopropanol (IP).

During the sintering process, the bonding of the metal particles occurred through the formation of a sintered neck area between the particles. However, sufficient neck formation required the reduction of the surface oxides during the early stages of the sintering [27]. In order to study the oxidation behavior of the SS particles at high temperature, thermogravimetry (TGA) experiments were performed, and consisted of heating a small bed of metal particles (10 mg) to 1000 °C with a heating rate of 10 °C min^{-1} in air, N_2, and N_2:H_2 (95:5 *v/v* %) (Figure 3a–c). These gases were chosen in order to mimic the environment of the furnace during the sintering of the green HFs. In particular, SS alloys become susceptible to corrosion and oxidation when exposed to a high temperature, typically referred to as the sensitization temperature [28,29]. In air and N_2 atmospheres, the sensitization temperatures of the SS_{10} and SS_{20} were found to be 310 °C and 500 °C, respectively, and as high as 650 °C for the SS_{44}. The TGA curves of the SS_{44} in air revealed the presence of residual organic contaminants, probably resulting from the fabrication process and associated with the slight weight loss from 200 to 400 °C (Figure 3c).

The weight gains of the three powders were found to occur at higher temperatures in reducing N_2:H_2 atmosphere with sensitization temperatures above 650 °C for the SS_{10} and SS_{20} particles and above 750 °C for the SS_{44} particles. The finest SS powders (SS_{10} and SS_{20}) showed the greatest weight gain upon reaching the 1000 °C mark due to the higher surface area to volume ratio of the particles in comparison to the larger SS_{44} particles. The SS_{20} and SS_{44} powders were found to be more prone to oxidation as compared to the SS_{10} powder due to their as-received oxidized surface state, which was analysed by EDS analysis. The oxygen content was 1.7 and 1.5 ± 0.1 wt % for the SS_{20} and SS_{44} powders, respectively, and was below the detection limit for the SS_{10} powder. Similarly, the carbon content of the SS_{20} and SS_{44} powders was found to be higher; 7.4 and 8.5 ± 0.3 wt %, respectively. Furthermore, the high carbon content of the SS_{44} particles could also explain the lower weight gain at high temperature. The weight gain was associated with the formation and growth of stable surface metal oxide compounds across their surface [30,31]. The weight gain of the particles observed in N_2

and N_2:H_2 atmospheres was due to the uptake of nitrogen atoms by the SS materials, leading to the formation of nitride species [32–34]. The kinetics of oxide formation and growth were shown to be greatly increased in air; however, the presence of oxygen did not seem to influence the sensitization temperature of the different particles.

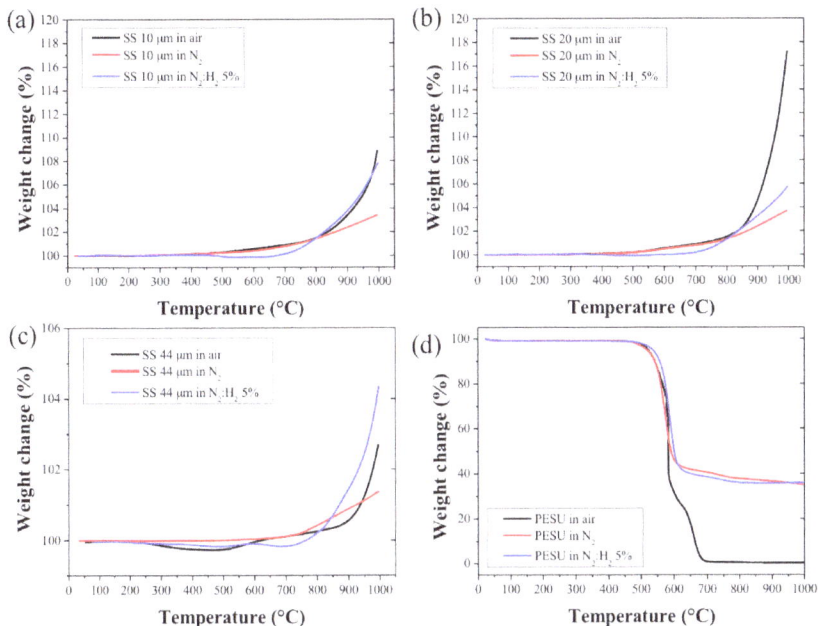

Figure 3. Thermogravimetry (TGA) thermal profiles measured in air, inert, and reducing atmosphere of (**a**): 10 µm SS powder, (**b**): 20 µm SS powder, (**c**): 44 µm SS powder and (**d**): PESU polymer binder.

The thermal characterization of the polymer binder is also of particular importance for the fabrication of metal HFs, since the polymer burn-out should occur below the sensitization temperatures of the metal particles in order to avoid any reaction and oxidation with the metal particles. In addition, the polymer matrix should be maintained at a lower temperature to ensure the structural integrity of the HF during the polymer de-binding step. Figure 3d shows the thermal decomposition of PESU in air, inert (N_2), and reducing (N_2:H_2 5%) atmosphere. The decomposition of PESU occurred abruptly at 600 °C below the temperature corresponding to the sensitization temperature of the SS particles. The TGA profile of the green SS_{10} HF in air atmosphere is presented in Figure S1, and shows a similar decomposition of the PESU matrix during the de-binding step at 600 °C. However, a weight gain was observed during the thermal treatment, suggesting that sensitization could not be avoided during the de-binding step. The thermal treatment of the green HF was therefore performed in reducing atmosphere in order to prevent the oxidation of the particles, promote metal-to-metal contact, and form neck area between the metal particles [27,31]. However, the thermal decomposition of PESU in inert and reducing atmosphere (Figure 3d) revealed that up to 35 wt % of organic residues remained. The sintered HFs were therefore thoroughly cleaned first with ethanol and distilled water before any characterization tests.

2.2. Green and Sintered HF Morphologies

The dopes comprised of SS particles and PESU polymer binder were then extruded to form the green HF material. The SEM images of the green HFs are shown in Figure 4. The average outer diameter of the green HFs made of SS_{10} particles was found to be 1.23 ± 0.03 mm, and to exhibit circular cross-sectional morphology with a uniform distribution of SS_{10} particles within the PESU matrix. The outer diameters of the green $SS_{10/20}$ and $SS_{10/44}$ HFs, 2.24 ± 0.05 and 2.17 ± 0.45 mm respectively, were found to be larger than the SS_{10} HFs due to a shorter air gap, resulting in less stretch and elongation stress on the forming HFs [3]. The ellipsoidal shape and irregular circularity of the green $SS_{10/20}$ and $SS_{10/44}$ HFs were attributed to aggregates and clusters of metal particles of different sizes, which may have formed in the dope during the degas step prior to spinning. The circularity of the green HFs could be further improved by increasing the bore flow rate in order to break the aggregates in the dope solution. The SEM images reveal the presence of macro-voids and finger-like structures, starting from both sides of the HFs exposed to the bore and external coagulation liquid and going toward the inner core of the HFs. This structure is more pronounced for the $SS_{10/20}$ and $SS_{10/44}$ HFs due to the presence of clusters of metal particles inside the continuous PESU phase.

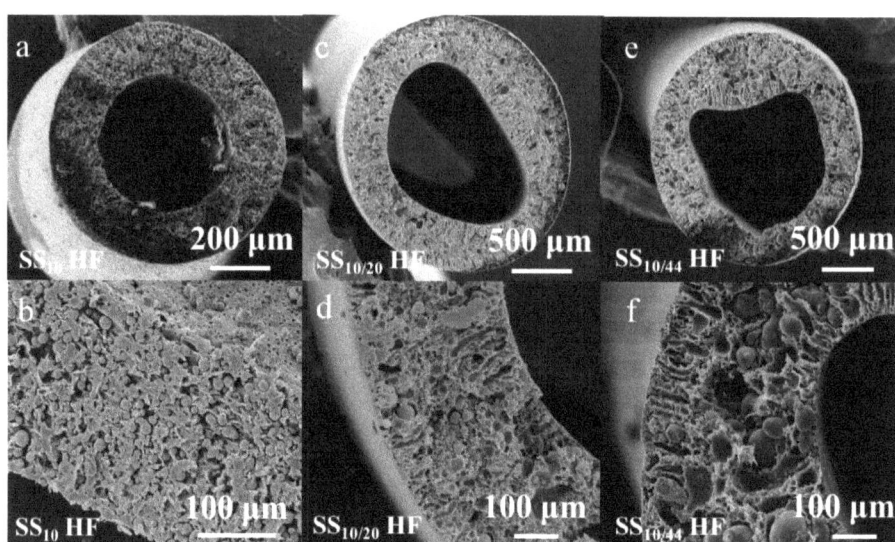

Figure 4. Cross-sectional SEM images of (**a,b**): the SS_{10} green hollow-fibers (HFs); (**c,d**): the $SS_{10/20}$ HFs; (**e,f**): the $SS_{10/44}$ HFs.

As seen in Figure 5, the sintered HFs retained their original lumen shapes upon thermal treatment. The HFs made of SS_{10} and $SS_{10/20}$ particles exhibited smooth surfaces, whereas the macro-voids present across the wall of the green HFs also remained after the thermal treatment. On the other hand, the HFs made of $SS_{10/44}$ particles presented extremely rough surfaces, with some of the largest particles pointing out of the structure. This geometry may be due to the shear displacement of the largest particles across the matrix of small particles during the sintering [35]. As opposed to the SS_{10} and $SS_{10/20}$ HFs, the cross-sectional SEM images of the $SS_{10/44}$ HFs revealed a denser structure. The original macro-voids observed in the green HFs were also absent after sintering, as expected due to the high diffusion rate of the smaller particles, which were able to fill the macro-voids during the thermal treatment. The shape of the metal particles present in the SS_{10} and $SS_{10/20}$ green HFs became indistinguishable when sintered at 1100 °C, indicating that the sintering process was completed, which was not the case for the $SS_{10/44}$ HFs where the shapes of the particles were still well-defined (Figure 5(f1,f2)). The SEM images of the sintered HFs across the whole range of sintering temperature and time are displayed in Figures S2–S4.

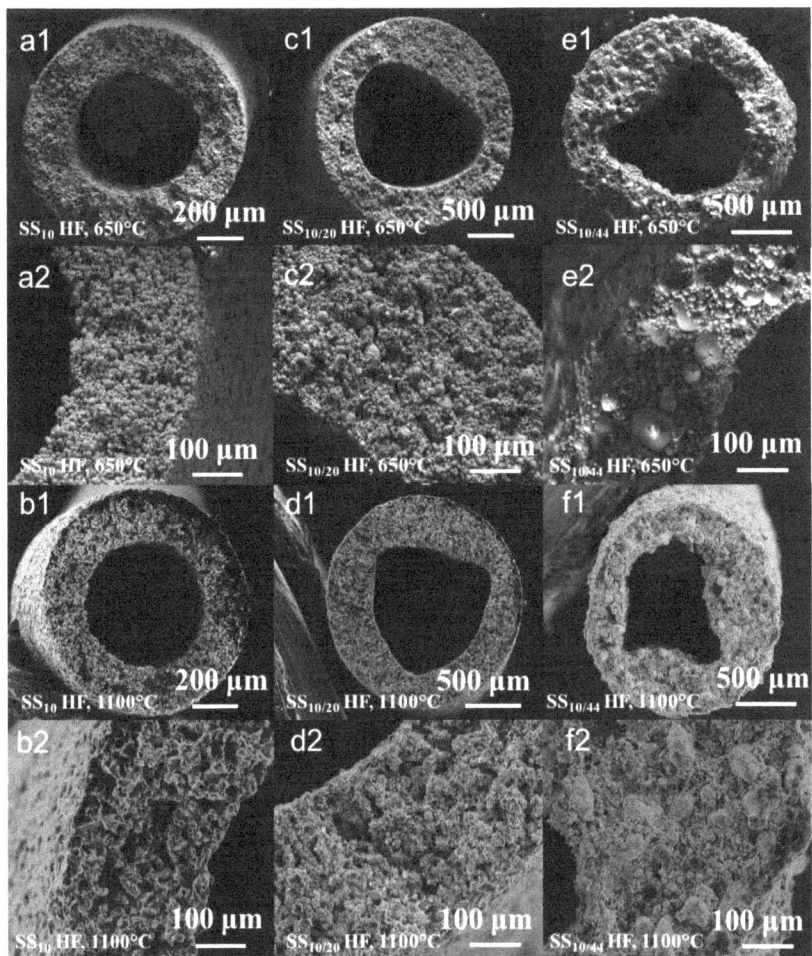

Figure 5. Cross-sectional SEM images of the sintered SS_{10} HFs for 90 min at 650 °C (**a1,a2**) and 1100 °C (**b1,b2**); sintered $SS_{10/20}$. HFs for 90 min at 650 °C (**c1,c2**) and 1100 °C (**d1,d2**); sintered $SS_{10/44}$. HFs for 90 min at 650 °C (**e1,e2**) and 1100 °C (**f1,f2**).

2.3. Sintered HF Properties

The radial shrinkage was measured based on the SEM cross-sectional images, and the measurement results (%) are shown in Figure 6. The sintered HFs exhibited isotropic radial shrinkage strongly dependent on the sintering temperature. A slight increase in shrinkage values was noticed for the SS_{10} and $SS_{10/20}$ HFs when sintered for a longer duration (90 min) as shown in Figure 6a,b. The radial shrinkage of the SS_{10} HFs remained below 10% up to 900 °C, which corresponded to the first stage of the sintering process only. On the other hand, the $SS_{10/20}$ and $SS_{10/44}$ HFs showed greater radial shrinkage (>15%) across the range of sintering temperature. This effect was particularly noticeable for the $SS_{10/44}$ HFs due to the difference in size and shapes of the metal particles, which promoted the densification of the structures across the whole range of sintering temperature [24].

As observed in cross-sectional SEM images (Figure 5(a1,c1,e1)), the denser structures resulting from the thermal treatment at a lower temperature can be explained by the incomplete degradation of the polymer during the de-binding step. This observation is supported by the EDX chemical analysis and SEM images (Figure S5 and Figure 5), which revealed that the de-binding step was incomplete at the lowest temperatures (650 and 700 °C), and that a large amount of carbon remained in the structure. The densification of the structure at a lower temperature is therefore not sufficient to induce the collective sintering of the metal particles due to the residual carbon hampering particle-to-particle contacts and particle neck formation. This effect may be alleviated by using a different polymer binder with a lower decomposition temperature or by prolonging the duration of the de-binding phase.

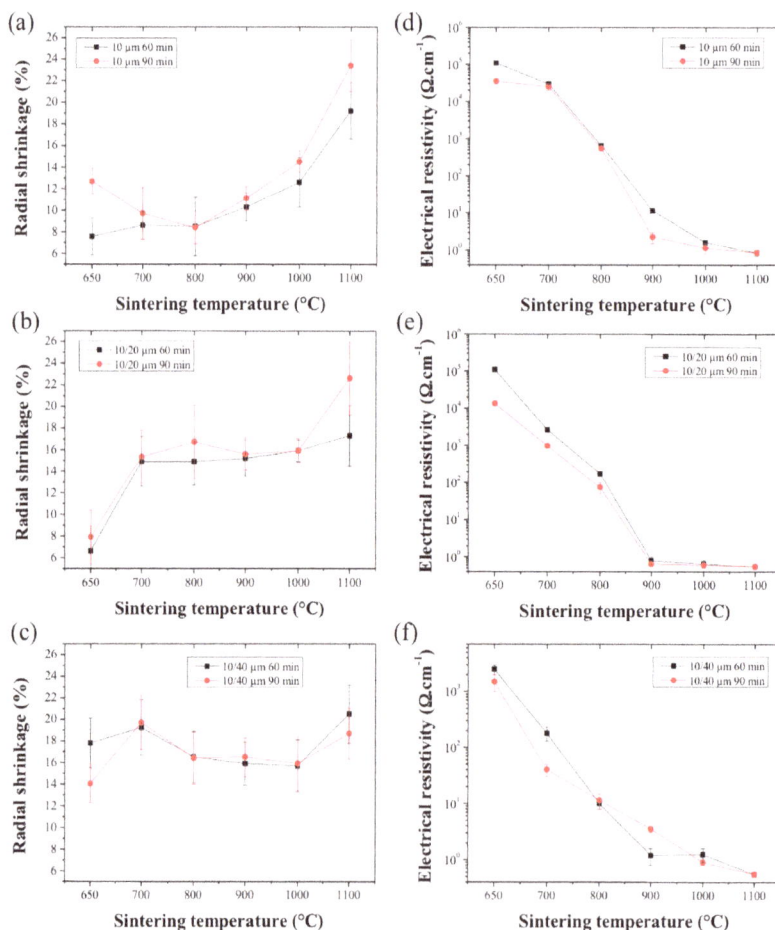

Figure 6. Radial shrinkage from green to sintered HFs as a function of the sintering time and temperature. (**a**): SS_{10} HFs, (**b**): $SS_{10/20}$ HFs and (**c**): $SS_{10/44}$ HFs. (**d**–**f**) are the electrical resistance of the SS HFs as a function of the sintering temperature and time for the SS_{10}, $SS_{10/20}$ and $SS_{10/44}$ HFs.

The electrical resistance measurements (Figure 6d–f) of the HFs sintered at a temperature lower than 900 °C indicated that the densification of the structure is not sufficient to create a significant conductive pathway across the inter-particle neck area. The enhanced mechanical strength of the $SS_{10/20}$ and $SS_{10/44}$ HFs at lower temperatures can be explained from the increased initial neck growth of the smallest particles to the largest particle. The initial neck growth between the smallest and largest metal particles is able to form a scaffold maintaining the structural integrity of the HFs even at the lowest sintering temperatures. Across all of the HF samples, the electrical resistance was drastically reduced from 900 °C of sintering temperature, indicating that the neck areas were able to form a continuous and conductive scaffold across the HFs [20,21]. No significant differences were, however, found between the HFs sintered for 60 or 90 min, suggesting that a percolation threshold had been reached by that time.

The mechanical properties of the sintered SS HFs are therefore highly related to the dope composition and to the thermal treatment conditions, including the sintering temperature, duration, and atmosphere. The mechanical strength of the SS_{10} HFs was found to be acceptable for handling and testing only when sintered above 900 °C (Figure 7a,b) with a maximal flexural stress of 23.9 ± 1.2 and 17.2 ± 0.9 N mm^{-2} when sintered for 60 and 90 min, respectively. On the other hand, the $SS_{10/20}$ and $SS_{10/44}$ HFs exhibited sufficient mechanical strength right from the lowest sintering temperature (650 °C), with a value above 37 N mm^{-2} when sintered for 60 or 90 min, and values slightly lower for the $SS_{10/20}$ HFs when sintered for 90 min (<40 N mm^{-2}). In these conditions, the extended sintering time was therefore unable to make up for the lower sintering temperature and to achieve a significant degree of bonding between the metal particles. The SS_{10} and $SS_{10/20}$ HFs exhibit exceptional mechanical strength and flexibility when sintered above 1000 °C. The maximal flexural stress of the SS_{10} HFs sintered at 1000 °C exceeded 200 and 300 N mm^{-2} for a sintering time of 60 and 90 min, respectively, while the $SS_{10/20}$ HFs showed values twice as high as the SS_{10} HFs. SS_{10} HFs sintered at 1100 °C showed a further mechanical strength improvement, with values increased by up to 5 and 10 times depending on the sintering time (Figure 7a,b). However, the $SS_{10/20}$ HFs sintered at 1100 °C for 60 or 90 min did not show any significant mechanical strength improvement. This lack of improvement indicates that the densification of the structure is at a maximum [24]. The mechanical strength of the $SS_{10/44}$ HFs sintered at the same temperatures were, however, found to be significantly lower (<60 N mm^{-2}) compared to the SS_{10} and $SS_{10/20}$ HFs. Although the densification of the $SS_{10/44}$ system occurred at lower temperatures and reached a maximum sooner, the lack of coalescence between the large and the small particles formed a more rigid and denser structure. The difference in shape and size between the two SS powders promoted the densification of the systems, with the small particles filling the pores between the large particles. Therefore, it may be preferential to decrease the size ratio difference between the particles as well as to work with uniform spherical particles. The SS_{10} and $SS_{10/20}$ HFs fabricated in this work exhibited similar mechanical performance when compared to previous studies available in the literature; however, the mechanical properties of the $SS_{10/44}$ were found to be inferior [20,34,36].

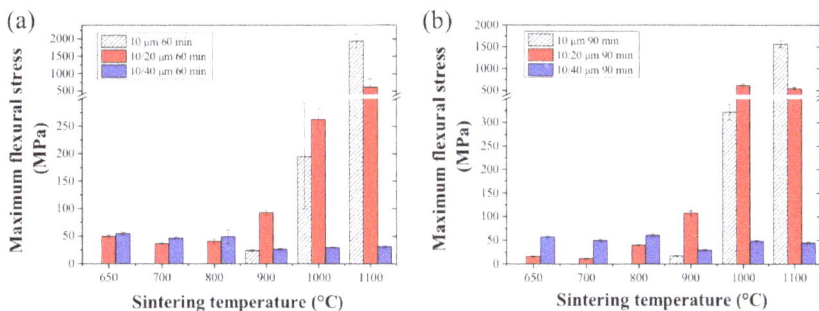

Figure 7. Maximal flexural stress of the SS HFs as a function of the sintering temperature with (**a**): HFs sintered for 60 min and (**b**): HFs sintered for 90 min.

2.4. Pore Size Distribution and Water Permeance

The average pore sizes of the sintered HF membranes, which correspond to the average dimension of the inter-particle domains, were measured by capillary flow porometry. Capillary flow porometry techniques have the advantage to be able to measure the open pores across materials, and is particularly suited to micron-ranged pore distributions [37]. The average pore sizes of the sintered HFs are presented in Figure 8a,b. The SS_{10} and $SS_{10/20}$ HFs sintered at 900 °C presented a similar average pore size—around 5 μm—for a sintering time of 60 min, and around 3 μm when sintered for 90 min. The same trend was observed for a sintering temperature of 1000 °C, with similar average pore sizes between the SS_{10} and $SS_{10/20}$ HFs, and smaller values when sintered for 90 min (Figure 8b). However, the $SS_{10/20}$ HFs sintered at 1100 °C presented larger pore sizes as compared to the SS_{10} HFs. The average pore size of the SS_{10} HFs sintered at 1100 °C for 90 min was reduced to 0.5 μm as opposed to the average pore size of 2.4 ± 0.3 μm found in the $SS_{10/20}$ HFs. The average pore sizes of the $SS_{10/44}$ HFs were found to be below 1 μm for a sintering temperature ranging from 900 to 1100 °C without significant effect on the sintering time. The smaller average pore size of the $SS_{10/44}$ HFs was related to the greater size difference between the two series of particles, which led to a denser structure with smaller pores [24]. A general aspect of sintering is that as the sintering time and temperature increase, the coalescence between the particles is promoted [38]. However, in the case of $SS_{10/20}$ HFs, the maximum density was limited due to the larger particles within the system. As opposed to the SS_{10} HFs, the increase in sintering temperature and time therefore promoted the creation of larger pores in the $SS_{10/20}$ HF structure [38]. On the other hand, the SS_{10} HFs could potentially form dense materials by further intensification of the sintering conditions [18].

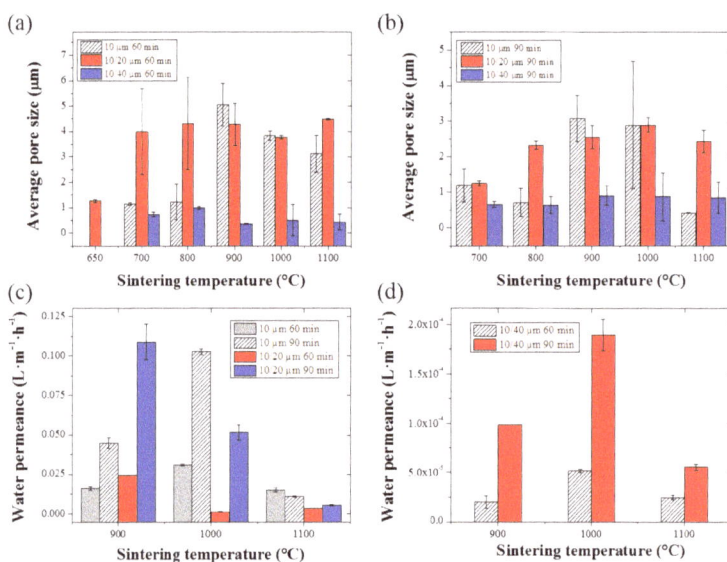

Figure 8. Average pore size measurements of the SS HFs as a function of the sintering temperature, for (**a**): 60 min and (**b**): 90 min sintering time. Water permeance values of (**c**): the SS_{10} and $SS_{10/20}$ HFs as a function of the sintering temperature and time. (**d**): Pure water fluxes of the $SS_{10/44}$ HFs as a function of the sintering temperature and time.

Pure water fluxes were measured across a range of pressures varying from 0.2 to 1 bar. The pure water flux measurements exhibited a linear relationship with respect to the applied pressure, and are shown in the supplementary materials (Figure S6). Membrane permeances as a function of the sintering temperature and time are shown in Figure 8c,d. The $SS_{10/20}$ HFs sintered at 900 °C for 90 min exhibited the greatest water permeance among all of the SS HFs (Figure 8c). The $SS_{10/20}$ HFs water permeance values were shown to progressively diminish with respect to the sintering temperature due to the densification process. The highest water permeance, corresponding to the SS_{10} HFs, was attained at a sintering temperature of 1000 °C and for a sintering time of 90 min, likely due to an increased open porosity across the structure. The SS_{10} HFs water permeance drastically diminished at the sintering temperature of 1100 °C, which was therefore considered to be close to the full densification of the material. The $SS_{10/44}$ HFs also presented the lowest water permeance due to their denser structures and smaller pore size distributions, as previously reported (Figure 8d).

3. Materials and Methods

3.1. Metal Particles and Chemicals

The 10 μm 316L stainless steel (SS_{10}) was purchased from Sandvik Osprey Ltd., Neath, UK, while the 20 μm (SS_{20}) and 325 mesh size (SS_{44}, 44 μm) 316L SS powders were sourced from Huarui Group Ltd., Hangzhou, China. Poly(ethersulfone) (PESU, Ultrason®, BASF, Tarragona, Spain) was dried in a vacuum oven at 120 °C for 6 h before being used as the polymer binder. N-Methyl-2-pyrrolidone (NMP, ACS reagent ≥99.0%, Sigma-Aldrich, Madrid, Spain) was chosen as a solvent for PESU and used as received. Distilled water (DI) and isopropanol (IP, ACS reagent ≥99.5%, Sigma-Aldrich, Madrid, Spain) were used during the green HFs drying procedure and for the determination of the metal particle size distributions.

3.2. Preparation of the Green HFs

Spinning dopes were prepared by first dissolving the polymer binder into NMP for more than 24 h on a roller tube mixer. Subsequently, the metal particles were added in multiple steps and finally the mixture was stirred overnight. The spinning dope mixture was then loaded into the spinning pump and degassed by applying a vacuum 24 h before the date of the spinning experiment. The compositions of the spinning dopes were as follows: 70 wt % of metal particles, 7.5 wt % of polymer binder, and 22.5 wt % of NMP. In the spinning experiments, where a mixture of SS particles of different size distributions was used, the blend of SS particles comprised 60 wt % of SS_{10} powder and 40 wt % of either SS_{20} or SS_{44} powder.

The spinning setup allowed the spinning of green HFs of various geometries by controlling different process parameters such as air gap (cm) and dope and bore flow rates (mL h^{-1}). The temperature was maintained at 25 °C through heating bands wrap around the spinneret assembly and spinning line. The spinning conditions are listed in the Supporting Information section's Table S3. The spinneret used had an inner and outer diameter of 0.7 and 1.3 mm, respectively.

3.3. Drying and Thermal Treatments of Green HFs

The green HFs were firstly kept in a DI bath for 24 h followed by an IP bath for another 24 h in order to exchange the NMP solvent. Finally, the green HFs were dried at room temperature for 24 h. Once dried, thermal treatments were performed in a tubular furnace (model GSL-1100 X, MTI Corporation, Richmond, CA, USA) in order to remove the PESU polymer binder and sinter the SS particles.

The polymer was removed at 600 °C for 60 min followed by a sintering process with temperatures ranging from 650 to 1100 °C for 60 or 90 min at a heating rate of 5 °C min^{-1} under a reducing atmosphere nitrogen/hydrogen mixture (N_2:H_2, 85:15 *v/v* %). The gas was first flown 30 min before the thermal treatment in order to purge the air from the furnace, and was maintained at 1 dm^3·min^{-1} during the thermal treatment.

3.4. Materials Characterization

The morphology of the metal particles and the green and sintered HFs was characterized by Scanning Electron Microscopy (SEM, JEOL Neoscope, JCM-5000, Peabody, MA, USA). The elemental distributions on the surface of the samples were evaluated by an Electron Dispersive X Ray Spectroscopy (EDS) analysis with an Oxford detector on a JEOL JSM 7800F model. SEM imaging (surface and cross-section) was performed at 10 keV and a 10 mm working distance, while surface elemental mapping was performed at 20 keV and 10 mm working distance.

The particle size distribution of the metal powders dispersed in water or isopropanol were analysed by Dynamic Light Scattering (DLS) using a Mastersizer 2000 (Malvern Instrument, Worcester, UK). The aspect ratio of the metal particles was determined using ImageJ software (version 1.50i). The specific surface area of the metal powders was measured by the BET method using a TriStar 3000 instrument (Micromeritics, Norcross, GA, USA). The average absolute and relative densities of the metal powders were determined with a 50 mL pycnometer bottle using ethanol as the wetting liquid. The temperature-dependence of the wetting fluid was taken into account. Thermogravimetry (TGA) experiments were performed under air, N_2, and N_2:H_2 mixture (95:5 *v/v* %) using a Q50 TGA (TA instrument, New Castle, DE, USA). During the TGA experiments, the gas flow was maintained at 60 $cm^3 \cdot min^{-1}$, and the heating rate was fixed at 1, 5, or 10 $^\circ C \cdot min^{-1}$.

The pore size and pore size distribution were measured using a capillary flow porometer (Porometer 3GZH Quantachrome Instruments, Boynton Beach, FL, USA) after wetting the sintered HFs with Porofil® (Quantachrome Instruments) wetting solution. The electrical resistivity of the green and sintered HFs was determined using an electrical resistivity cell with a variable resistor. Three-point bending tests were performed on a DMA Q800 (TA instrument, USA) in order to characterize the mechanical resistance of the HFs. The stress rate was set at 0.5 $N \cdot min^{-1}$, and the temperature keep at 25 $^\circ C$. The bending stress or flexural stress was calculated using the following equation:

$$\sigma = \frac{8FKD_0}{\pi(D_0^4 - D_i^4)} \tag{1}$$

where σ is the bending stress (MPa), F is the measured maximal flexure load (N), K is the support span (mm), and D_0 and D_i are the outer and inner diameters of the HFs (mm), respectively.

Pure water permeation tests were performed in a lab-scale polyurethane module. The metal HF membranes with an effective length of 40 mm were first cleaned with ethanol, flushed thoroughly with distilled water, and dried at room temperature. The metal HF membranes were then placed into the module and the distilled feed water was circulated onto the membranes and forced through the membrane walls under pressure varying from 0.2 to 1 bar. The pure water flux of the metal HF membranes was calculated using the equation:

$$J = \frac{Q}{A_m} = \frac{Q}{n\pi D_{HF}l_{eff}} \tag{2}$$

where J is the membrane pure water flux (L $m^{-2} \cdot h^{-1}$), Q is water flux rate (L h^{-1}), A_m is the effective membrane surface area (m^2), n is the number of HFs, D_{HF} the outer diameter, and l_{eff} is the effective length of the HFs.

4. Conclusions

Macro-porous SS HFs were successfully fabricated by the dry-wet spinning of a dope comprised of SS metal powders of different sizes and a polymer binder. A systematic study of the effect of the sintering parameters on the starting materials and sintered HFs was performed. Both the mechanical strength and water permeance of the HFs made of different sized SS particles were improved at a low (900 $^\circ C$) sintering temperature due to the increased neck density offered by the multi-modal particles' distribution. At a sintering temperature higher than 1000 $^\circ C$, SS particle repacking during the thermal treatment gave rise to a denser structure with a lower water permeance. The porous sintered HFs were highly electrically conductive and able to sustain high filtration pressures, making them valuable membranes for micro-filtration applications. Particularly, the SS HF membranes and filters developed in this study presented appropriate pore size distributions for bacteria and yeast rejection toward beverage sterilization applications. In addition, the optimization of the mixed particles' aspect and size ratio could lead to the development of composite HFs with enhanced mechanical properties and defined pore geometry and porosity for filtration applications, or that act as an electrically conductive diffuser in electro-catalytic reactors.

Supplementary Materials: The following are available online at http://www.mdpi.com/2077-0375/7/3/40/s1, Figure S1: Thermal treatment and TGA profile of the green SS_{10} HF in air atmosphere with a heating rate of $10^\circ C.min^{-1}$, Figure S2: SEM images of the green and sintered SS_{10} HFs at temperature ranging from 650 to 1100°C for 60 or 90 min, Figure S3: SEM images of the green and sintered $SS_{10/20}$ HFs at temperature ranging from 650 to 1100°C for 60 or 90 min, Figure S4: SEM images of the green and sintered $SS_{10/44}$ HFs at temperature ranging from 650 to 1100°C for 60 or 90 min, Figure S5: Carbon and oxygen elemental analysis of the SS_{10} HFs (a and d), $SS_{10/20}$ HFs (b and e) and $SS_{10/44}$ HFs (c and f), Figure S6: Pure water fluxes as a function of the feed pressure: (a), (b) and (c): SS_{10} HFs; (d), (e) and (f): $SS_{10/20}$ HFs; g: $SS_{10/44}$ HFs; Table S1: Size distribution of stainless steel particles, Table S2: Metal powder properties: surface area, total BET surface area, average absolute density and relative density determined versus average bulk density of 316L grade stainless steel (8 g.cm^{-3}), Table S3: Spinning conditions.

Acknowledgments: Francois-Marie Allioux would like to thank the Institute for Frontier Materials, Deakin University, Victoria, Australia, for funding his Ph.D scholarship and his Travel Award, which enable him to work at TECNALIA, San Sebastian-Donostia, Spain, and thanks AINSE Ltd for providing financial assistance (PGRA Award-30290). Ludovic F. Dumée acknowledges Deakin University for his Alfred Deakin Postdoctoral Fellowship. This research did not receive any specific grant from funding agencies in the public, commercial, or not-for-profit sectors.

Author Contributions: The manuscript was written through contributions of all authors.

Conflicts of Interest: The authors declare no conflict of interest.

References

1. Li, N.N.; Fane, A.G.; Ho, W.S.W.; Matsuura, T. *Advanced Membrane Technology and Applications*; John Wiley & Sons: Hoboken, NJ, USA, 2011.
2. Mat, N.C.; Lou, Y.; Lipscomb, G.G. Hollow fiber membrane modules. *Curr. Opin. Chem. Eng.* **2014**, *4*, 18–24. [CrossRef]
3. Feng, C.; Khulbe, K.C.; Matsuura, T.; Ismail, A.F. Recent progresses in polymeric hollow fiber membrane preparation, characterization and applications. *Sep. Purif. Technol.* **2013**, *111*, 43–71, Repeated with 10. [CrossRef]
4. Baker, R.W. *Membrane Technology*; Wiley Online Library: Hoboken, NJ, USA, 2000.
5. Luiten-Olieman, M. *Inorganic porous Hollow Fiber Membranes: With Tunable Small Radial Dimensions*; Gildeprint Drukkerijen: Enschede, The Netherlands, 2012.
6. Wickramasinghe, S.; Semmens, M.J.; Cussler, E. Mass transfer in various hollow fiber geometries. *J. Membr. Sci.* **1992**, *69*, 235–250. [CrossRef]
7. Peng, N.; Widjojo, N.; Sukitpaneenit, P.; Teoh, M.M.; Lipscomb, G.G.; Chung, T.-S.; Lai, J.-Y. Evolution of polymeric hollow fibers as sustainable technologies: Past, present, and future. *Prog. Polym. Sci.* **2012**, *37*, 1401–1424. [CrossRef]
8. Kas, R.; Hummadi, K.K.; Kortlever, R.; de Wit, P.; Milbrat, A.; Luiten-Olieman, M.W.J.; Benes, N.E.; Koper, M.T.M.; Mul, G. Three-dimensional porous hollow fibre copper electrodes for efficient and high-rate electrochemical carbon dioxide reduction. *Nat. Commun.* **2016**, *7*, 10748. [CrossRef] [PubMed]
9. Lai, C.Y.; Groth, A.; Gray, S.; Duke, M. Enhanced abrasion resistant PVDF/nanoclay hollow fibre composite membranes for water treatment. *J. Membr. Sci.* **2014**, *449*, 146–157. [CrossRef]
10. Irfan, M.; Idris, A. Overview of PES biocompatible/hemodialysis membranes: PES–blood interactions and modification techniques. *Mater. Sci. Eng.* **2015**, *56*, 574–592. [CrossRef] [PubMed]
11. Dumee, L.F.; He, L.; Lin, B.; Ailloux, F.-M.; Lemoine, J.-B.; Velleman, L.; She, F.; Duke, M.C.; Orbell, J.D.; Erskine, G.; et al. The fabrication and surface functionalization of porous metal frameworks—A review. *J. Mater. Chem. A* **2013**, *1*, 15185–15206. [CrossRef]
12. Dumée, L.F.; He, L.; Wang, Z.; Sheath, P.; Xiong, J.; Feng, C.; Tan, M.Y.; She, F.; Duke, M.; Gray, S. Growth of nano-textured graphene coatings across highly porous stainless steel supports towards corrosion resistant coatings. *Carbon* **2015**, *87*, 395–408. [CrossRef]

13. Gitis, V.; Rothenberg, G. *Ceramic Membranes: New Opportunities and Practical Applications*; John Wiley & Sons: Hoboken, NJ, USA, 2016.
14. Cassano, A.; Rastogi, N.K.; Basile, A. 18-Membrane technologies for water treatment and reuse in the food and beverage industries. In *Advances in Membrane Technologies for Water Treatment*; Woodhead Publishing: Oxford, UK, 2015; pp. 551–580.
15. Luiten-Olieman, M.W.J.; Winnubst, L.; Nijmeijer, A.; Wessling, M.; Benes, N.E. Porous stainless steel hollow fiber membranes via dry–wet spinning. *J. Membr. Sci.* **2011**, *370*, 124–130. [CrossRef]
16. David, O.; Gendel, Y.; Wessling, M. Tubular macro-porous titanium membranes. *J. Membr. Sci.* **2014**, *461*, 139–145. [CrossRef]
17. Lee, S.-M.; Choi, I.-H.; Myung, S.-W.; Park, J.-Y.; Kim, I.-C.; Kim, W.-N.; Lee, K.-H. Preparation and characterization of nickel hollow fiber membrane. *Desalination* **2008**, *233*, 32–39. [CrossRef]
18. Meng, B.; Tan, X.; Meng, X.; Qiao, S.; Liu, S. Porous and dense Ni hollow fibre membranes. *J. Alloys Compd.* **2009**, *470*, 461–464. [CrossRef]
19. Fernandez, E.; Helmi, A.; Coenen, K.; Melendez, J.; Viviente, J.L.; Tanaka, D.A.P.; van Sint Annaland, M.; Gallucci, F. Development of thin Pd–Ag supported membranes for fluidized bed membrane reactors including WGS related gases. *Int. J. Hydrogen Energy* **2015**, *40*, 3506–3519. [CrossRef]
20. Schmeda-Lopez, D.R.; Smart, S.; Nunes, E.H.M.; Vasconcelos, D.; Vasconcelos, W.L.; Bram, M.; Meulenberg, W.A.; da Costa, J.C.D. Stainless steel hollow fibres—Sintering, morphology and mechanical properties. *Sep. Purif. Technol.* **2015**, *147*, 379–387. [CrossRef]
21. Luiten-Olieman, M.W.J.; et al. Towards a generic method for inorganic porous hollow fibers preparation with shrinkage-controlled small radial dimensions, applied to Al$_2$O$_3$, Ni, SiC, stainless steel, and YSZ. *J. Membr. Sci.* **2012**, *407–408*, 155–163. [CrossRef]
22. German, R.M. Chapter Four—Measurement Tools and Experimental Observations. In *Sintering: From Empirical Observations to Scientific Principles*; Butterworth-Heinemann: Boston, MA, USA, 2014; pp. 71–130.
23. Dumée, L.F.; She, F.; Duke, M.; Gray, S.; Hodgson, P.; Kong, L. Fabrication of meso-porous sintered metal thin films by selective etching of silica based sacrificial template. *Nanomaterials* **2014**, *4*, 686–699. [CrossRef] [PubMed]
24. German, R.M. Chapter Eleven—Mixed Powders and Composites. In *Sintering: From Empirical Observations to Scientific Principles*; Butterworth-Heinemann: Boston, MA, USA, 2014; pp. 335–385.
25. Bouvard, D.; Lange, F.F. Relation between percolation and particle coordination in binary powder mixtures. *Acta Metall. Mater.* **1991**, *39*, 3083–3090. [CrossRef]
26. Spierings, A.; Levy, G. Comparison of density of stainless steel 316L parts produced with selective laser melting using different powder grades. In Proceedings of the Annual International Solid Freeform Fabrication Symposium, Austin, TX, USA, 3–5 August 2009.
27. Higgins, P.; Munir, Z. Influence of Surface Oxide Layer on Sintering Process. of Lead. *Powder Metall.* **1978**, *21*, 188–194. [CrossRef]
28. Lima, A.S.; Nascimento, A.M.; Abreu, H.F.G.; de Lima-Neto, P. Sensitization evaluation of the austenitic stainless steel AISI 304L, 316L, 321 and 347. *J. Mater. Sci.* **2005**, *40*, 139–144. [CrossRef]
29. German, R.M. Sintering theory and practice. *Solar-Terr. Phys.* **1996**, 568.
30. Tarabay, J.; Peres, V.; Pijolat, M. Oxidation of Stainless Steel Powder. *Oxid. Met.* **2013**, *80*, 311–322. [CrossRef]
31. Bergman, O. Influence of Oxygen Partial Pressure in Sintering Atmosphere on Properties of Cr–Mo Prealloyed Powder Metallurgy Steel. *Powder Metall.* **2013**, *50*, 243–249. [CrossRef]
32. Nayar, H.S.; Wasiczko, B. Nitrogen absorption control during sintering of stainless steel parts. *Met. Powder Rep.* **1990**, *45*, 611–614. [CrossRef]
33. Karlsson, H.; Nyborg, L.; Frykholm, R. Surface Reactions During Fabrication and Sintering of 410 L Stainless Steel Powder. *Powder Metall. Prog.* **2005**, *5*, 220–233.
34. Rui, W.; Zhang, C.; Cai, C.; Gu, X. Effects of sintering atmospheres on properties of stainless steel porous hollow fiber membranes. *J. Membr. Sci.* **2015**, *489*, 90–97. [CrossRef]
35. Bordia, R.K.; Raj, R. Sintering Behavior of Ceramic Films Constrained by a Rigid Substrate. *J. Am. Ceram. Soc.* **1985**, *68*, 287–292. [CrossRef]
36. Luiten-Olieman, M.W.J.; Raaijmakers, M.J.T.; Winnubst, L.; Wessling, M.; Nijmeijer, A.; Benes, N.E. Porous stainless steel hollow fibers with shrinkage-controlled small radial dimensions. *Scr. Mater.* **2011**, *65*, 25–28. [CrossRef]

37. Dumee, L.; Velleman, L.; Sears, K.; Hill, M.; Schutz, J.; Finn, N.; Duke, M.; Gray, S. Control of porosity and pore size of metal reinforced carbon nanotube membranes. *Membranes* **2010**, *1*, 25–36. [CrossRef] [PubMed]
38. Kang, S.-J.L. *Sintering: Densification, Grain Growth and Microstructure*; Butterworth-Heinemann: Oxford, UK, 2004.

membranes

MDPI

Article

Nanofiltration and Tight Ultrafiltration Membranes for Natural Organic Matter Removal—Contribution of Fouling and Concentration Polarization to Filtration Resistance

Joerg Winter [1], Benoit Barbeau [2] and Pierre Bérubé [1,*]

[1]   Department of Civil Engineering, The University of British Columbia, 6250 Applied Science Lane, Vancouver, BC V6T1Z4, Canada; jwinter@civil.ubc.ca

[2]   Department of Civil, Geological and Mining Engineering, École Polytechnique de Montréal, Montréal, QC H3T 1J4, Canada; benoit.barbeau@polymtl.ca

*   Correspondence: berube@civil.ubc.ca; Tel.: +1-604-822-5665

Received: 14 March 2017; Accepted: 26 June 2017; Published: 2 July 2017

Abstract: Nanofiltration (NF) and tight ultrafiltration (tight UF) membranes are a viable treatment option for high quality drinking water production from sources with high concentrations of contaminants. To date, there is limited knowledge regarding the contribution of concentration polarization (CP) and fouling to the increase in resistance during filtration of natural organic matter (NOM) with NF and tight UF. Filtration tests were conducted with NF and tight UF membranes with molecular weight cut offs (MWCOs) of 300, 2000 and 8000 Da, and model raw waters containing different constituents of NOM. When filtering model raw waters containing high concentrations of polysaccharides (i.e., higher molecular weight NOM), the increase in resistance was dominated by fouling. When filtering model raw waters containing humic substances (i.e., lower molecular weight NOM), the increase in filtration resistance was dominated by CP. The results indicate that low MWCO membranes are better suited for NOM removal, because most of the NOM in surface waters consist mainly of humic substances, which were only effectively rejected by the lower MWCO membranes. However, when humic substances are effectively rejected, CP can become extensive, leading to a significant increase in filtration resistance by the formation of a cake/gel layer at the membrane surface. For this reason, cross-flow operation, which reduces CP, is recommended.

Keywords: nanofiltration; tight ultrafiltration; concentration polarization; fouling; natural organic matter

---

1. Introduction

Over the past 20 years, membrane filtration has been increasingly implemented in water purification processes [1]. This increase has mainly been driven by the decreasing costs of membrane systems, and increasingly stringent drinking water quality regulations. In drinking water treatment applications, ultrafiltration membranes (UF) can effectively remove particulate contaminants, such as protozoa and bacteria, from raw water sources. However, effective removal of natural organic matter (NOM) and viruses is generally limited [2,3]. This is because the molecular weight cut off (MWCO) of UF membranes typically used in drinking water treatment applications is relatively large (i.e., greater than 100,000 Da). Membranes with a MWCO of less than 10,000 Da are required to effectively remove NOM and all pathogens [4–8]. The present study investigated the use of membranes with MWCOs ranging from 300 to 8000 Da for the removal of NOM. In this range, membranes are commonly referred to as nanofiltration (NF) membranes (i.e., with a typical MWCO range of 200–1000 Da) and tight UF membranes (i.e., with a typical MWCO range of 1000 to 10,000 Da). The removal of NOM from source waters is of importance in drinking water treatment because it can cause color, taste, and odor issues, increase chlorine demand, as well as contribute to disinfection by-product formation and microbial regrowth in distribution systems [9–14]. Because NF and tight UF membranes can achieve sufficient primary disinfection (i.e., >4-log removal of all pathogens) and extensive removal of NOM, they can provide effective and comprehensive drinking water treatment in a single step. A simple, single-step approach is of particular interest for small/remote communities, because of the limited financial and technical resources generally available to implement and operate water treatment systems.

The use of NF and tight UF membranes in drinking water treatment applications is still limited. This is mainly because spiral wound configurations, which generally require extensive pre-treatment and a high trans-membrane pressure, have historically been used for NF and tight UF membrane applications. In addition, fouling control measures, such as backwashing or surface scouring, cannot be applied to spiral wound configurations. Recent developments in NF and tight UF membrane configurations, such as hollow fiber configurations, promise to address some of these limitations [15–18]. Frank et al. have also reported that a substantially higher permeability can be maintained using hollow fiber configurations [16]. However, other studies have reported that hollow fiber configurations are less effective in mitigating concentration polarization (CP) than spiral wound configurations [19,20]. Because CP and fouling are still major challenges in the application of NF and tight UF membranes for drinking water treatment, insight into CP and fouling is necessary to develop recommendations for the design and operation of NF and tight UF membranes, and to enable greater adoption of this advanced treatment technology.

Fouling occurs as material that is retained at the membrane surface or inside the membrane pores, accumulates, and increases resistance to the permeate flow. Material accumulates when the permeation drag—the force that transports potential foulants towards the membrane—is greater than forces acting in the opposite direction (i.e., away from the membrane) [21,22]. NOM is generally considered to be a main contributor to membrane fouling in drinking water treatment applications [4,23–27]. The extent of fouling is not necessarily proportional to the total amount of NOM retained, but is rather governed by the retention of specific NOM fractions [28–31]. The NOM fractions that have been reported to be mainly relevant to membrane fouling are biopolymers and humic substances, which are mainly of aquatic and terrestrial origin, respectively [32–34]. For UF membranes, the biopolymer fraction of NOM generally contributes the most to fouling [35–38]. Fouling due to humic substances is generally not as extensive, although it is more difficult to control hydraulically (e.g., by backwashing) [25,35]. The extent and the hydraulic reversibility of UF fouling can also be affected by other constituents in raw waters, notably calcium [25,35,39]. Calcium can form bridges between NOM molecules, between the membrane surface and NOM, as well as contribute to aggregation of NOM by charge destabilization [25,35,39]. Biopolymers have also been reported to substantially contribute to fouling of NF and tight UF membranes [28,38,40]. Humic substances, although effectively rejected, have not been reported to substantially contribute to fouling of NF membranes [41]. The limited contribution of humic substances to fouling has been attributed to charge repulsion effects, which enhance the back transport of humic substances in cross-flow systems [41,42]. The extent to which electrostatic repulsion contributes to fouling control has been demonstrated to increase with the ratio of cross-flow velocity to permeate flux [39]. Therefore, the effect of electrostatic repulsion on fouling due to humic substances, is likely to be substantially less pronounced in dead-end systems. The reported effects of calcium on fouling in NF and tight UF membranes systems have been inconsistent, and likely depend on the concentration in the solution being filtered [29,39,43,44].

CP, resulting from the accumulation of dissolved material rejected by the membrane, can also increase the resistance to permeate flow [45–47]. Unlike the situation with fouling, material does not deposit on the membrane surface or inside the membrane pores, but accumulates in proximity to the membrane surface. CP is characterized by an equilibrium between the convective transport of material towards the membrane, and the diffusive transport of retained material away from the membrane; a steady state that is expected to develop within the first minutes of filtration [29,45,48]. CP can affect the resistance to permeate flow by increasing the viscosity, the osmotic pressure of the solution being filtered, and the back-diffusion of retained solutes [45]. If CP effects become extensive, solutes at the membrane surface can reach a critical concentration (i.e., solubility limit), beyond which they form a cake/gel layer which fouls the membrane surface [26,46–48]. In literature, this critical concentration is also referred to as the "gel concentration" [46,47]. If a material reaches the critical concentration, the back-diffusion of material is limited by the formation of a cake/gel layer. Under such conditions, the convective transport towards the membrane becomes greater than the diffusive transport away from the membrane, and fouling occurs, which increases the resistance to permeate flow throughout the filtration phase. According to the Stokes–Einstein equation, the diffusivity and the diffusive transport of material depends on its size. NOM ranges in size from a few hundred Daltons (low molecular weight acids and neutrals) to over 20,000 Da (biopolymers) [39,49–51]. The size of NOM also generally increases with ionic strength, and decreases with pH of a solution, due to changes in the shape of NOM from uncoiled to coiled [52]. CP due to the retention of salts in Reverse Osmosis (RO) and NF systems has been extensively investigated [19,53–56]. However, only a few studies have investigated CP due to the retention of NOM in NF membranes. These studies have indicated that NOM can result in CP, and therefore, affect system performance [26,57]. Also, comprehensive knowledge on the effect of (i) the type of NOM (e.g., polysaccharides and humic substances), and (ii) the membrane MWCO on the contribution of CP to the total increase in filtration resistance, is not currently available. This is a knowledge gap that limits the ability to develop recommendations for the optimal design and operation of NF and tight membrane systems for drinking water treatment.

The objective of the present study was to quantify the contribution of fouling and CP to the total increase in resistance during filtration of model raw waters containing polysaccharides, humic substances, and a mixture of both (as typical constituents of NOM in natural surface waters), and to identify conditions under which an extensive increase in filtration resistance occurs. Since fouling and CP are not only expected to depend on the raw water characteristics, but also on the membrane's selectivity, the impact of the MWCO of a membrane on CP and fouling was also investigated. As previously discussed, CP is expected to develop rapidly within the first minutes of filtration [39,45]. However, this has never been investigated for NF and tight UF membranes in drinking water applications.

2. Material and Methods

2.1. Experimental Approach

Model raw waters of various compositions were considered. The biopolymer and humic material content of the raw waters were modeled using polysaccharide alginate derived from brown algae (Sodium Alginate, Sigma Aldrich, Oakville, ON , Canada) and Suwannee River NOM (SRNOM) 2R101N (International Humic Substances Society, St. Paul, MN, USA), respectively. Model raw waters with three different NOM compositions were considered: (i) alginate, (ii) SRNOM, and (iii) mixtures of SRNOM and alginate with a carbon mass ratio of 4:1. The ratio of 4:1 in the mixtures of SRNOM and alginate was selected based on humic substances to biopolymers ratios obtained from size exclusion chromatography (SEC) analyses of local surface water (Jericho Pond, Vancouver, BC, Canada) and is also consistent with the ratios reported in literature, which indicates that humic substances account for 70–80% of the NOM in natural waters [32]. All three model raw water compositions were tested at two dissolved organic carbon (DOC). concentrations: 5 mg/L, and 10 mg/L.

To prepare the model raw waters, NOM surrogates (i.e., the polysaccharide alginate and/or SRNOM) were dissolved in Milli-Q laboratory water (MilliporeSigma, Temecula, CA, USA). Sodium bicarbonate at a concentration of 1 mM was added as a buffer. This concentration of bicarbonate also corresponds to an alkalinity that is within a range typical of that present in natural source waters [58]. The retention of alkalinity during filtration ranged from approximately 2% to 40% (depending on the membrane MWCO), and the impact of alkalinity rejection on CP and resistance during filtration was calculated [28], and found to be negligible relative to the overall increases in resistance observed in the present study (results not shown), and hence, is not further discussed. The pH was then adjusted to 7.0 ($\pm$ 0.1), using sodium hydroxide or hydrochloric acid, as needed. Prior to all filtration tests, all model raw water solutions were filtered through a 0.45 μm nitrocellulose filter. Milli-Q laboratory water was used for clean water flux tests prior- and post-filtration of model raw waters. The pH of the Milli-Q laboratory water used for clean water flux tests was also adjusted to 7.0 ($\pm$0.1), using sodium hydroxide.

The experimental setup, illustrated in Figure 1, consisted of feed vessels, membrane cells, bleed lines and permeate collection systems. Two feed vessels were used, one containing clean water and the other containing a model raw water. The pressure applied to both vessels was equal, enabling the feed to the membrane to be switched between model raw water and clean water without causing any pressure variation in the membrane cells. Three CF042 membrane cells (Sterlitech, Seattle, WA, USA) were used in parallel. These accommodated flat sheet membrane coupons, 39.2 mm wide and 85.5 mm long. Flat sheet, thin film composite polyamide (PA) membranes with nominal MWCOs of approximately 300 Da (DK series, GE Osmonics, Trevose, PA, USA), 2000 Da (GH series, GE Osmonics) and 8000 Da (GM series, GE Osmonics) were considered as a representative range for NF–tight UF range membranes. The intrinsic membrane resistances for the 300 Da, 2000 Da and 8000 Da MWCO membranes were $9.0 \times 10^{13} \pm 1.1 \times 10^{13}$ m^{-1}, $8.1 \times 10^{13} \pm 1.1 \times 10^{13}$ m^{-1} and $2.2 \times 10^{13} \pm 2.0 \times 10^{12}$ m^{-1}, respectively. The transmembrane pressure applied to the vessels was kept constant at 40.0 $\pm$ 1.0 psi (i.e., 2.8 $\pm$ 0.1 bar). The tests were conducted at room temperature (23 $\pm$ 1 °C). The membrane cells were operated in a pseudo dead-end mode with a continuous bleed that introduced a very low and constant cross-flow of less than 0.006 m/min along the membrane surface. The bleed line allowed the liquid to be purged from the membrane cell when the feed was switched, while still providing pseudo dead-end operation. Based on preliminary tests, this cross-flow velocity had a negligible effect on fouling and concentration polarization (results not presented). All filtration tests were conducted in triplicate in three sequential phases:

1. Pre-clean water filtration phase (for a minimum of 3 h).
2. Filtration phase with model raw waters (for approximately 3 h).
3. Post-clean water filtration phase (for a minimum of 45 min).

The filtration tests were conducted in these three sequential phases to quantify the contribution of CP and fouling to the increase in the resistance to permeate flow, as discussed in more detail in Section 3.2.

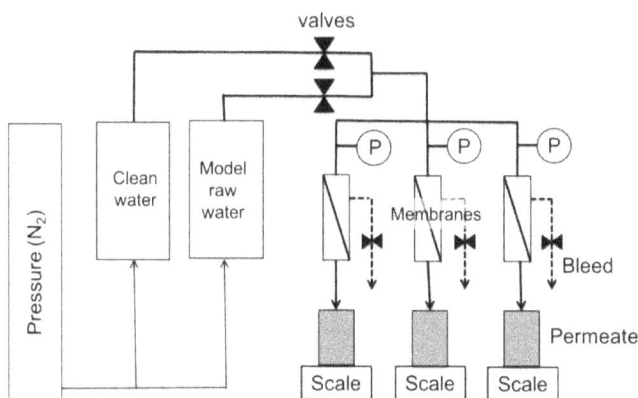

Figure 1. Experimental setup.

Immediately prior to the start of the filtration phase with model raw water, the drain valve was opened and the content of the flow cell was purged, ensuring that the concentration of NOM in the flow cell at the start of the filtration of model raw water, was equivalent to that of the model raw water. The drain was then partially closed to generate a constant low cross-flow velocity of 0.006 m/min.

2.2. Data Evaluation and Sample Analyses

The rate of fouling was quantified based on fouling coefficients obtained from standard filtration laws fitted to the experimental data collected during the filtration of model raw water [59]. Based on the minimum R^2 value for all conditions investigated in the present study, the permeate flux could be best modeled by assuming that fouling was predominantly due to the formation of a cake layer on the membrane surface (see Equation (1); results not presented). Therefore, the results are presented in terms of a cake fouling coefficient (k_c) (see Table 1).

$$J = \frac{J_0'}{1 + k_c \times \frac{\mu}{\Delta P} \times V \times J_0'} \left[\text{m}^3 \, \text{m}^{-2} \, \text{s}^{-1} \right] \tag{1}$$

In Equation (1), J represents the permeate flux [$\text{m}^3 \, \text{m}^{-2} \, \text{s}^{-1}$], V the specific volume of water filtered [$\text{m}^3 \, \text{m}^{-2}$], μ the dynamic viscosity [Pa s] and ΔP the trans-membrane pressure [Pa]. The initial flux J_0' was defined as the measured clean water flux (J_0), minus any rapid decrease observed in the flux within the first few minutes of a filtration test. Because CP is expected to develop within the first minutes of filtration, the difference between J_0 and J_0' was assumed to be due to CP [39,45]. The estimated cake fouling coefficient k_c [$\text{m}^{-1} \, \text{m}^{-3} \, \text{m}^2$] quantifies the increase in resistance [m^{-1}] per specific volume filtered [$\text{m}^3 \, \text{m}^{-2}$].

Size exclusion chromatography (SEC) was performed using High Performance Liquid Chromatography (HPLC) (Perkin Elmer, Burnaby, BC, Canada) with DOC detection for the analysis of NOM in the feed and permeate. The method used was adopted from Huber et al. [51]. A TSK HW-50S column (Tosoh, Tokyo, Japan) was used as the stationary phase, and a phosphate buffer (2.5 g/L KH_2PO_4 + 1.5 g/L $Na_2HPO_4 \cdot H_2O$) was used as the mobile phase. The sample injection volume and flow rate were 1 mL and 1 mL/min, respectively. A GE Sievers 900 Turbo Portable TOC Analyzer (GE Sievers, Boulder, CO, USA) with a sampling rate of 4 s and a detection range of 0.2 mg C/L to 10 mg C/L was used as the DOC detector.

3. Results

3.1. NOM Rejection

The amount of material rejected by the membranes was defined as the difference between material present in the feed (i.e., the model raw water) and material present in the permeate. As illustrated in Figure 2, the membranes with MWCOs of 300 Da and 2000 Da could effectively reject all of the alginate present in model raw waters, and the membrane with a MWCO of 8000 Da could reject most of the alginate present in model raw waters. This was expected because the molecular weight of alginate is greater than 10,000 Da [39,51] and the MWCOs of all the membranes considered are lower than 10,000 Da (i.e., ranging from 300 Da to 8000 Da). Only a small amount of the larger constituents in SRNOM was rejected by the membrane with a MWCO of 8000 Da. The extent of the rejection increased as the MWCO of the membranes decreased, with all of the organic material being rejected by the membrane with a MWCO of 300 Da. Again, this was expected, because most of the constituents of SRNOM have a molecular weight in between 300 Da and 8000 Da [49,50].

Note that NOM in most surface waters consists mainly of humic substances (i.e., lower molecular weight NOM) [32]. Because effective NOM removal is one of the main motivations behind implementing NF and tight UF membranes in drinking water treatment, membranes with lower MWCOs (i.e., <2000 Da) are recommended for this application.

a)

b)

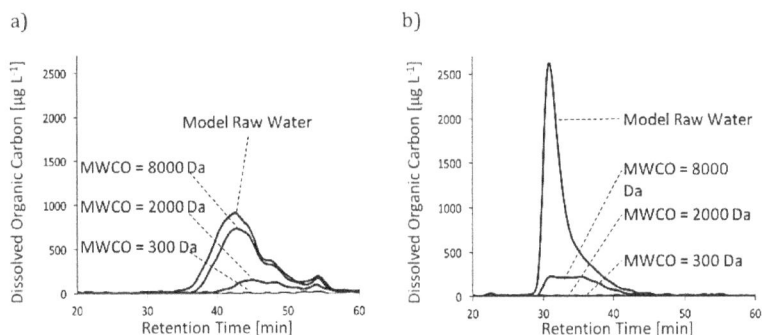

Figure 2. Typical size exclusion chromatograms of model raw water and permeate samples: (**a**) model raw water containing Suwannee River natural organic matter (SRNOM) at 10 mg/L; (**b**) model raw water containing alginate at 10 mg/L. Dalton values correspond to the molecular weight cut offs (MWCOs) of the different membranes; note that the retention time of the chromatograms is inversely proportional to the log of the molecular weight.

3.2. NOM Fouling and Concentration Polarization

Typical results from the filtration tests are presented in Figure 3. For all conditions investigated, the permeate flux remained constant during the pre-clean water filtration phase, then decreased over time (i.e., volume filtered) when filtering model raw waters, and increased during the post-clean water filtration phase.

As previously discussed, CP is characterized by an equilibrium between the convective transport of material towards the membrane, and the diffusive transport of retained material away from the membrane. Steady state is expected to be established within minutes once filtration of model raw water begins [39,45]. Hence, the rapid initial decrease in the permeate flux when filtering model raw waters was attributed to CP, and the subsequent slower longer term decrease in permeate flux was attributed to fouling.

Fouling occurs if the permeation drag is greater than the forces acting on potential foulants away from the membrane [22]. Because the permeation drag remains constant when switching from the model raw water filtration phase to the post-clean water filtration phase, no back transport and no recovery in resistance is expected during post-clean water filtration. However, a recovery in resistance due to CP is expected during post-clean water filtration. This is because a concentration gradient of retained material is still present, for a relatively short period of time, after switching from the model raw water filtration phase to the post-clean water filtration phase. During this time, the convective transport of material towards the membrane is zero (i.e., clean water filtration), while the diffusive transport of retained material away from the membrane surface is greater than zero, resulting in a reduction in resistance (i.e. recovery in resistance). Note, that if CP becomes extensive, and a cake/gel layer is formed, the recovery in resistance during post-clean water resistance will be limited [26,46–48]. Therefore, although a high recovery during post-clean water filtration suggests that CP likely dominates the increase in resistance due to permeate flow, other indicators of fouling and CP must be considered when interpreting the results.

As previously discussed, for all conditions investigated, the permeate flux could be best modeled assuming fouling was predominantly due to the formation of a cake layer on the membrane surface (see Section 2.2). The fit of the cake fouling model (Equation (1)) to typical data is presented in Figure 3. For all conditions investigated, the extent of fouling was quantified with respect to the fouling coefficient, as discussed in Section 3.2.1, and the contributions of CP and fouling to the increase in filtration resistance was analyzed as discussed in Section 3.2.2.

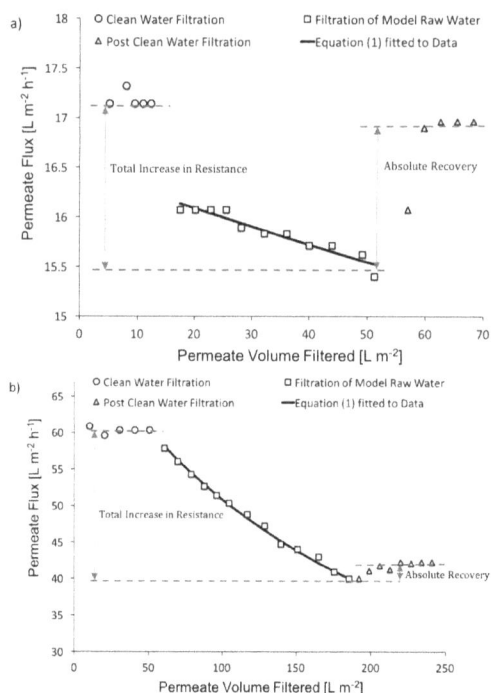

Figure 3. Typical results from filtration tests: (**a**) MWCO = 300 Da and SRNOM at 5 mg/L of DOC, (**b**) MWCO = 8000 Da and alginate at 5 mg/L of DOC.

3.2.1. Contribution of Fouling to the Total Increase in Filtration Resistance

The fouling coefficients (k_c) for the different experimental conditions investigated are summarized in Table 1. When filtering model raw waters containing alginate, the fouling coefficients were similar for the membranes with MWCOs of 300 and 2000 Da and only slightly lower (and not statistically different) for the membrane with the MWCO of 8000 Da. These results are consistent with those presented in Section 3.1, where the rejection of alginate was similar when filtering with membranes with MWCOs of 300 Da and 2000 Da MWCO, and slightly lower when filtering with the membrane with a MWCO of 8000 Da. Considering the substantially different initial permeate fluxes for the membranes with different MWCOs, and the similar fouling coefficients observed for all membranes, it can be concluded that the fouling coefficient was independent of the initial permeate flux. Further, the fouling coefficient was proportional to the concentration of alginate in the model raw water being filtered (i.e., the fouling coefficient doubled when the concentration of alginate in the raw water was twice as high), suggesting that for all membranes, fouling was similar and predominantly due to convective transport and the accumulation of a cake layer with similar characteristics for all membranes. These results are also consistent with the fact that the molecular weight of alginate is greater than the MWCO of all membranes considered (see Section 3.1).

When filtering model raw waters containing SRNOM, the fouling coefficient was substantially higher for the membrane with a MWCO of 300 Da than for the membranes with MWCOs of 2000 and 8000 Da. This is again consistent with the higher rejection of SRNOM by the membrane with a MWCO of 300 Da, than the membranes with MWCOs of 2000 and 8000 Da. It should be noted that the fouling coefficient was similar for the membranes with MWCOs of 2000 and 8000 Da, even though the rejection of organic material present in SRNOM was greater for the membrane with a MWCO of 2000 Da, than for the membrane with a MWCO of 8000 Da. For the membrane with a MWCO of 8000 Da, the fouling coefficient was proportional to the concentration of SRNOM in the model raw water. However, for the membranes with MWCOs of 300 and 2000 Da, the fouling coefficient was not proportional to the concentration of SRNOM in the model raw water. The deviation from proportionality increased as the MWCO of the membrane decreased. These results indicate that, in contrast to filtering model raw water containing alginate, when filtering model raw water containing SRNOM (i.e., using membranes with low MWCOs), fouling was not only governed by the convective transport of material towards the membrane, but also likely by the formation of a CP-induced cake/gel layer at the membrane surface, as discussed in Section 3.2.2.

When filtering model raw waters containing a mixture of SRNOM and alginate, the fouling behavior was similar to that of model raw water containing SRNOM, suggesting that the higher percentage of low molecular weight NOM (i.e., 80% low molecular weight NOM versus 20% high molecular weight NOM) governed the fouling behavior.

Table 1. Fouling coefficient k_c for different filtration tests (ranges correspond to observed minimum/maximum values).

NOM Constituents	NOM Concentration [mg/L]	Fouling Coefficient k_c [10^{11} m^{-1} m^{-3} m^2] Average (Range)		
		MWCO = 300 Da	MWCO = 2000 Da	MWCO = 8000 Da
SRNOM	5	0.7 (0.5–0.9)	0.1 (0–0.1)	0.1 (0.1–0.1)
	10	3.5 (3.5–3.5)	0.6 (0–1.1)	0.2 (0.2–0.2)
Alginate	5	1.1 (0.8–1.6)	0.8 (0.5–1.1)	0.8 (0.7–0.9)
	10	2.1 (2.0–2.2)	2.1 (1.6–2.4)	1.6 (1.5–1.7)
SRNOM + Alginate	4 + 1	0.7 (0.4–1.0)	0.2 (0.1–0.3)	0.2 (0.2–0.2)
	8 + 2	5.1 (4.7–5.5)	1.2 (0.7–2.0)	0.5 (0.5–0.5)

3.2.2. Contribution of CP to the Total Increase in Filtration Resistance

The contribution of CP to the total increase in resistance during filtration of the model raw waters was quantified by the relative recovery. The relative recovery was defined as the ratio of the absolute recovery to the total increase in resistance during filtration of model raw water (Figure 3). The total increase in resistance was calculated based on the difference between the permeate flux during a pre-clean water filtration phase, and at the end of a filtration of model raw water phase. The absolute recovery was calculated as the difference between the permeate flux at the end of a filtration of model raw water phase, and at the end of a post-clean water filtration phase. Figure 4 illustrates the relative recoveries for the different conditions investigated.

The relative recovery generally increased as the MWCO of the membrane decreased. Also, the relative recovery generally increased when the concentration of NOM in the raw water increased. When filtering model raw water containing alginate, the relative recovery was generally low, ranging from approximately 10% to 40%. Also, the fouling coefficient increased proportionally with the concentration of alginate in the model raw water (see Section 3.2.1). This, combined with the low relative recoveries, suggest that the increase in resistance to permeate flow was dominated by fouling when filtering model raw waters containing alginate, at all of the considered concentrations (i.e., 5 mg L^{-1} and 10 mg L^{-1}). However, when filtering model raw waters containing SRNOM, the relative recovery was generally higher, ranging from approximately 50% to 100%. The greater relative recoveries when filtering model raw waters containing SRNOM, rather than alginate, was attributed to the differences in the diffusion coefficients of these types of NOM. Diffusion coefficients of 2.2×10^{-10} to 3.8×10^{-10} m^2 s^{-1} have been reported for humic substances [29], which account for most of the organic material in SRNOM, while that for alginate is in the order of 9.4×10^{-11} m^2 s^{-1}, an estimate based on Dextran T-70 [29]. The diffusive transport of retained material from the membrane back into solution is expected to increase relative to the molecular diffusion [29]. Also, the permeation drag is expected to be greater for alginate than for SRNOM because of the larger size of alginate compared to SRNOM [22]. The relative recovery for the membrane with a MWCO of 300 Da was lower when the concentration of SRNOM in the model raw water increased (i.e., from 5 to 10 mg/L). Under these conditions, the concentration at the membrane surface is expected to be high, and likely resulted in the formation of a cake/gel layer at the membrane surface [26,46–48]. Once formed, the resistance offered by the cake/gel layer could not be reversed during post-clean water filtration, which is consistent with previous research where humic substances fouling was observed to be irreversible [25,26,29,35].

When filtering model raw waters containing a mixture of SRNOM and alginate, the relative recovery was similar to that of model raw waters containing SRNOM, again suggesting that the higher percentage of low molecular weight NOM (i.e., 80% low molecular weight NOM versus 20% high molecular weight NOM) was mainly responsible for the observed changes in permeate flux.

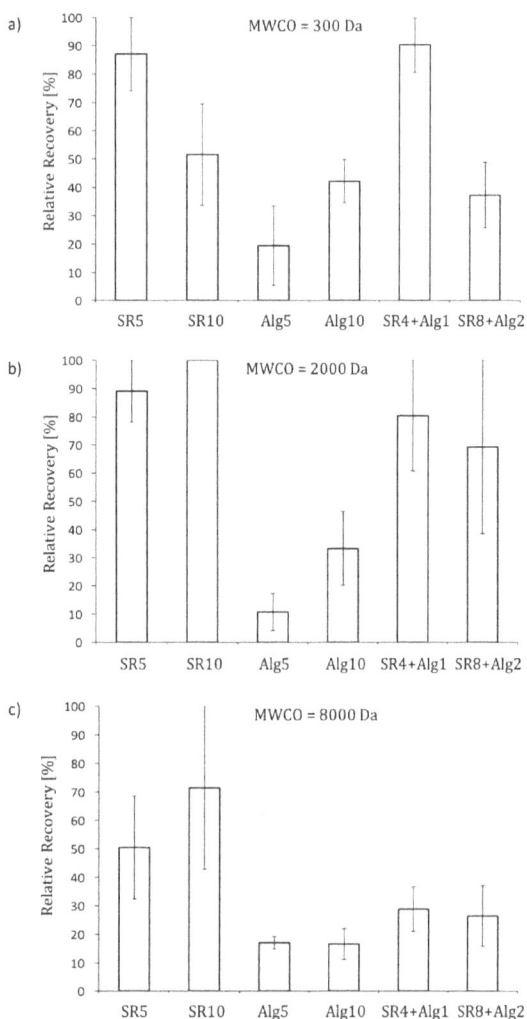

Figure 4. Relative recovery during post-clean water filtration (**a**) MWCO = 300 Da; (**b**) MWCO = 2000 Da; (**c**) MWCO = 8000 Da; SR5: SRNOM at 5 mg/L; SR10: SRNOM at 10 mg/L; Alg5: Alginate at 5 mg/L; Alg10: Alginate at 10 mg/L; SR4 + Alg1: SRNOM at 4 mg/L + alginate at 1 mg/L; SR8 + Alg2: SRNOM at 8 mg/L + alginate at 2 mg/L; error bars correspond to observed minimum/maximum values).

As illustrated in Figure 5, when filtering model raw waters containing SRNOM or a mixture of SRNOM and alginate, a linear correlation was observed between the initial rapid increase in resistance during model raw water filtration and the absolute recovery during post-clean water filtration ($R^2 = 0.97$). When filtering model raw waters containing alginate, no correlation was observed ($R^2 = 0.04$, results not shown). The correlation supports the assumption that the initial rapid increase in resistance is due to CP rather than fouling, that CP develops rapidly within the first minutes of filtration when filtering raw waters containing NOM, and confirms that the relative recovery observed when filtering model raw waters containing SRNOM can be attributed to CP. In addition, these results confirm that the total increase in filtration resistance is mainly impacted by CP when filtering model raw waters containing SRNOM.

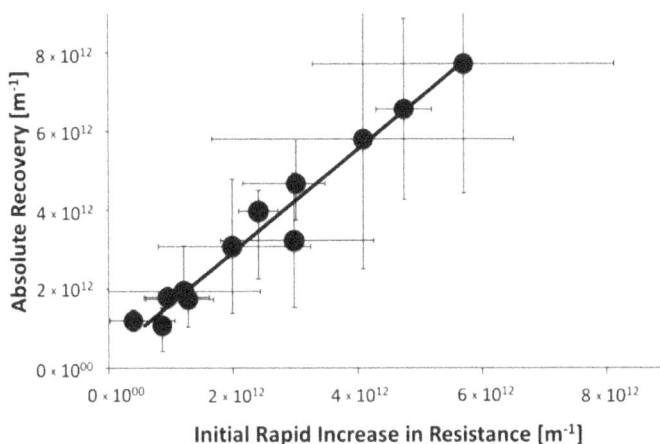

Figure 5. Correlation between Absolute Recovery and Initial Rapid Increase in Resistance (model raw waters containing SRNOM and a mixture of SRNOM + alginate; solid line: linear regression, $R^2 = 0.97$, slope = 1.31 ± 0.35, intercept = 3×10^{11}; $p = 0.05$; error bars correspond to minimum/maximum values).

3.2.3. Overall Impact of Fouling and CP to the Increase in Resistance to Permeate Flow

Overall, the results suggest that when filtering model raw waters containing alginate, the total increases in the filtration resistance is dominated by fouling, while the total increase in filtration resistance is dominated by CP when filtering model raw waters containing SRNOM (i.e., humic substances, lower molecular weight NOM). The impact of CP is greatest for membranes with lower MWCOs (i.e., <2000 Da), which are recommended for the effective removal of NOM in drinking water treatment applications. Filtration using membranes with higher MWCOs (e.g., 8000 Da) is not prone to CP because the lower molecular weight NOM, responsible for CP, is not effectively retained.

It should be noted that when the impact of CP is extensive (i.e., filtration of raw waters containing mainly humic substances/low molecular weight NOM with low MWCO membranes), the increase in resistance is much greater than when filtration was dominated by fouling. For this reason, and because NOM in surface waters consist predominantly of humic substances [32], operation with control measures to continuously limit CP, such as cross-flow operation, should be considered for drinking water treatment using NF and tight UF range membranes. Dead-end operation, which is typically used for drinking water production by UF, is not recommended because it cannot continuously limit CP.

Future research will investigate the impact of control measures (e.g., cross-flow) and different membrane configurations (e.g., hollow fiber versus spiral wound membrane configurations) to mitigate CP and fouling in NF and tight UF membrane systems applied to the treatment of drinking water.

4. Conclusions

The following four main conclusions can be made, based on the results of the present study:

1. Since NOM in most surface waters consists mainly of humic substances (i.e., lower molecular weight NOM), and because effective NOM removal is likely to be one of the main motivations for implementing NF and tight UF range membranes in drinking water treatment, membranes with lower MWCOs (i.e., <2000 Da) are recommended for this application.

2. When filtering raw waters containing high concentrations of alginate (i.e., polysaccharides, high molecular weight NOM), the total increase in the filtration resistance is mainly dominated by fouling; when filtering raw waters containing predominantly humic substances (i.e., lower molecular weight NOM), the total increase in filtration resistance is mainly dominated by CP. The impact of CP is greatest for membranes with low MWCOs (i.e., <2000 Da).

3. When filtering raw waters containing high concentrations of humic substances (i.e., 10 mg L^{-1}) with the membrane with the lowest MWCO (i.e., 300 Da), the impact of CP became extensive and the increase in resistance was greater than that observed for any other experimental condition considered, likely due to the formation of a CP-induced cake/gel layer.

4. Operations with control measures to continuously limit CP, such as cross-flow operation, should be considered in drinking water applications using NF and tight UF range membranes. Dead-end operation, which is typically used for drinking water production using UF, is not recommended because it cannot continuously limit CP.

Acknowledgments: The authors wish to thank IC-Impacts for providing funding for this project.

Author Contributions: Joerg Winter, Benoit Barbeau and Pierre Bérubé conceived and designed the experiments; Joerg Winter performed the experiments; Joerg Winter analyzed the data with the input by Benoit Barbeau and Pierre Bérubé; Joerg Winter wrote the present manuscript.

Conflicts of Interest: The authors declare no conflict of interest.

References

1. Gao, W.; Liang, H.; Ma, J.; Man, M.; Chen, Z.L.; Han, Z.S.; Li, G.B. Membrane fouling control in ultrafiltration technology for drinking water water production: A review. *Desalination* **2011**, *272*, 1–8. [CrossRef]
2. ElHadidy, A.M.; Peldszus, S.; van Dyke, M.I. An evaluation of virus removal mechanisms by ultrafiltration membranes using MS2 and φX174 bacteriophage. *Sep. Purif. Technol.* **2013**, *120*, 215–223. [CrossRef]
3. Song, H.; Shao, J.; He, Y.; Hou, J.; Chao, W. Natural organic matter removal and flux decline with charged ultrafiltration and nanofiltration membranes. *J. Membr. Sci.* **2011**, *376*, 179–187. [CrossRef]
4. Lee, N.; Amy, G.; Croué, J.-P.; Buisson, H. Identification and understanding of fouling in low-pressure membrane (MF/UF) filtration by natural organic matter (NOM). *Water Res.* **2004**, *38*, 4511–4523. [CrossRef] [PubMed]
5. Lee, S.; Lee, K.; Wan, W.M.; Choi, Y. Comparison of membrane permeability and a fouling mechanism by pre-ozonation followed by membrane filtration and residual ozone in membrane cells. *Desalination* **2005**, *178*, 287–294. [CrossRef]
6. Odegaard, H.; Osterhus, S.; Melin, E.; Eikebrokk, B. NOM removal technologies—Norwegian experiences. *Drink. Water Eng. Sci.* **2010**, *3*, 1–9. [CrossRef]
7. Patterson, C.; Anderson, A.; Sinha, R.; Muhammad, N.; Pearson, D. Nanofiltration Membranes for Removal of Color and pathogens in Small Drinking Water Sources. *J. Environ. Eng.* **2012**, *138*, 48–57. [CrossRef]
8. Sadmani, A.H.M.; Anwar, A.; Robert, C.; Bagley, D.M. Impact of natural water colloids and cations on the rejection of pharmaceutically active and endocrine disrupting compounds by nanofiltration. *J. Membr. Sci.* **2014**, *450*, 271–281. [CrossRef]
9. Chin, A.; Bérubé, P. Removal of disinfection by-product precursors with ozone-UV advanced oxidation process. *Water Res.* **2005**, *39*, 2136–2144. [CrossRef] [PubMed]
10. Croue, J.-P.; Korshin, G.; Mark, B.M. *Characterization of Natural Organic Matter in Drinking Water*; AWWA Research Foundation (AWWARF): Denver, CO, USA, 1999.

11. Hozalski, R.M.; Bouwer, E.J.; Goel, S. Removal of natural organic matter from drinking water supplies by ozone-biofiltration. *Water Sci. Technol.* **1999**, *40*, 157–163. [CrossRef]

12. Rittmann, B.E.; Stilwell, D. Modelling biological processes in water treatment: The integrated biofiltration model. *J. Water Supply Res. Technol. AQUA* **2002**, *51*, 1–14.

13. Sarathy, S.; Mohseni, M. The fate of natural organic matter during UV/H₂O₂ advanced oxidation of drinking water. *Can. J. Civ. Eng.* **2009**, *36*, 160–169. [CrossRef]

14. Wang, G.-S.; Hsieh, S.-T. Monitoring natural organic matter in water with scanning spectrophotometer. *Environ. Int.* **2001**, *26*, 205–212. [CrossRef]

15. Bequet, S.; Remigy, J.C.; Rouch, J.C. From ultrafiltration to nanofiltration hollow fiber membranes: A continuous UV-photografting process. *Desalination* **2002**, *144*, 9–14. [CrossRef]

16. Frank, M.; Bargeman, G.; Zwijnenburg, A. Capillary hollow fiber nanofiltration membranes. *Sep. Purif. Technol.* **2001**, *22*, 499–506. [CrossRef]

17. Spruck, M.; Hoefer, G.; Fili, G.; Gleinser, D.; Ruech, A.; Schmidt-Baldassari, M.; Rupprich, M. Preparation and characterization of composite multichannel capillary membranes on the way to nanofiltration. *Desalination* **2013**, *314*, 28–33. [CrossRef]

18. Verberk, J.Q.J.C.; van Dijk, J.C. Air Sparging in Capillary Nanofiltration. *J. Membr. Sci.* **2006**, *284*, 339–351. [CrossRef]

19. Geraldes, V.; Semião, V.; de Pinho, M.N. Flow management in nanofiltration spiral wound modules with ladder-type spacers. *J. Membr Sci.* **2002**, *203*, 87–102. [CrossRef]

20. Thorsen, T.; Flogstat, H. Nanofiltration in Drinking Water Treatment—Literature Review. Available online: https://www.techneau.org/fileadmin/files/Publications/Publications/Deliverables/D5.3.4b.pdf (accessed on 13 March 2017).

21. Song, L.; Elimelech, M. Particle deposition onto a permeable surface in laminar flow. *J. Colloid Interface Sci.* **1995**, *173*, 165–180. [CrossRef]

22. Belfort, G.; Davis, R.H.; Zydney, A.L. The behavior of suspensions and macromolecular solutions in cross-flow microfiltration. *J. Membr. Sci.* **1994**, *96*, 1–58. [CrossRef]

23. Fonseca, A.C.; Summers, R.S.; Greenberg, A.R.; Hermandez, M.T. Extra-cellular Polysaccharides, soluble microbial products, and natural organic matter impact on nanofiltration flux decline. *Environ. Sci. Technol.* **2007**, *41*, 2491–2497. [CrossRef] [PubMed]

24. Huang, H.; Lee, N.; Young, T.; Amy, G.; Lozier, J.C.; Jacangelo, J.G. Natural organic matter fouling of low-pressure, hollow-fiber membranes: Effects of NOM source and hydrodynamic conditions. *Water Res.* **2007**, *41*, 3823–3832. [CrossRef] [PubMed]

25. Jermann, D.; Pronk, W.; Kägi, R.; Halbeisen, M.; Boller, M. Influence of interactions between NOM and particles on UF fouling mechanism. *Water Res.* **2008**, *42*, 3870–3878. [CrossRef] [PubMed]

26. Schäfer, A.I.; Fane, A.G.; Waite, T.D. Nanofiltration of natural organic matter: Removal, fouling and influence of multivalent ions. *Desalination* **1998**, *118*, 109–122. [CrossRef]

27. Zheng, X.; Ernst, M.; Jekel, M. Identification and quantification of major organic foulants in treated domestic wastewater affecting filterability in dead-end ultrafiltration. *Water Res.* **2009**, *43*, 238–244. [CrossRef] [PubMed]

28. Her, N.; Amy, G.; Park, H.-R.; Song, M. Characterizing algogenic organic matter (AOM) and evaluating associated NF membrane fouling. *Water Res.* **2004**, *38*, 1427–1438. [CrossRef] [PubMed]

29. Schäfer, A.I. *Natural Organics Removal Using Membranes—Principles, Performance and Cost*; Technomic Publishing: Lancaster, PA, USA, 2007.

30. Zularisam, A.W.; Ismail, A.F.; Salim, M.R.; Sakinah, M.; Ozaki, H. The effects of natural organic matter (NOM) fractions on fouling characteristics and flux recovery of ultrafiltration membranes. *Desalination* **2007**, *212*, 191–208. [CrossRef]

31. Winter, J.; Uhl, W.; Bérubé, P.R. Integrated Oxidation Membrane Filtration Process—NOM Rejection and Membrane Fouling. *Water Res.* **2016**, *104*, 418–424. [CrossRef] [PubMed]

32. Zumstein, J.; Buffle, J. Circulation of Pedogenic and Aquagenic Organic Matter in an Eutrophic Lake. *Water Res.* **1989**, *23*, 229–239. [CrossRef]

33. Amy, G.; Her, N. Size exclusion chromatography (SEC) with multiple detectors: A powerful tool in treatment process selection and performance. *Water Sci. Technol.* **2004**, *4*, 19–24.

34. Biber, M.V.; Guelacar, F.O.; Buffle, J. Seasonal Variations in Principal Groups of Organic Matter in a Eutrophic Lake using Pyrolysis/GC/MS. *Environ. Sci. Technol.* **1996**, *30*, 3501–3507. [CrossRef]

35. Jermann, D.; Pronk, W.; Meylan, S.; Boller, M. Interplay of different NOM fouling mechanisms during ultrafiltration of drinking water production. *Water Res.* **2007**, *41*, 1713–1722. [CrossRef] [PubMed]

36. Laabs, C.; Amy, G.; Jekel, M. Understanding the Size and character of Fouling-Causing Substances from Effluent Organic Matter (EfOM) in Low Pressure Membrane filtration. *Environ. Sci. Technol.* **2006**, *40*, 4495–4499. [CrossRef] [PubMed]

37. Wray, H.E.; Andrews, R.C.; Bérubé, P.R. Surface shear stress and membrane fouling when considering natural water matrices. *Desalination* **2013**, *330*, 22–27. [CrossRef]

38. Jarusutthirak, C.; Amy, G.; Croue, J.-P. Fouling characteristics of wastewater effluent organic matter (EfOM) isolates on NF and UF membranes. *Desalination* **2002**, *145*, 247–255. [CrossRef]

39. Seidel, A.; Elimelech, M. Coupling between chemical and physical interactions in natural organic matter (NOM) fouling of nanofiltration membranes: Implications for fouling control. *J. Membr. Sci.* **2002**, *203*, 245–255. [CrossRef]

40. Sari, M.A.; Chellam, S. Surface water nanofiltration incorporating (electro) coagulation-microfiltration pretreatment: Fouling control and membrane characterization. *J. Membr. Sci.* **2013**, *437*, 249–256. [CrossRef]

41. Cho, J.; Amy, G.; Pellegrino, J. Membrane filtration of natural organic matter: Initial comparison of rejection and flux decline characteristics with ultrafiltration and nanofiltration membranes. *Water Res.* **1999**, *33*, 2517–2526. [CrossRef]

42. Her, N.; Amy, G.; Plottu-Pecheux, A.; Yoon, Y. Identification of nanofiltration foulants. *Water Res.* **2007**, *41*, 3936–3947. [CrossRef] [PubMed]

43. Listiarini, K.; Chun, W.; Sun, D.D.; Leckie, J. Fouling mechanism and resistance analyses of systems containing sodium alginate, calcium, alum and their combination in dead-end fouling of nanofiltration membranes. *J. Membr. Sci.* **2009**, *344*, 244–251. [CrossRef]

44. Mahlangu, T.O.; Thwala, J.M.; Mamba, B.B.; D'Haese, A.; Verliefde, A.R.D. Factors governing combined fouling by organic and colloidal foulants in cross-flow nanofiltration. *J. Membr. Sci.* **2015**, *49*, 53–62. [CrossRef]

45. Schäfer, A.I.; Fane, A.G.; Waite, T.D. *Nanofiltration: Principles and Applications*; Elsevier: Oxford, UK; New York, NY, USA, 2005.

46. Mulder, M. *Basic Principles of Membrane Technology*, 2nd ed.; Kluwer Academic: Dordrecht, The Netherlands, 1996.

47. Van den Berg, G.B.; Smolders, C.A. Flux Decline in Ultrafiltration Processes. *Desalination* **1990**, *77*, 101–133. [CrossRef]

48. Elimelech, M.; Bhattacharjee, S. A novel approach for modeling concentration polarization in crossflow membrane filtration based on the equivalence of osmotic pressure model and filtration theory. *J. Membr. Sci.* **1998**, *145*, 223–241. [CrossRef]

49. Chin, Y.-P.; Aiken, G.; O'Loughlin, E. Molecular Weight, Polydispersity, and Spectroscopic Properties of Aquatic Humic Substances. *Environ. Sci. Technol.* **1994**, *28*, 1853–1858. [CrossRef] [PubMed]

50. Her, N.; Amy, G.; Foss, D.; Cho, J.; Yoon, Y.; Kosenka, P. Optimization of Method for Detecting and Characterizing NOM by HPLC-Size Exclusion Chromatography with UV and On-Line DOC detection. *Environ. Sci. Technol.* **2002**, *36*, 1069–1076. [CrossRef] [PubMed]

51. Huber, S.A.; Balz, A.; Abert, M.; Pronk, W. Characterisation of aquatic humic and non-humic matter with size-exclusion chromatograpgy-organic carbon detection-organic nitrogen detection (LC-OCD-OND). *Water Res.* **2011**, *45*, 879–885. [CrossRef] [PubMed]

52. Cornel, P.K.; Summers, S.R.; Roberts, P.V. Diffusion of Humic Acid in Dilute Aqueous Solution. *J. Colloid Interface Sci.* **1986**, *110*, 149–164. [CrossRef]

53. Bhattacharjee, S.; Chen, J.C.; Elimelech, M. Coupled model of concentration polarization and pore transport in cross-flow nanofiltration. *AIChE J.* **2001**, *47*, 2733–2745. [CrossRef]

54. Brian, P.L.T. Concentration Polarization in Reverse Osmosis Desalination with Variable Flux and Incomplete Salt Rejection. *Ind. Eng. Chem. Fundam.* **1965**, *4*, 439–445. [CrossRef]

55. Hoek, E.M.V.; Elimelech, M. Cake-Enhanced Concentration Polarization—A New Fouling Mechanism for Salt-Rejecting Membranes. *Environ. Sci. Technol.* **2003**, *37*, 5581–5588. [CrossRef] [PubMed]

56. Rautenbach, R.; Albrecht, R. *Membrane Processes*; John Wiley & Sons: New York, NY, USA, 1989.

57. Yoon, Y.; Amy, G.; Cho, J.; Her, N. Effects of retained natural organic matter (NOM) on NOM rejection and membrane flux decline with nanofiltration and ultrafiltration. *Desalination* **2005**, *173*, 209–221. [CrossRef]

58. Hincks, S.S.; Mackie, G.L. Effect of pH, calcium, alkalinity, hardness, and chlorophyll on the survival, growth, and reproductive success of zebra mussel (Dreissena polymarpha) in Ontario lakes. *Can. J. Fish. Aquat. Sci.* **1997**, *54*, 2049–2057. [CrossRef]

59. Hermia, J. Constant Pressure Blocking Filtration Laws—Application to Power-Law Non-Newtonian Fluids. *Chem. Eng. Res. Des.* **1982**, *60*, 183–187.

membranes

MDPI

Article

Poly(vinylbenzylchloride) Based Anion-Exchange Blend Membranes (AEBMs): Influence of PEG Additive on Conductivity and Stability

Jochen A. Kerres [1,2,*] and Henning M. Krieg [2]

[1] Institute of Chemical Process Engineering, University of Stuttgart, 70199 Stuttgart, Germany
[2] Faculty of Natural Science, North-West University, Focus Area: Chemical Resource Beneficiation, Potchefstroom 2520, South Africa; henning.krieg@nwu.ac.za
* Correspondence: jochen.kerres@icvt.uni-stuttgart.de; Tel.: +49-711-68585-244

Academic Editor: Tongwen Xu
Received: 4 May 2017; Accepted: 7 June 2017; Published: 16 June 2017

Abstract: In view of the many possible applications such as fuel cells and electrolysers, recent interest in novel anion exchange membranes (AEMs) has increased significantly. However, their low conductivity and chemical stability limits their current suitability. In this study, the synthesis and characterization of several three- and four-component anion exchange blend membranes (AEBMs) is described, where the compositions have been systematically varied to study the influence of the AEBM's composition on the anion conductivities as well as chemical and thermal stabilities under strongly alkaline conditions. It was shown that the epoxide-functionalized poly(ethylene glycol)s that were introduced into the four-component AEBMs resulted in increased conductivity as well as a marked improvement in the stability of the AEBMs in an alkaline environment. In addition, the thermal stability of the novel AEBMs was excellent showing the suitability of these membranes for several electrochemical applications.

Keywords: anion-exchange blend membrane; poly(vinylbenzylchloride); polybenzimidazole; ionic cross-link; Cl⁻ conductivity; impedance spectroscopy; TGA-FTIR coupling

1. Introduction

Over the past decades, the research interest in anion exchange membranes (AEMs) for electrochemical conversion processes has increased significantly with possible applications of AEMs in alkaline polymer electrolyte fuel cells (APEFCs) [1], alkaline polymer electrolyte electrolysis (APEE) [2], redox flow batteries (RFBS) [3], reverse electrodialysis (RED) [4] and bio-electrochemical systems, including microbial fuel cells (MFC) [5] and enzymatic fuel cells [6]. In addition, AEMs are used in electrodialysis (ED) [7] and Donnan Dialysis [8]—or diffusion dialysis (DD) [9]. A significant advantage of using AEMs in electrochemical conversion processes such as fuel cells or electrolysis is that no noble metal catalysts (consisting of platinum group metals (PGM)) are required for the electrocatalytic reactions at the electrodes, as was shown by Piana et al. [1], which implies that AEM containing membrane electrode assemblies (MEAs) are potentially much cheaper than cation exchange membranes (CEM) containing MEAs. However, AEMs have, compared to CEM, the following disadvantages:

- The ionic conductivity of AEMs is significantly lower than that of CEMs despite having comparable ion exchange capacity (IEC), which is partly due to the fact that most AEM types have a hydrocarbon backbone, which is significantly less hydrophobic than, for example, the perfluorinated polymer main chain of the perfluorinated membranes of the Nafion® (Fayetteville, NC, USA) type, leading to a smaller separation between ionic groups and the polymer backbone and, therefore, to lower ionic conductivities due to the lower local density of the anion-exchange groups, particularly since in most AEM types the cationic head groups are connected to the polymer backbone only via a CH_2 (benzylic) bridge [10] which hinders clustering of the anion-exchange groups.

- If the AEMs are exchanged specifically with OH^- counterions, their chemical stability is limited when used in APEFCs or APEEs, since the OH^- counterion can degrade the cationic head group [11] or even the polymer backbone [12].

Most current research aims at minimizing these disadvantages of AEMs thereby improving their properties. The starting polymers for commonly used AEM polymers are aromatic group-containing polymers, such as polystyrene, polyphenylene ether, or other aromatic polyethers such as polyethersulfones or polyether ketones which are substituted with methyl groups. The first step for the preparation of an AEM is the introduction of halomethyl groups into the polymer. This halomethylation is achieved by either: (1) bromomethylation with hydrogen halide, formaldehyde and a Lewis acid such as $ZnCl_2$ or $AlCl_3$ (Blanc reaction [13,14]); or (2) bromination of the $-CH_3$ side groups of aromatic polymers using *N*-bromo-succinimide (NBS) and a radical starter as the bromination agent (Wohl–Ziegler reaction [15]). Since the Blanc reaction includes the strongly carcinogenic intermediate *bis*(chloromethyl) ether, the Wohl–Ziegler reaction is currently preferably used for the synthesis of halomethylated aromatic polymers. Literature examples of the preparation of bromomethylated aromatic polymers by the Wohl–Ziegler reaction are the bromomethylation of poly(phenylene oxide) [16], or methylated poly(ether sulfone) [17]. The conversion of the $-CH_2Hal$ group (Hal = Cl, Br) to an anion exchange group can be attained by the reaction with a tertiary amine such as trimethylamine [14], pyridine [18], pentamethylguanidine [19] or an *N*-alkylated (*benz*)imidazole [20]. The conductivity of AEMs can be increased by increasing the distance between the polymer backbone and the cationic head groups thereby obtaining a greater local concentration of ion-conducting groups. Phase-segregated AEMs with improved ionic conductivity can be achieved by preparing linear (multi) block copolymers composed of hydrophobic and ion-containing blocks [21] or graft copolymers with an anion-exchange graft side chain [22] (for example, by the grafting of vinyl benzyl side chains onto an e^--irradiated ETFE (ethylene-tetrafluoroethylene copolymer) film and by quaternizing the chloromethylated side chains with trimethylamine [23]). Another promising approach to improve the distance between the backbone and cationic group, to facilitate the clustering of anion-exchange groups to increase the anion conductivity, is to place the anion-exchange groups either within or at the end of an aliphatic side chain or both. For example, Dang and Jannasch have recently prepared novel PPO AEMs by a four-step procedure [24]: First, PPO was lithiated at the methyl groups, followed by quenching with an excess of a dihalogenoalkane, e.g., dibromohexane. The formed PPO-AEMs with the halomethyl group at the end of the aliphatic side chain were then reacted with an excess of a hexanediamine, e.g. N,N,N′,N′-tetramethyldiaminohexane, affording AEMs with the cationic group between the first and the second aliphatic group. The tertiary amino group remaining at the end of the side chain was then quaternized with CH_3I, ending up with AEMs containing two ammonium groups in the side chain. These AEMs showed the expected high anion conductivity and, surprisingly, excellent chemical stabilities in an alkaline environment which was ascribed to the absence of chemically labile benzylic links in these AEMs. Hickner et al. used a similar multication approach for the synthesis of side-chain AEMs based on PPO. Similar to those of the Jannasch group, the membranes were highly conductive (up to 80 mS·cm^{-1}) and remarkably stable (15%–50% conductivity loss after 20 days in 1 M NaOH at 80 °C) [25]. In order to achieve an improvement of the chemical stability of AEMs, the combination of the anion exchange groups and the polymer main chain must be studied as the stability of the anion exchange always also depends on the polymeric main chain. For example, it was shown that a benzyltrimethylammonium (BTMA) head group bound to a poly(styrene) backbone shows slightly better chemical stability in an alkaline medium (0.6 M KOH, 80 °C) than BTMA bound to PPO and a much higher chemical stability than BTMA bound to a poly(phenylene ether sulfone (PES)) backbone [12]. It is obviously not easy to predict which polymer backbone is stable, which can be seen from the above example, since all three polymers

(PSt, PPO and PES) contain aromatic groups of comparable electron density. However, it has been found that the alkali stability of the cationic head groups of AEMs can be significantly improved by sterically shielding the anion-exchange groups from nucleophilic OH^- counterion attack. In a study by Thomas et al., two different polybenzimidazolium- ($PBIm^+$) based AEMs were studied in terms of their stability in an alkaline environment [26]. One of the $PBIm^+$-AEMs contained methyl groups adjacent to the dimethylbenzimidazolium cation, the other not. While the sterically hindered $PBIm^+$-AEM showed high stability in 2 M KOH, the unhindered one was rapidly degraded. The high stability of the hindered $PBIm^+$-AEM was explained as follows: at the sterically hindered $PBIm^+$-AEM, the OH^- group cannot attack the imidazolium ring. At the unhindered $PBIm^+$-AEM, the OH^- group can attack the imidazolium ring leading to ring opening [27]. From these results, it is clear that shielding the anion-exchange groups from a possible OH^- attack by introducing bulky groups in the vicinity of the anion-exchange group is a promising concept. Other groups have also synthesized AEMs where the cationic group is surrounded by bulky functional groups. For example, Liu et al. synthesized sterically highly hindered PPO AEMs containing 1,4,5-trimethyl-2-(2,4,6-trimethoxyphenyl) imidazolium head groups, which possessed excellent alkaline stabilities (no decrease of ion exchange capacity (IEC) after 25 h storage in 1 M KOH at 80 °C [28]). Contrary, dimethylimidazolium group-modified PPO showed a strong IEC decrease (approximately 50% in 2 M KOH at 60 °C) after nine days [29]. Other strategies to reduce the chemical degradation of AEMs include:

- search for alternative fixed cations;
- chemical and/or physical cross-linking; and
- embedding of the anion-exchange polymers in an inert polymer matrix.

As an alternative to the most commonly used trialkylammonium cationic groups, the already mentioned pentamethylguanidinium group (PMG) has been introduced. However, it was found that the PMG cations are only chemically stable when they are resonance-stabilized, i.e., the positive charge of the PMG-cation is delocalized, which is the case when the PMG group is attached to an aromatic group which should preferably be electron-deficient, as shown by Kim et al. [30,31]. Another example of a cationic functional group chemically stabilized by steric hindrance is the tris (2,4,6-trimethoxyphenyl) phosphonium cation, which was bound to polyvinylbenzyl-graft chains [32], exhibiting no degradation after 75 h of storage in 1 N NaOH at 60 °C. In a study by Zha et al., a positively charged *bis*(terpyridine) ruthenium (II) complex was attached to a norbornene polymer [33]. This AEM exhibited excellent stability in alkaline environment as immersion of the polymer in 1 N NaOH at room temperature showed no degradation even after half a year. Another option of stabilizing AEMs is crosslinking. He et al. reported the synthesis of mechanically robust PPO-based AEMs which have been covalently cross-linked (via quaternization) in a multistage process with tertiary diamines and vinyl benzyl chloride [34]. In a study by Cheng et al. chloromethylated PSU was cross-linked (via quaternization) with the new N-basic difunctional reagent guanidimidazol. These novel crosslinked polymers had better stability to alkali than similar AEMs, which have been quaternized with 1-methylimidazole without cross-linking [35].

In our group, bromomethylated PPO was quaternized with the diamine DABCO and 1.4-diiodobutane and embedded in the matrix polymer PVDF. Mechanically and chemically stable covalently crosslinked AEMs were obtained, which exhibited no degradation (IEC and conductivity) even after 10 days of storage in 1 N KOH at 90 °C, while yielding a high performance in direct methanol fuel cells (DMFC) (4 M MeOH and 5 M KOH) [36]. In another study methylated PBIOO obtained from Fuma-Tech (Bietigheim-Bissingen, Germany), which had been prepared using a novel non-carcinogenic reagent, was blended with sulfonated PSU and covalently crosslinked (quaternized) using DABCO and 1.4-diiodobutane [37]. In a DMFC using non-platinum catalysts (anode: 6% $Pd/CeO_2/C$, cathode: 4% FeCo/C) these AEMs at 80 °C (anode Feed 4 M MeOH + 5 M KOH) performed comparable to that of a commercial Tokuyama-AEM (maximum power density 120 mW/cm^2). In another study from our group, ionic and covalently crosslinked AEM blends were synthesized with bromomethylated PPO or a bromomethylated and partially fluorinated arylene main-chain polymer and a partially fluorinated PBI (F_6PBI) as the mechanically and chemically stable matrix and sulfonated polyethersulfone sPPSU as the ionic macromolecular cross-linker. The halomethylated blend component was quaternized with N-methylmorpholine (NMM) [38]. Due to the interaction of the sulfonate groups of the sulfonated polymer and the basic N-methylmorpholinium cations, ionic crosslinking sites were formed, leading to an improved mechanical and chemical stability of the AEM blends. Covalent cross-links in the membrane matrix were formed during evaporation of the solvent by the reaction of a minor portion of the $-CH_2Cl$ groups of the PVBCl with some of the N–H-functional groups of the PBI blend components [39]. The alkaline stability of the membranes was investigated in 1 M KOH at 90 °C over a period of 10 days and compared to a commercial AEM of Tokuyama (A201). The most stable of the manufactured AEM blends lost about 40% of their initial Cl^- conductivity, while the commercial A201 possessed only 21% of the initial conductivity after this time period. Similar AEM blends were synthesized in a further study where NBS-brominated PPO was blended with PBIOO or F_6PBI as the matrix polymer, while sPPSU was added as ionic cross-linker to the blend of brominated PPO and F_6PBI. The quaternization of bromomethylated PPO leading to the anion-exchange groups was attained with 1-methylimidazole or 1-ethyl-3-methylimidazole [40]. After the stability tests (1 M KOH, 90 °C, 10 days), the blend membrane of 1-methylimidazole-quaternized PPO, F_6PBI and sPPSU showed a conductivity of 69% of the original conductivity, while blends from PBIOO and PPO quaternized with the two imidazoles mentioned above exhibited a residual ion conductivity which was 31–43% of the original value.

A further interesting new approach was suggested by Hickner et al. where they synthesized mechanically strong and chemically stable AEMs from rigid-flexible semi-interpenetrating networks with trimethylammonium PPO as the rigid compound and a flexible network built up from poly(ethylene glycol)s (PEGs). These membranes possessed, probably due to the high hydrophilicity of PEG, high OH^- conductivities (up to 75 mS·cm^{-1}), and good alkaline stabilities (conductivity decreased 25%–61% after 30 days at 80 °C in 1 M NaOH) [41].

In this study, several of the approaches described above were combined resulting in two types of novel anion-exchange blend membranes (AEBMs). Poly(vinylbenzylchloride) (PVBCl), was used as the AEM precursor, as this polymer is known to form stable AEMs [12], while different sterically hindered tertiary N-basic compounds were used for the quaternization of the PVBCl. The two novel AEBMs consisted of:

(a)   Three-component blends (after quaternization of the chlormethyl groups with the tertiary N-basic compounds) consisting of PVBCl, a sterically hindered tertiary N base, the polybenzimidazoles PBIOO or F_6PBI, and a nonfluorinated and a partially fluorinated aromatic sulfonated polyether as ionic cross-linker (Figure 1a);

(b)   Four-component blends (after quaternization of the chlormethyl groups with the tertiary N-basic compounds) consisting of PVBCl, a sterically hindered tertiary N base, PBIOO, a nonfluorinated aromatic sulfonated polyethersulfone polymer as ionic cross-linker, and poly(ethylenglycol)s (PEGs) with different chain lengths and epoxide groups at the end of the PEG chain for anchoring in the PBI matrix by reacting the epoxy groups with the N–H groups of the imidazole moiety (Figure 1b). The reaction of the epoxy end groups of the PEG with the N-H groups of the PBI blend component is illustrated in Figure 2.

In this study, the synthesis and characterization of several three- and four-component AEBMs is described, where the compositions were systematically varied to study the influence of the AEMBs composition on the resulting properties. AEBMs prepared from nonfluorinated blend components (PBIOO and sPPSU) were compared to those containing partially fluorinated polymeric compounds (F_6PBI and sFPE (FPE = fluorinated polyether)). In the framework of the investigations, a particular focus was to study the influence of the AEBM composition on the anion conductivities and chemical and thermal stabilities under strongly alkaline conditions.

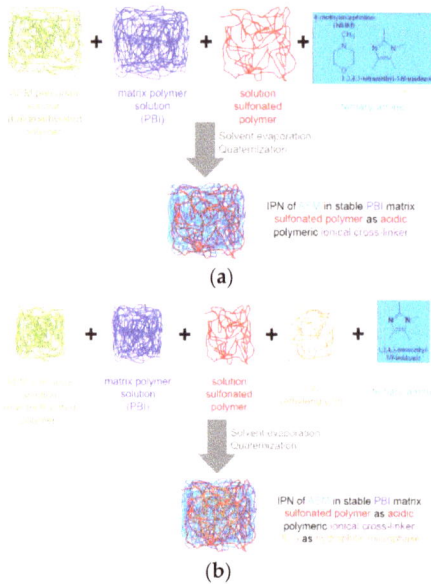

Figure 1. Preparation schemes for the: (**a**) three-component; and (**b**) four-component AEBMs.

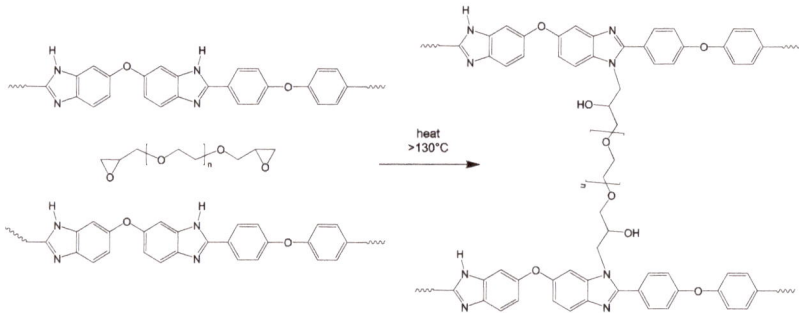

Figure 2. Reaction of epoxide end groups of PEGs with the imidazole groups of PBI.

Figure 3 shows the polymers used for the preparation of the AEBMs.

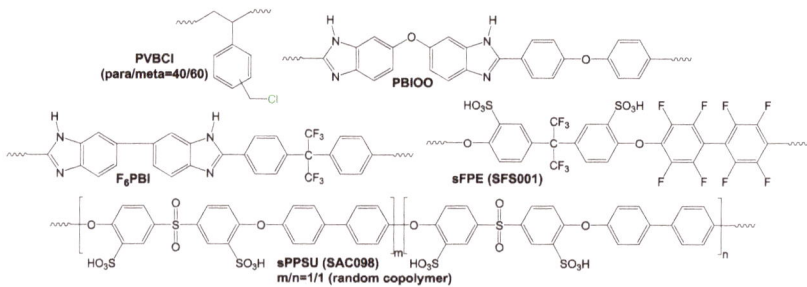

Figure 3. Polymers used for the AEBMs.

2. Results and Discussion

2.1. Membrane Properties

2.1.1. Brief Discussion of the Physico-Chemical Properties of the AEBMs

In this study, three- and four-component AEBMs using poly(vinylbenzylchloride) as the anion-exchange precursor polymer were prepared, where the composition was systematically varied in terms of the blend component type and mass ratio. Most membranes were prepared using TMIm as the tertiary N compound as this sterically hindered methylated imidazole was found to be more stable in alkaline environments than other imidazoles that were used in an earlier study [40]. For comparative purposes, two AEBMs quaternized with PMP (2256) and NMM (2257), respectively, were also added to this study.

Before varying the composition of the blend membranes, the extent of quaternization of the CH_2Cl groups of poly(vinylbenzylchloride) (PVBCl) with TMIm was investigated by Attenuated Total Reflection Fourier-Transform Infrared Spectroscopy (ATR-FTIR). According to the literature, the stretching vibrations of the CH_2Cl group of PVBCl are found in the vicinity of 675, 709 and 1267 cm^{-1} [42]. Since the band around 1267 cm^{-1} is very strong, this band was selected as indication of unreacted CH_2Cl groups in the AEBMs. In Figure 4, the ATR-FTIR spectra of several of the AEBMs are compared to that of PVBCl (in this study, the CH_2Cl band was found at 1279 cm^{-1}), clearly indicating that no unreacted CH_2Cl groups were present in the AEBMs.

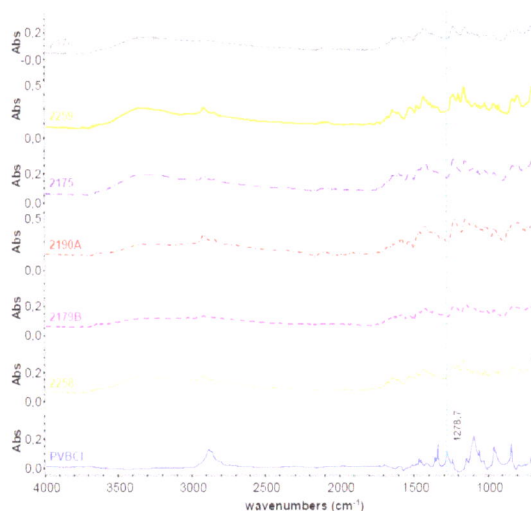

Figure 4. FTIR spectra of PVBCl and several of the prepared AEBMs.

The membrane compositions were varied in terms of: (a) the nonfluorinated/partially fluorinated content (for the nonfluorinated AEBMs, PBIOO was used as matrix polymer and sulfonated PPSU as the ionic macromolecular cross-linker, and for the partially fluorinated AEBMs, F_6PBI as matrix polymer and SFPE001 as the ionic macromolecular cross-linker (Table 1, Figure 3); (b) the different anion-exchange polymer contents leading to different IECs of the AEBMs; and (c) two different types of PEGs (PEG-diglycidylether with M_W = 500 Da or 6000 Da) and their amounts in the AEBMs resulting in varying hydrophilicity of the blends. The purpose of these variations was to investigate their impact on important membrane properties such as membrane conductivity, alkaline stability and thermal stability.

In this section, the membrane properties are discussed by first presenting an overview of important membrane properties such as the IEC values, conductivities and water uptake values obtained for the various membranes, followed by a more detailed discussion of the membrane conductivities (Section 2.1.2) and the thermal stabilities using thermogravimetry (TGA) (Section 2.1.3). It was of particular interest during this study to determine how the membrane properties such as conductivity and thermal stability would change during alkaline treatment as both alkaline stability and anion conductivity are the most important prerequisites for the suitability of AEMs for alkaline membrane fuel cells or alkaline membrane electrolysis (see Introduction). TGA was used to determine the alkaline stability of the AEBMs since alkaline degradation would most probably lead to a change in the shape of the TGA traces by anion-exchange group splitting-off and/or backbone degradation due to the attack of the OH^- ions.

In Table 1, the most important AEBM data including IEC, conductivity (measured in a 1 N NaCl solution using the in-plane setup (see Section 3.3.2)) and water uptake are listed. As stated previously, the –3C and –4C in the membrane number refer to the three- and four-component AEBMs, respectively, NF refers to AEBMs composed of nonfluorinated blend components, while F stands for AEBMs containing the partially fluorinated F_6PBI and SFPE001.

Table 1. Physico-chemical properties of three- and four-component AEBMs.

Membrane (No.)	IEC_{calc} (meq OH^-/g)	IEC_{exp} (meq OH^-/g)	IEC_{exp} * (meq OH^-/g)	Weight Loss ** (%)	σ_{Cl}^- (S/cm)	σ_{Cl}^-* (S/cm)	WU (%, RT)	WU (%, 90 °C)
2256_3C_NF	2.3	2.9	3.17	2.4	8.2	1.1	49	51
2257_3C_NF	2.3	0.39	0.45	7.3	0.17	0.12	7	8
2179B_3C_NF	2.3	2.5	2.64	5.8	10.7	15.9	63	100
2223_3C_NF	2.0	2.97	3.15	2.5	10.8	5.4	56	79
2177_4C_NF	2.3	2.2	2.44	3.4	14.0	10.6	36	42
2175_4C_NF	2.3	2.92	2.96	2.4/0 §	29.3	72.7	367	n.m.
2176_4C_NF	2.3	2.79	2.84	2.6/3 §	21.6	69.9	370	436
2190A_4C_NF	2.0	2.1	2.73	4.1	14.3	16.3	67	120
2261_4C_NF	2.3	2.2	n.a.	0.6	42.8	32.2	132	212
2262_4C_NF	2.6	n.a.	3.53	2.1	72.9	56.9	242	377
2264_4C_NF	2.0	3.07	2.93	0	14.0	11.1	71	90
2265_4C_NF	1.7	3.07	3.04	0	4.9	1.3	41	47
2279_4C_NF	1.4	2.89	3.04	0	0.97	0.1	28	28
2267_4C_NF	2.3	2.92	2.84	0	51.7	60.5	290	371
2246_4C_F	2.0	2.23	2.38 ***	2.1	3.9	4.4 ***	29	34
2258_4C_F	2.3	2.42	2.53	1.9	26.4	16.1	77	91
2259_4C_F	2.6	2.41	2.74	0.9	45.4	46.8	157	258

* After 10 days in 10% KOH at 90 °C; ** after extraction with DMAc at 90 °C; *** after 30 days in 10% KOH at 90 °C; § weight loss after 10 days of KOH storage.

The characterization results presented in Table 1 are briefly discussed below.

(a) Influence of quaternized cationic groups:

The TMIm-based AEBMs had significantly higher Cl^- conductivities than the NMM- and PMP-based AEBMs. From the low measured IEC of the PMP-based AEBM, it is clear that only a minor portion of the anion-exchange groups was formed under the applied membrane formation conditions which can be traced back to the high bulkiness of PMP.

(b) Calculated vs. measured IECs:

In nearly all cases the measured were higher than the calculated IECs, particularly after the 10 days KOH test (with the exception of the PMP-based 2257 as explained above). In two studies, Aili et al., when investigating the interaction of poly(2,2'-(m-phenylene)-5,5'-bibenzimidazole membranes with KOH [43,44], observed that KOH deprotonates a part of the imidazole-N–H groups, forming imidazole-N$^-$–K$^+$ ion pairs. Subsequently, the N$^-$ anions can form ionic cross-links by electrostatic interaction with the cations of the anion-exchange blend component according to:

$$\text{PBI-N}^- \text{-K}^+ + \text{Poly-N}^+ \text{-Cl}^- \rightarrow \text{PBI-N}^- \text{-}^+ \text{N-Poly} + \text{K}^+ \text{-Cl}^-$$

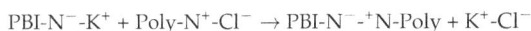

When determining the IEC, the AEBM that is in the OH$^-$ form is immersed in an excess of HCl solution which not only neutralizes the OH$^-$ counter ions but also splits the N$^-$–$^+$N ionic cross-links which consumes additional HCl, leading to the increased measured IEC.

(c) Cl$^-$ conductivities of the three- and four-component AEBMs:

The conductivities of the PEG-containing four-component AEBMs are higher than those of the three-component AEBMs, which clearly highlights the positive effect of the hydrophilic PEG blend component on the conductivity. It is particularly noteworthy that already small amounts of PEG (8–10%) lead to a marked conductivity rise. These results confirm literature stating that the introduction of a PEG component into the AEMs increases the anion-conductivity [41,45]. When comparing the conductivities of membranes 2179B, 2177 and 2176, it is clear that the Cl$^-$ conductivity increased with increasing PEG content in the AEBM (see also Figure 5).

(d) Dependence of Cl$^-$ conductivity on the anion-exchange polymer content of the four-component AEBMs:

As expected and reported in the literature (e.g., [46]), the conductivity of the AEBMs increased with increasing anion-exchange polymer content of the blend membranes. The dependence of the Cl$^-$ conductivity on the anion-exchange polymer content for AEBMs composed of PVBCl, TMIm, PBIOO and sPPSU is displayed in Figure 6. Interestingly, the decrease of the Cl$^-$ conductivity when treated with KOH became less with increasing anion-exchange polymer content, indicating the high OH$^-$ stability of the anion-exchange polymers containing PVBCl and TMIm.

(e) Conductivities of partially and non-fluorinated AEBMs with similar calculated IECs:

As can be seen in Table 1, the conductivities of nonfluorinated AEBMs were higher than those of partially fluorinated AEBMs (examples: 2264 (nonfluorinated)/2246 (partially fluorinated; 2261 (nonfluorinated)/2258 (partially fluorinated); 2262 (nonfluorinated)/2259 (partially fluorinated)).

(f) Water uptake:

In general, the water content of the AEBMs increased with increasing Cl$^-$ conductivity.

(g) Membrane weight loss during extraction:

The weight loss remained below 7% for all investigated membranes indicating complete cross-linking within the AEBM blend components. The small weight decrease during extraction is due to the leaching of residual low-molecular tertiary N compounds (they were introduced in 100% excess, as mentioned before) and residual solvents. For two of the membranes (2175 and 2176, see Table 2), the extracted amounts before and after KOH treatment were nearly the same, suggesting alkaline stability of the AEBMs and particularly of the covalent cross-links formed by reaction of the PEGs' end groups with the PBI-N–H-groups.

(h) Cl$^-$ conductivities before and after KOH treatment:

In all cases, the conductivity declined with KOH storage except for the PBIOO-based AEBMs (2175, 2176 and 2190A), where the Cl$^-$ conductivity was higher after 10 days of KOH storage. Particularly for 2175 and 2176, the conductivity rise after KOH treatment was significant. The reason for this is unclear, but it can be speculated that these membranes might contain residual KOH physically bound to the PBI blend component by electrostatic interaction (see (b)) which was not completely washed out during the membrane rinsing after the KOH test. Another possible explanation is the possible rearrangement of the microstructure of the membranes during the 10 days KOH treatment. To further investigate this phenomenon, the impedance of some of the membranes were measured using the Scribner Impedance facility which allows the determination of the Cl⁻ conductivity in a vapor state as a function of temperature under constant relative humidity (the amount of relative humidity chosen was 90%). Moreover, the 2176 membrane was subjected to a 30-day exposure to KOH (1 M). The impedance of this membrane was measured after 20 and 30 days of KOH treatment. The results of the detailed impedance investigations of selected AEBMs are presented in Section 2.1.2.

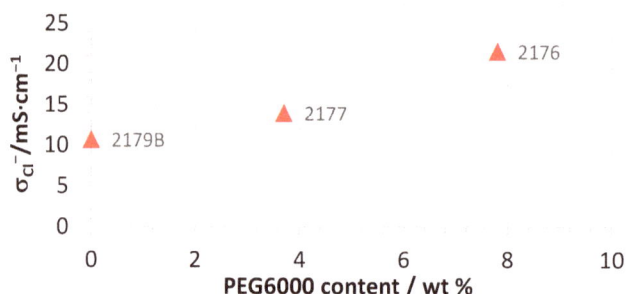

Figure 5. Dependence of Cl⁻ conductivity on the PEG 500 content (measured at room temperature in 1 M NaCl solution) for AEBMs composed of PVBCl, TMIm, PBIOO and sPPSU.

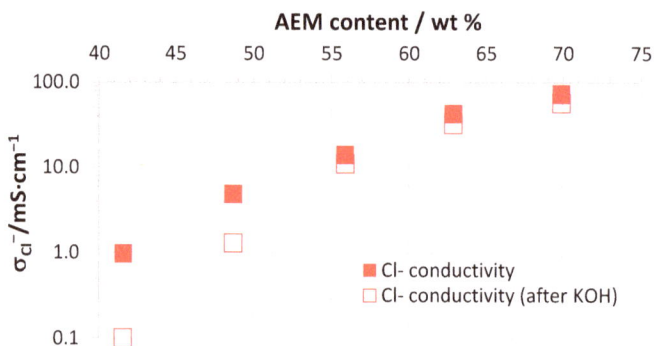

Figure 6. Dependence of Cl⁻ conductivity on the anion-exchange polymer content (measured at room temperature in 1 M NaCl solution) for AEBMs composed of PVBCl, TMIm, PBIOO and sPPSU.

2.1.2. Membrane Conductivities of the AEBMs as a function of Temperature

A selection of membranes were investigated in terms of Cl⁻ conductivity as a function of temperature using the Scribner Membrane Test System (Southern Pines, NC, USA) at a relative humidity of 90% and a temperature range of 30–90 °C. The following comparisons were done:

- Membranes with the same calculated IECs (IEC_{calc} = 2.3 meq/g), but with or without PEGs (NF-AEBM 2179B without PEG, and 2176 with PEG, see Figure 7).
- Membranes with different calculated IECs, but the same PBI:PEG molar ratios (NF-AEBMs 2176 and 2190A, see Figure 8).
- Membranes with the same composition, but different PEG molecular weights (NF-AEBM 2175 with PEG500, and NF-AEBM 2176 with PEG6000, see Figure 9)
- Conductivities of selected NF-AEBMs and two F-AEBMs before and after KOH treatment (2176 (Figure 10a), 2190A (Figure 10b), 2177 (Figure 10c), 2179B (Figure 10d), 2258 (Figure 10e) and 2259 (Figure 10f)).

- Conductivities of NF-AEBMs in comparison with F-AEBMs (membranes 2264/2246, 2261/2258 and 2262/2259 with the calculated IECs$_{calc}$ 2.0, 2.3 and 2.6, respectively, where the IEC variation was obtained by varying the TMIm-quaternized PVBCl content in the blend (Figure 11).

Figure 7 shows that the conductivity was roughly doubled from the membrane without PEG (2179B) to that containing 7.8% PEG (2176) reaching conductivity values of nearly 80 mS·cm^{-1} at 90 °C.

In Figure 8, the strong influence of the IEC on the conductivity is demonstrated. With an IEC$_{calc}$ increase of 0.3 meq/g at 30 °C, the Cl$^-$ conductivity was more than doubled. Moreover, due to the higher water uptake of the 2176 membrane (because of its higher IEC), the conductivity increase with temperature was much more pronounced compared to that of the 2190A membrane.

Figure 9 presents the Cl$^-$ conductivity vs. T curves of the two membranes 2175 and 2176 with the same composition, but containing PEGs with different molecular weights (2175: PEG 500, 2176: PEG 6000). In Figure 9, it is clear that the molecular weight of the PEG does not have an impact on the conductivity.

In Figure 10, the Cl$^-$ conductivity vs. T curves of several of the investigated membranes are compared before and after 10 days of KOH treatment. In the case of the 2176 membrane, the plot was obtained after 30 days of KOH treatment.

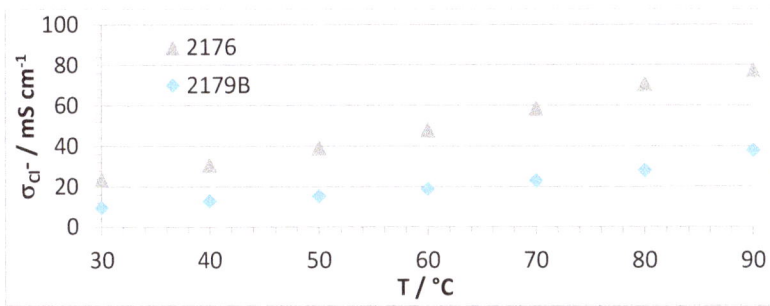

Figure 7. Cl$^-$ conductivities of AEBMs with the same calculated IEC, but without PEG (2179B) and with PEG (2176) in dependence of temperature @r.h. of 90% (used PEG: PEG 6000).

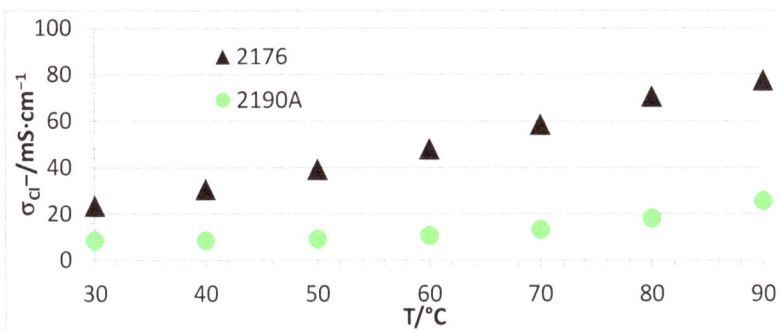

Figure 8. Cl$^-$ conductivities of AEBMs with different calculated IEC, but same molar ratio between PBI and PEG component in dependence of temperature @r.h. of 90% (used PEG: PEG 6000; calculated IEC of 2176: 2.3 meq OH$^-$/g, and calculated IEC of 2190A: 2.0 meq OH$^-$/g).

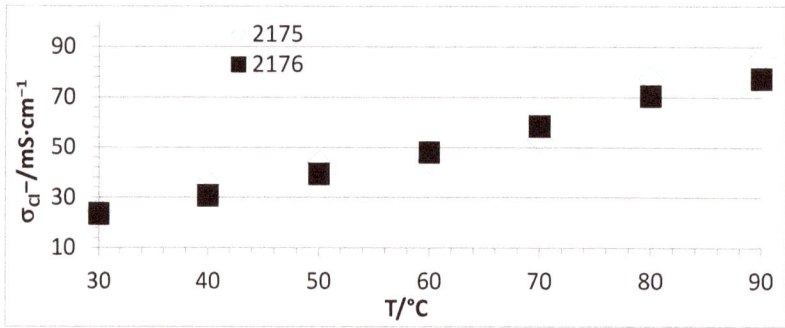

Figure 9. Cl⁻ conductivity vs. T curves of two membranes, 2175 and 2176, with the same composition, but PEGs with different molecular weights (2175: PEG 500, 2176: PEG 6000).

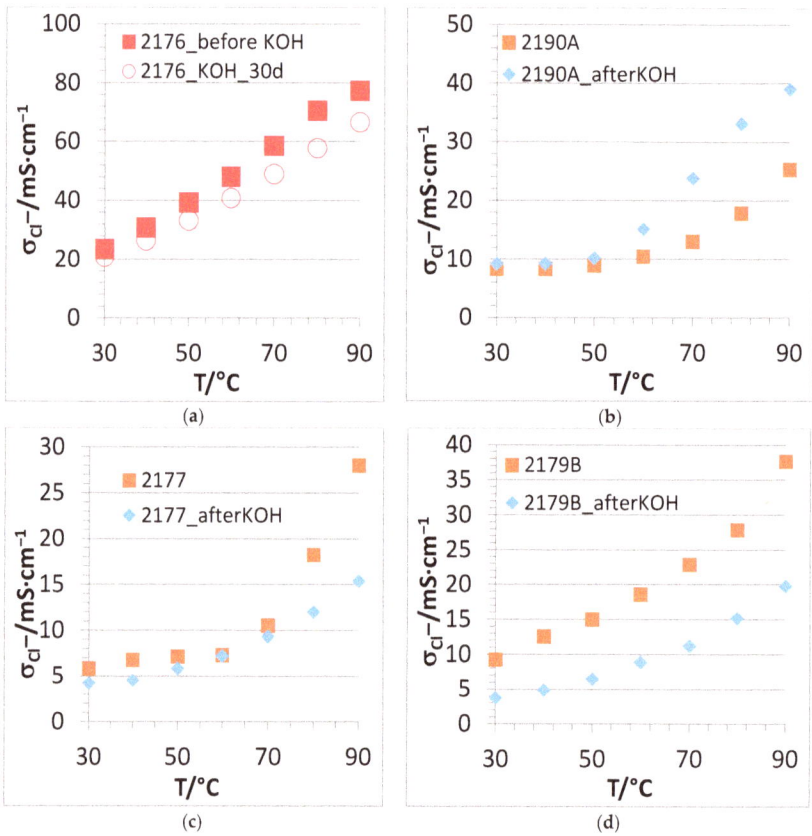

Figure 10. *Cont.*

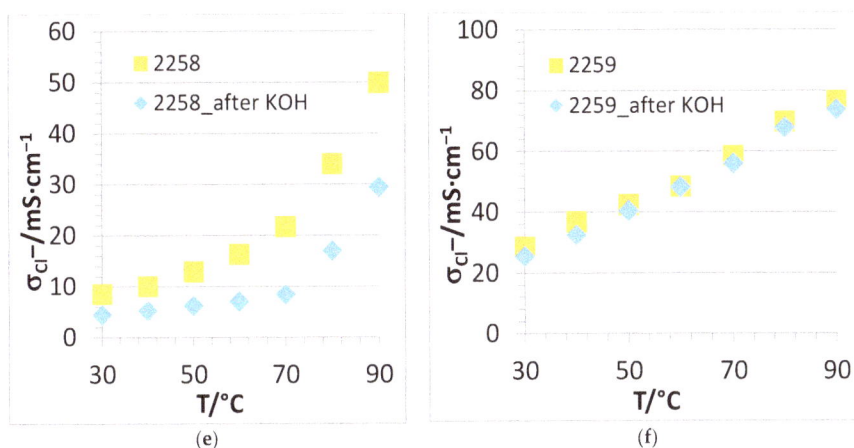

Figure 10. Comparison of Cl⁻ conductivities of several PBIOO-based AEBMs before and after 10 days of KOH treatment (2176: after 30 d KOH treatment) as a function of temperature @90% r.h.: (**a**) membrane 2176; (**b**) membrane 2190A; (**c**) membrane 2177; (**d**) membrane 2179B; (**e**) membrane 2258; and (**f**) membrane 2259.

The fact that the 2176 membrane lost only 14% of Cl⁻ conductivity after a 30 days KOH treatment (Figure 10a) confirms the excellent alkaline stability of this membrane, which is among the best alkaline stability values reported for anion-exchange membranes to date. When comparing the before/after KOH conductivity results of the 2190A membrane, which comprises a lower calculated (and experimental) IEC than 2176, it is clear (Figure 10b) that this membrane possessed a higher conductivity after 10 days, which was even more pronounced at higher temperatures. This behavior cannot currently be satisfactorily explained, but it could be that low-molecular hydrophilic residues in the membrane, which had not been washed out completely from the membrane during the washing procedure after KOH immersion (particularly KOH which is able to deprotonate the N–H group of the PBI blend compound forming a salt pair Im-N⁻K⁺), might also have contributed to the observed increased conductivity. A similar increase of conductivity with KOH treatment time has been reported for another type of PEG-containing AEM [45].

In contrast, 2177 (with roughly half the PEG content of 2176 (Figure 10c)), and 2179B (without PEG (Figure 10d)) showed lower Cl⁻ conductivities after 10 days of KOH treatment. This behavior is a strong indication that the PEG blend component stabilizes the membrane in an alkaline environment, with this effect increasing with increasing PEG content. However, the origin of this behavior is, as already stated, still unclear and requires a more in-depth investigation in ongoing studies. When comparing the two partially fluorinated AEBMs 2258 (Figure 10e) and 2259 (Figure 10f), it is clear that the membrane possessing the higher anion-exchange polymer content (2259) is more stable in KOH than that having the lower anion-exchange polymer content (2258) as would have been expected.

Figure 11 compares the conductivities of NF- and F-four-component AEBMs for the three calculated IECs of 2.0, 2.3, and 2.6 meq OH⁻/g. It is clear that, for all three different anion-exchange polymer contents, the nonfluorinated AEBMs had higher conductivities than the partially fluorinated AEBMs. This can be ascribed to the higher hydrophobicity of the F-AEBMs leading to lower water uptake values causing lower Cl⁻ conductivities in comparison to the nonfluorinated AEBMs.

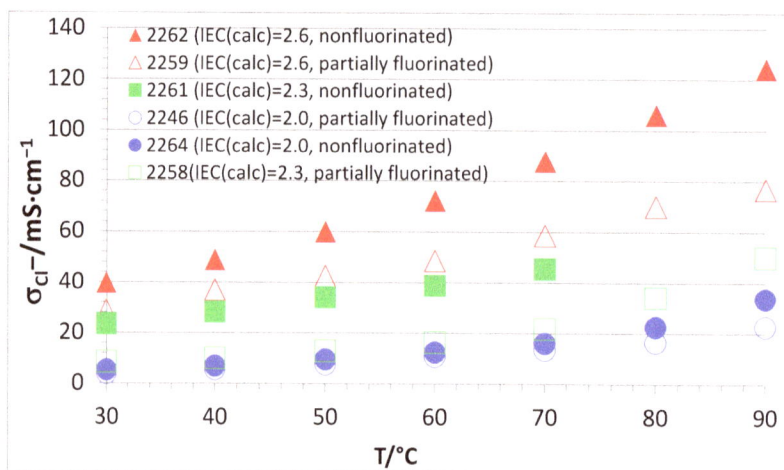

Figure 11. Cl⁻ conductivities of nonfluorinated and partially fluorinated AEBMs.

2.1.3. Thermal Stability of the AEBMs

A high thermal stability is an important prerequisite for the application of ionomer membranes to those electrochemical applications with working temperatures above room temperature such as fuel cells and electrolysers which should be operated in the T range > 60–80 °C to benefit from improved electrode kinetics at elevated temperatures. The thermal stability of the AEBMs was determined using thermogravimetry (TGA). Of particular interest was whether the shapes of the TGA traces were altered after KOH treatment, indicating membrane degradation.

In Figure 12, the TGA traces of membranes 2176, 2177 and 2179B with similar calculated IEC values, but different PEG contents (see Figure 5), are depicted. It is clear that all displayed membranes possess high and comparable thermal stabilities.

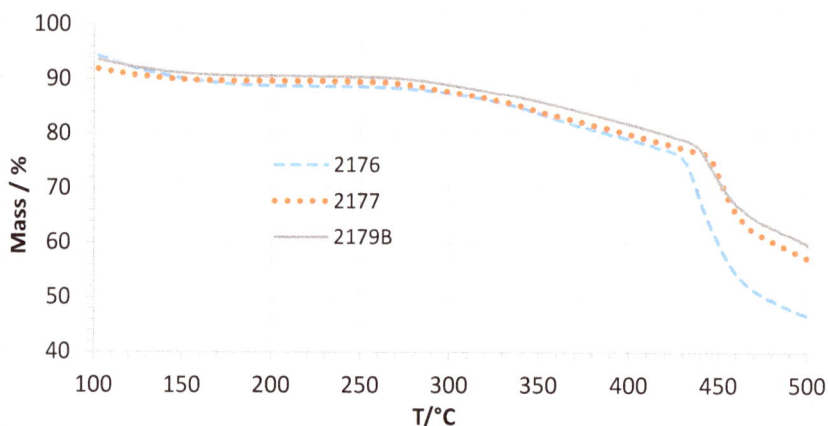

Figure 12. TGA traces of membranes 2176, 2177 and 2179B (different PEG 6000 content).

To further elucidate the TGA traces, coupling of the TGA with other spectroscopical techniques can be used. In our group, TGA-FTIR coupling experiments were done to identify the thermal degradation processes during heating of the membranes in the TGA measurement cell [47]. This will be exemplified with the example of membrane 2176 and its blend components.

In Figure 13, the TGA traces of the 2176 and its blend components are depicted. It can be seen that the blend membrane is more stable than the least stable of the blend components, PEG-diepoxide, which reflects the high chemical reactivity of the epoxide group and the relatively low thermal stability of the poly(ethylene glycol) macromolecular chains. This obviously also indicates the high thermal stability of the blend membrane. The shape of the TMIm TGA trace is similar to a vapor evaporation curve, which means that the TMIm is probably not thermally decomposed during the TGA experiment. This finding was confirmed by the TGA-FTIR coupling experiment where the series spectra showed the signature FTIR spectrum of TMIm.

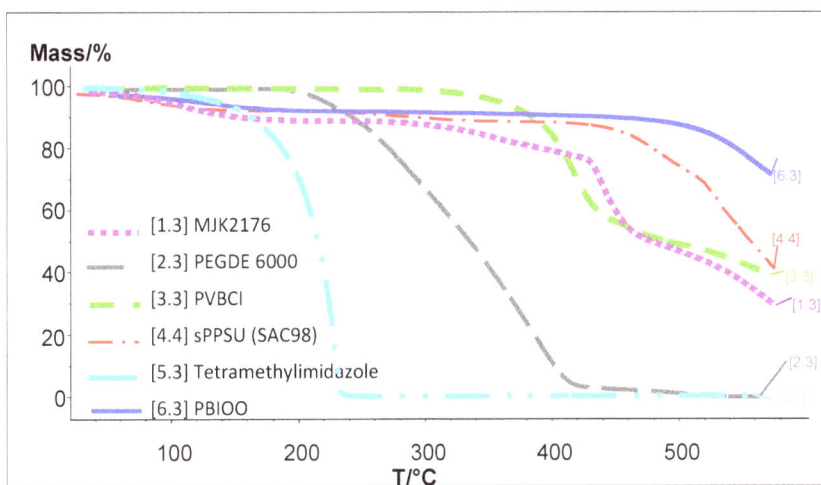

Figure 13. TGA traces of 2176 and its blend components.

Figure 14 shows the Gram–Schmidt traces of the 2176 and its blend components, again confirming that the PEG has the lowest thermal stability among the blend components. In Figure 14, the large peak after roughly 8 min of FTIR running time is the CO_2 marker signal which allows the assignment of temperatures to the passed time of the TGA-FTIR experiment, or, in other words, the assignment of the temperature to each FTIR spectrum of the gaseous degradation products. The low thermal stability of the PEG-diepoxide blend compound can be confirmed by its Gram–Schmidt trace, i.e., the large peak between 20 and 32 min. This large degradation peak has only a weak echo in the 2176 membrane, showing a slight raise in the 20–32 min. range. It seems that the thermal stability of the PEG blend component was increased by the proposed cross-linking reaction between the epoxide end groups and the imidazole N–H groups of the PBIOO blend component. The high thermal stability of the cross-linked AEBM can be explained not only by the covalent cross-linking reaction between the PEG-epoxide and the PBIOO–N–H, but also by ionic cross-linking between the sPPSU and the basic blend components, the TMIm-quaternized PVBCl and the PBIOO. Figure 15 depicts a more detailed Gram–Schmidt trace of 2176, where the correlation between temperature and FTIR running time is represented. In Figure 15 the large peak after roughly 8 min of FTIR running time is again the CO_2 marker signal.

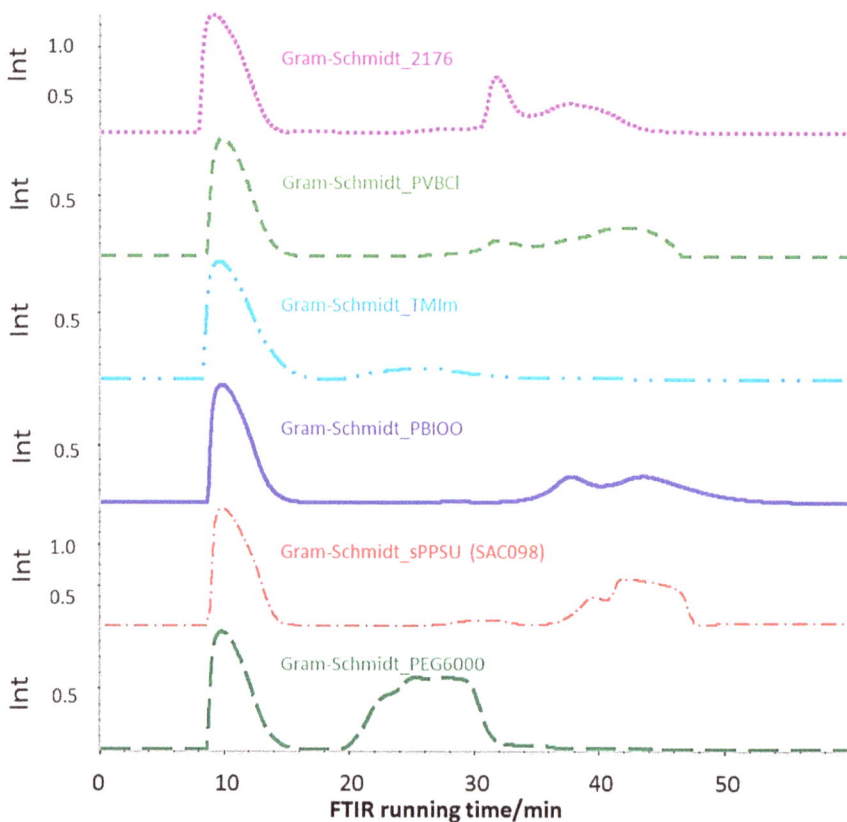

Figure 14. Gram–Schmidt traces of 2176 and its blend components.

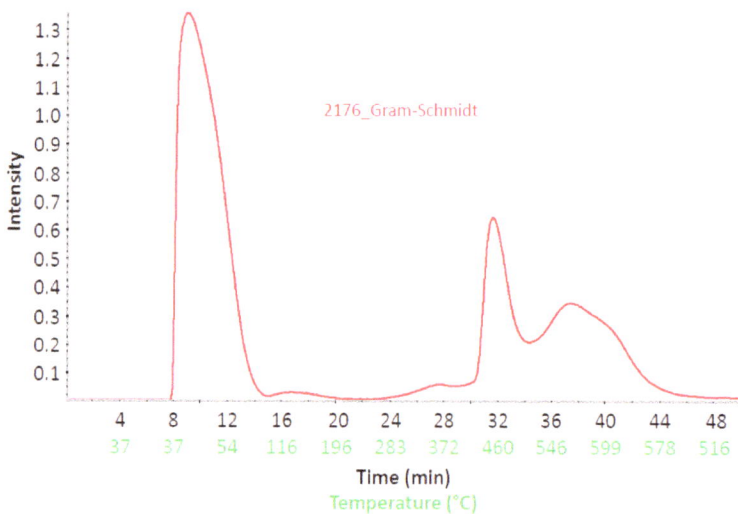

Figure 15. Gram–Schmidt trace of 2176 as intensity vs. FTIR running time and temperature.

The next step of the analysis of the TGA-FTIR coupling results is the assignment of TGA trace steps and Gram–Schmidt trace peaks to the FTIR spectra behind them. In Figure 16, the FTIR spectra of the TGA-FTIR experiments of 2176 assigned to T = 37, 267, 430, and 450 °C are presented. The first FTIR spectrum in Figure 16 displays the CO_2 band (2250–2400 cm^{-1}) from the CO_2 marker signal injected into the TGA cell. At T = 267 °C, the following bands were found:

- water evaporating from the membrane(4000 to 3400 cm^{-1})
- onset of CO_2 development indicating the beginning degradation of the blend membrane, probably originating from the PEG blend component (the TGA-FTIR coupling experiment of pure PEG-diepoxide also showed an onset of CO_2 formation in the same temperature range).

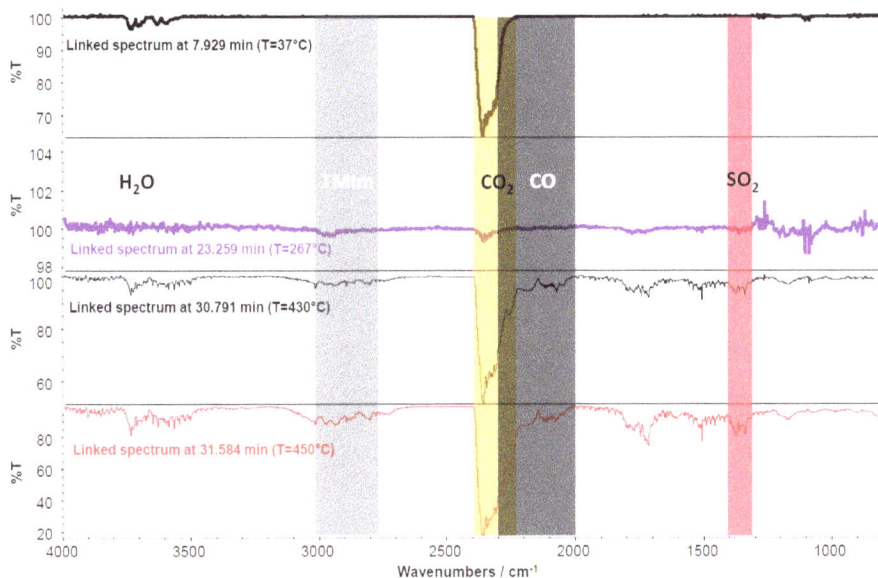

Figure 16. FTIR spectra of the TGA-FTIR experiment of 2176 assigned to T = 37, 267, 430, and 450 °C.

The FTIR spectrum of the TGA decomposition products at 430 °C indicates an ongoing degradation of the AEBM confirmed by CO_2 and CO development (CO band from 2000 to 2300 cm^{-1}). Moreover, at this temperature, a SO_2 band (1300–1400 cm^{-1}) appeared, indicating the onset of the splitting-off of the sulfonate groups of the sulfonated PPSU blend component [48]. In addition, a broad band appears between 2800 and 3000 cm^{-1}, which can be assigned to the TMIm being split off from the quaternized anion-exchange polymer blend component. Water vapor is also observed, probably originating from the thermal decomposition of the AEBM (keeping in mind that the TGA experiments were performed under an oxygen-enriched air which facilitates the thermal oxidation of the organic constituents of the AEBM finally leading to CO_2/CO and H_2O formation). The bands between 1700 and 1800 cm^{-1} and between 1100 and 1200 cm^{-1} observed at this temperature probably originated from the decomposition of the PEG blend component as had been observed for the pure PEG-diepoxide.

The FTIR spectrum of 2176 at 450 °C (where the maximum intensity of the Gram–Schmidt trace was observed) confirms the ongoing degradation of the AEBM. In Figure 17, the weight gain vs. TGA running time (Figure 17a) and weight gain vs. temperature (Figure 17b) TGA traces are presented where the onset of CO_2 formation, the onset of SO_2 formation, and the maximum of the Gram–Schmidt trace are highlighted.

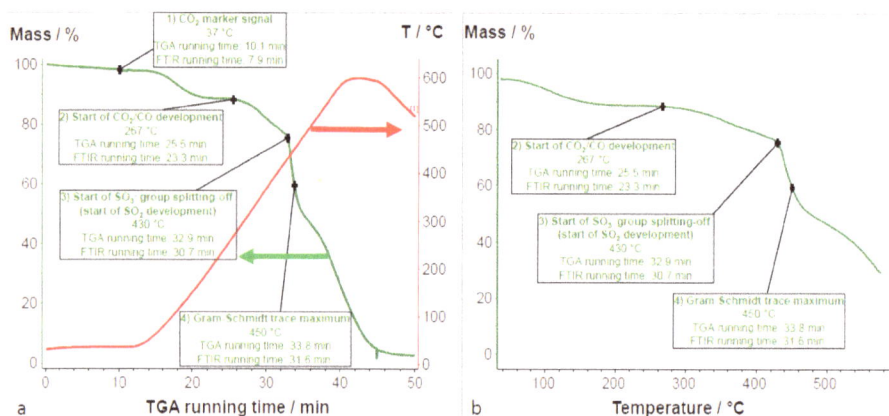

Figure 17. Weight gain vs. TGA running time (**a**) and weight gain vs. temperature (**b**) TGA traces with highlighted points.

The first step of the TGA trace (from room temperature to 200 °C) was due to water vapor evaporation from the water-swollen membrane, as confirmed by the FTIR spectra in this temperature range. Interestingly enough, CO_2 development was only observed at the beginning of the second stage (from 260 to 430 °C) of the TG vs. T TGA trace. While the second step of the TGA trace was mainly caused by the degradation of the PEG blend component, the weight loss within this step was higher than the PEG content of the membrane (PEG content of 2176 was 7.8%, while the weight loss during the second step was 13.5%). This implies that not only the PEG was degraded during the second step, but also other constituents of the membrane which contributed to the formation of CO_2 and CO.

At the end of the second stage at 430 °C, the onset of both SO_2 and TMIm was observed, indicating the onset of the splitting-off of ionically-cross-linking anion-exchange and cation-exchange groups which intensified during the third step of the TGA trace. At 450 °C, the maximum intensity of the Gram–Schmidt trace was observed for 2176. The complex FTIR spectrum at this temperature indicates the simultaneous existence of different thermal degradation processes of both the functional groups and polymer backbones of the blend components. Since the alkaline stability of anion-exchange membranes is an important prerequisite for their application in alkaline fuel cells, the comparison of the TGA traces and the TGA-FTIR coupling experiment results is of interest, as the chemical degradation of the AEBMs by the KOH treatment would result in a change of the shape of both the TGA traces and the TGA-FTIR coupling results, where an OH attack might lead to partial splitting-off of anion-exchange groups and possibly also to partial degradation of the polymer backbone of the blend components. In Figure 18a,b, the TGA traces and the Gram–Schmidt traces of 2176 before and after 30 days of KOH treatment are presented.

Figure 18 shows only small differences in both TGA and Gram–Schmidt traces in the *T* range up to 420–430 °C, which implies that the structure of 2176 basically remained intact during the KOH treatment. This is also supported by the conductivity results obtained with 2176, which decreased by only 14%.

The conductivity results of the AEBMs suggest that the chemical stability of the AEBMs against KOH-induced degradation of the anion-exchange groups increased with increasing anion-exchange polymer content (see Figure 5), which should also reflect in the TGA traces. As an example, the TGA traces of 2262 (calculated IEC = 2.6 meq/g) and 2279 (calculated IEC = 1.4 meq/g) before and after KOH treatment are presented in Figure 19.

From the TGA traces of 2262, it can be seen that the height of the third step of the TGA traces between 420 and 460 °C, where the largest part of the anion-exchange groups was split off (see the analysis of the TGA-FTIR results of 2176), was nearly the same before and after KOH treatment, confirming the significant alkaline stability that had been discussed in Table 1 (conductivity results of 2262). In the case of 2279, the third stage of degradation is much less pronounced after KOH treatment than before, indicating a substantial degradation of anion-exchange groups during the KOH immersion time. This was again confirmed by the 90% conductivity decrease of conductivity by the KOH treatment.

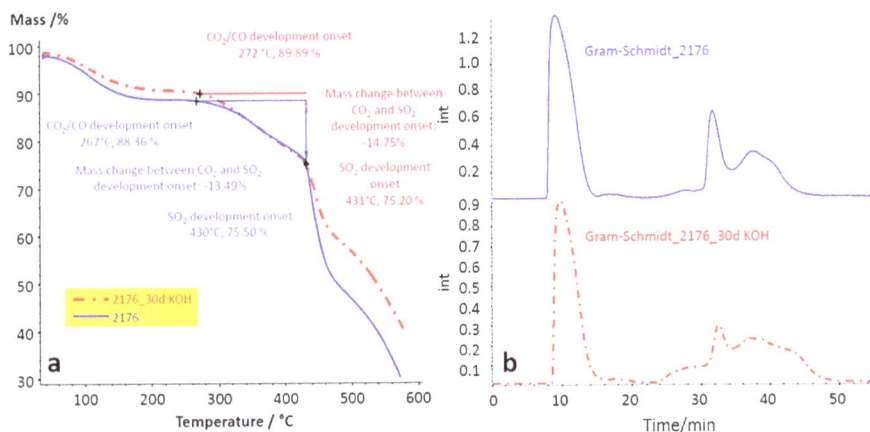

Figure 18. TGA traces (**a**) and Gram–Schmidt traces (**b**) of the 2176 before and after 30 days of KOH treatment.

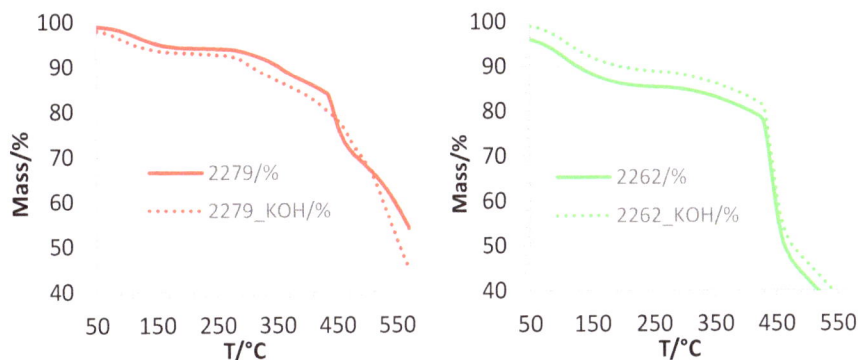

Figure 19. TGA traces of the AEBMs 2262 and 2279 before and after KOH treatment (10 days).

Subsequently, the influence of the molecular weight of the applied PEG-diepoxide on the thermal stability of the AEBMs was determined. In Figure 20, the TGA traces of 2175 (containing PEG500) and 2176 (containing PEG6000) are presented. It is clear that the molecular weight of the applied PEG did not influence the thermal stability.

Finally, the TGAs of nonfluorinated and partially fluorinated AEBMs was compared using the nonfluorinated 2261 and the partially fluorinated 2258 (Figure 21). One can see that the thermal stability of both membranes was comparable. This implies that it is not necessarily advantageous to prepare partially fluorinated anion-exchange membranes, which has the further disadvantage that

C–F bonds are not very stable in an alkaline environment as F can be split off by an OH^- ion attack in both aliphatic and aromatic C–F bonds [48,49]

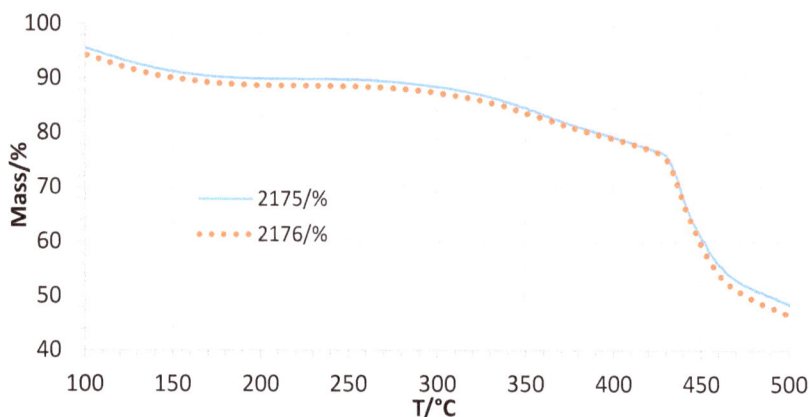

Figure 20. TGA traces of membranes 2175 and 2176 (different PEG molecular weight).

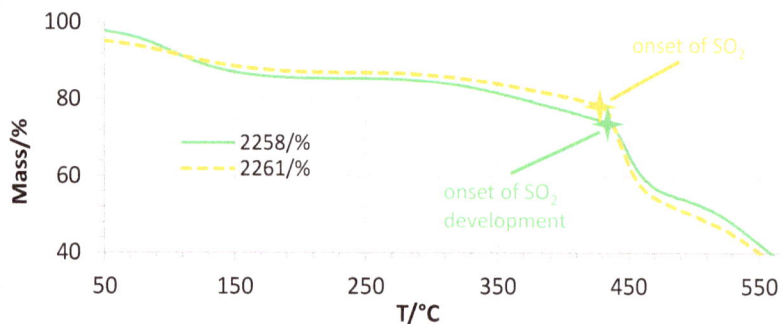

Figure 21. TGA traces of 2258 (partially fluorinated) and 2261 (nonfluorinated).

3. Materials and Methods

3.1. Materials

The solvents (DMAc and DMSO) were used as received. The sulfonated polymer (SPPSU098) was synthesized as described earlier [50]. PBIOO was purchased from Fuma-Tech (Bietigheim-Bissingen, Germany). N-methylmorpholine (NMM) and 1,2,4,5-tetramethylimidazole (TMIm) were purchased from TCI Chemicals, and 1,2,2,6,6-pentamethylpiperidine (pempidine, PMP) from Manchester Organics. PEG-diglycidyl ether (M_n = 500 Da, product No. 475696, and M_n = 6000 Da, product No. 731803, respectively) were obtained from Sigma-Aldrich Germany (Munich, Germany). Poly(vinylbenzyl chloride) (PVBCl), 60/40 mixture of 3- and 4- isomers, was purchased from Sigma-Aldrich, product number 182532. Following the information provided by Sigma-Aldrich, the average molecular weight of PVBCl was: average $M_n \approx$ 55,000 Da, average $M_w \approx$ 100,000 Da (determined by GPC/MALLS according to the Sigma-Aldrich website).

3.2. Membrane Preparation and Posttreatment

All blend components (except for the PEGs in the four-component AEBMs) were separately dissolved in DMAc or DMSO as 5%, 10%, 15% or 20% solutions, before being added in specific ratios to prepare the AEBMs (see Table 1 for composition of blends). For the four-component AEBMs, the different PEG-diglycidyl ethers (mol. wt.: 500 and 6000 Da) were added in pure form. To accelerate the dissolution rate of the 6000 Da PEG, the temperature was increased to 60 °C. After dissolution of the PEGs, the tertiary amines were added as 33.3% solutions in DMAc (TMIm) or in pure form (NMM, PMP) in 100% excess in terms of the amount of chloromethyl groups in the blend solution. After mixing, the blend solutions were cast onto glass plates and the solvent was evaporated in a convection oven at $T = 140$ °C for 2 h. After immersion in deionized water, the membranes came off the glass plates. The membranes were posttreated as follows:

1. The membranes were immersed in 10% ethanolic solutions of different tertiary amines. Most of the membranes were posttreated with 1,2,4,5-tetramethylimidazole (TMIm) to complete the quaternization reaction.
2. The membranes were immersed in 10% NaCl solution for 48 h at 90 °C.
3. The membranes were washed thoroughly with DI water and stored in DI water at 60 °C for 48 h.
4. The membranes were soaked in an aqueous 1 M KOH solution at 90 °C for 10 days. Some of the membranes were stored in a 1 M KOH solution at 90 °C for 20 and 30 days, respectively.
5. To determine the effectiveness of the reaction of the PEG's epoxy end groups with the N–H groups of the PBI, a 10 wt % DMSO solution of pure PBI (80 wt %) was mixed with the PEG (PEG500, 20 wt %), before casting a membrane from this mixture. After membrane formation, the blend membrane was extracted with DMAc to determine the extent of cross-linking within the blend. According to the results the DMAc residual after extraction was nearly 100%, proving the completeness of cross-linking of the PBI with the epoxide end groups which confirms the covalent cross-linking of PBI with bisphenol A-bisepoxide and other bisepoxide compounds [51].

The compositions of the AEBMs are listed in Table 2.

Table 2. Composition of blend components for AEBM preparation.

Membrane Type * (No.)	IEC$_{calc}$ (meq OH$^-$/g)	Type Amine	PVBCl-AEM (%)	PBI (%)	S-Polymer (%)	Type PEG (Da)	PEG (%)
2256_3C_NF	2.3	NMM	58.3	PBIOO/32.8	SPPSU **/8.9	-	-
2257_3C_NF	2.3	PMP	70.0	PBIOO/23.6	SPPSU/6.4	-	-
2179B_3C_NF	2.3	TMIm	63.5	PBIOO/28.7	SPPSU/7.8	-	-
2223_3C_NF	2.0	TMIm	54.9	PBIOO/38.4	SPPSU/6.8	-	-
2177_4C_NF	2.3	TMIm	62.9	PBIOO/26.0	SPPSU/7.4	6000	3.7
2175_4C_NF	2.3	TMIm	63.5	PBIOO/20.9	SPPSU/7.8	500	7.8
2176_4C_NF	2.3	TMIm	63.5	PBIOO/20.9	SPPSU/7.8	6000	7.8
2190A_4C_NF	2.0	TMIm	54.9	PBIOO/27.9	SPPSU/6.8	6000	10.4
2261_4C_NF	2.3	TMIm	62.9	PBIOO/24.1	SPPSU/7.4	500	5.6
2262_4C_NF	2.6	TMIm	69.9	PBIOO/19.5	SPPSU/6.0	500	4.5
2264_4C_NF	2.0	TMIm	55.9	PBIOO/28.6	SPPSU/8.8	500	6.6
2265_4C_NF	1.7	TMIm	48.7	PBIOO/33.4	SPPSU/10.3	500	7.7
2279_4C_NF	1.4	TMIm	41.6	PBIOO/41.6	SPPSU/11.7	500	8.8
2267_4C_NF	2.3	TMIm	65.2	PBIOO/18.2	SPPSU/12.3	500	4.2
2246_4C_F	2.0	TMIm	56.6	F$_6$PBI/28.98	SFPE ***/7.8	500	6.7
2258_4C_F	2.3	TMIm	63.5	F$_6$PBI/24.4	SFPE/6.5	500	5.6
2259_4C_F	2.6	TMIm	70.5	F$_6$PBI/19.7	SFPE/5.3	500	4.5

* Abbreviations: 3C_NF: three-component AEBMs, containing the nonfluorinated blend polymers PBIOO and sPPSU; 4C_NF: four-component AEBMs, containing the nonfluorinated blend polymers PBIOO and sPPSU; 4C_F: four-component AEBMs, containing the partially fluorinated blend polymers F$_6$PBI and SFPE001; ** SPPSU: sulfonated poly(phenylene sulfone); *** SFPE: sulfonated partially fluorinated aromatic polyether; for structures of the sulfonated polymers, see Figure 3.

3.3. Membrane Characterization

3.3.1. Ion-Exchange Capacity (IEC)

The IEC was determined as described in a recent paper of our group [38].

3.3.2. Ionic Conductivity

The ionic Cl^- conductivity of the membranes was determined via impedance spectroscopy (EIS). For the measurement, two different setups were used: for series impedance measurements an in-plane self-constructed impedance cell was used which allowed EIS measurements in a liquid state (1 N NaCl) at room temperature (25 °C). The cell was connected to a Zahner elektrik IM6 impedance spectrometer (Kronach, Germany). The impedances of the membranes were investigated in a frequency range of 200 Hz–8 MHz (amplitude: 5 mV). Details of the measurement are described in [38]. The ohmic resistance obtained from EIS measurement was converted into conductivity using the following equation:

$$\sigma = \frac{1}{R_{sp}} = \frac{d}{R \times A} \tag{1}$$

where, σ: Conductivity, $mS \cdot cm^{-1}$; R_{sp}: resistivity, $\Omega \cdot cm$; d: thickness of the membrane, cm; R: ohmic resistance, Ω; A: electrode area, cm^2.

The second impedance setup used for selected membranes was a Membrane Test System (MTS 740, Scribner Associates Inc., Southern Pines, NC, USA), which allows through-plane impedance measurements in a vapour phase as a function of temperature and relative humidity independently from each other, connected to a Solartron 1260 impedance analyser. Details of the measurements are described in a previous paper [40].

3.3.3. Gel Content

The gel content of the membranes, which is a measure for the extent of cross-linking within the AEBMs, was determined by extraction with 90 °C hot DMAc, as described in [40].

3.3.4. Water Uptake

The water uptake of the membranes was determined by immersion in water at the respective temperature until equilibration had been reached, as described in [40].

3.3.5. Thermal Stability

Characterization of the thermal stability of the membranes in the chloride form was performed by thermogravimetry (TGA, Netzsch, model STA 499C, Netzsch, Selb, Germany) at a heating rate of 20 °C·min^{-1} under O_2-enriched O_2/N_2 atmosphere (65%–70% O_2) which was coupled with a FTIR for FTIR analysis of the gaseous thermal degradation products.

3.3.6. Alkaline Stability

Due to the application of AEMs in electrochemical systems such as alkaline membrane fuel cells or alkaline membrane electrolysis where the AEMs are in the hydroxide form, the stability in strongly alkaline environments is the most important prerequisite for determining their suitability for these processes. Consequently, the stability of the AEBMs was investigated under harsh conditions, i.e., in 1 M KOH at a temperature of 90 °C for 10 days, or even in one case (membrane 2176) for up to 30 days. Membranes were characterized before and after the stability tests using Cl^- conductivity, IEC and thermal stability.

4. Conclusions

In this study, novel AEBMs based on poly(vinylbenzyl chloride) quaternized with different tertiary basic N compounds, PBIOO or F_6PBI as the stabilizing matrix polymer and a sulfonated poly(ethersulfone) or a partially fluorinated sulfonated aromatic polyether as ionic macromolecular cross-linker were synthesized and characterized. Epoxide-end-functionalized poly(ethylene glycol)s were introduced into the AEBMs as an additional hydrophilic microphase, leading to a significant increase in the chloride conductivity of the AEBMs. In addition, the introduction of the PEG also led to a marked improvement of the stability of the AEBMs in an alkaline environment. It seems that the presence of the PEG led to a microphase-separation of the AEBMs, which, with high probability, facilitated the clustering of the anion-exchange groups within the AEBM microstructure thereby contributing to the observed conductivity increase of the PEG-containing AEBMs. The thermal stability of the novel AEBMs was excellent. The favorable properties determined for the novel AEBMs make them promising candidates for several electrochemical applications such as redox-flow batteries, alkaline membrane fuel cells and alkaline membrane electrolysers.

Author Contributions: Jochen A. Kerres conceived and designed the experiments. Jochen A. Kerres performed the experiments. Jochen A. Kerres and Henning M. Krieg analysed the data. Jochen A. Kerres and Henning M. Krieg wrote the paper.

Conflicts of Interest: The authors declare no conflict of interest.

References

1. Piana, M.; Boccia, M.; Filpi, A.; Flammia, E.; Miller, H.A.; Orsini, M.; Salusti, F.; Santiccioli, S.; Ciardelli, F.; Pucci, A. H_2/air alkaline membrane fuel cell performance and durability, using novel ionomer and non-platinum group metal cathode catalyst. *J. Power Sources* **2010**, *195*, 5875–5881. [CrossRef]
2. Marini, S.; Salvi, P.; Nelli, P.; Pesenti, R.; Villa, M.; Berrettoni, M.; Zangari, G.; Kiros, Y. Advanced alkaline water electrolysis. *Electrochim. Acta* **2012**, *82*, 384–391. [CrossRef]
3. Schwenzer, B.; Zhang, J.; Kim, S.; Li, L.; Yang, Z. Membrane Development for Vanadium Redox Flow Batteries. *ChemSusChem* **2011**, *4*, 1388–1406. [CrossRef] [PubMed]
4. Logan, B.E.; Elimelech, M. Membrane-based processes for sustainable power generation using water. *Nature* **2012**, *488*, 313–319. [CrossRef] [PubMed]
5. Leong, J.X.; Daud, W.R.W.; Ghasemi, M.; Liew, K.B.; Ismail, M. Ion exchange membranes as separators in microbial fuel cells for bioenergy conversion: A comprehensive review *Renew. Sustain. Energy Rev.* **2013**, *28*, 575–587. [CrossRef]
6. Gellett, W.; Schumacher, J.; Kesmez, M.; Le, D.; Minteer, S.D. High Current Density Air-Breathing Laccase Biocathode. *J. Electrochem. Soc.* **2010**, *157*, B557–B562. [CrossRef]
7. Sandeaux, J.; Sandeaux, R.; Gavach, C. Competition between the electrotransports of acetate and chloride ions through a polymeric anion-exchange membrane. *J. Memb. Sci.* **1991**, *59*, 265–276. [CrossRef]
8. Kliber, S.; Wisniewski, J.A. Removal of bromate and associated anions from water by Donnan dialysis with anion-exchange membrane. *Desal. Water Treat.* **2011**, *35*, 158–163. [CrossRef]
9. Emmanuel, K.; Cheng, C.; Erigene, B.; Mondal, A.N.; Hossain, M.M.; Khan, M.I.; Afsar, N.U.; Liang, G.; Wu, L.; Xu, T. Imidazolium functionalized anion exchange membrane blended with PVA for acid recovery via diffusion dialysis process. *J. Memb. Sci.* **2016**, *497*, 209–215. [CrossRef]
10. Varcoe, J.R.; Atanassov, P.; Dekel, D.R.; Herring, A.M.; Hickner, M.A.; Kohl, P.A.; Kucernak, A.R.; Mustain, W.E.; Nijmeijer, K.; Scott, K.; Xu, T.; Zhuang, L. Anion-exchange membranes in electrochemical energy systems. *Energy Environ. Sci.* **2014**, *7*, 3135–3191. [CrossRef]
11. Slade, R.C.T.; Varcoe, J.R. Prospects for alkaline anion-exchange membranes in low temperature fuel cells. *Fuel Cells* **2005**, *5*, 187–200.
12. Nunez, S.A.; Hickner, M.A. Quantitative [H-1]NMR Analysis of Chemical Stabilities in Anion-Exchange Membranes. *ACS Macro Lett.* **2013**, *2*, 49–52. [CrossRef]
13. Fuson, R.C.; McKeever, C.H. Chloromethylation of Aromatic Compounds. In *Organic Reactions*; John Wiley & Sons: Hoboken, NJ, USA, 1942; pp. 63–90.

14. Zschocke, P.; Quellmalz, D. Novel ion-exchange membranes based on an aromatic polyethersulfone. *J. Memb. Sci.* **1985**, *22*, 325–332. [CrossRef]

15. Wohl, A. Brominised unsaturated compounds with N-bromo-acetamide, a report on the knowledge of the progression of chemical processes. *Ber. Dt. Chem. Ges.* **1919**, *52*, 51–63. [CrossRef]

16. Liska, J.; Borsig, E.; Tkac, I. A route to preparation of bromomethylated poly(2,6-dimethyl-1,4-phenylene oxide). *Macromol. Mat. Eng.* **1993**, *211*, 121–129.

17. Zhao, C.H.; Gong, Y.; Liu, Q.L.; Zhang, Q.G.; Zhu, A.M. Self-crosslinked anion exchange membranes by bromination of benzylmethyl-containing poly(sulfone)s for direct methanol fuel cells. *Int. J. Hydrogen Energy* **2012**, *37*, 11383–11393. [CrossRef]

18. Sata, T.; Yamane, Y.; Matsusaki, K. Preparation and properties of anion exchange membranes having pyridinium or pyridinium derivatives as anion exchange groups. *J. Polym. Chem. Part A Polym. Chem.* **1998**, *36*, 49–58. [CrossRef]

19. Wang, J.; Li, S.; Zhang, S. Novel Hydroxide-Conducting Polyelectrolyte Composed of an Poly(arylene ether sulfone) Containing Pendant Quaternary Guanidinium Groups for Alkaline Fuel Cell Applications. *Macromolecules* **2010**, *43*, 3890–3896. [CrossRef]

20. Zhang, F.; Zhang, H.; Qu, C. Imidazolium functionalized polysulfone anion exchange membrane for fuel cell application. *J. Mater. Chem.* **2011**, *21*, 12744–12752. [CrossRef]

21. Rebeck, N.T.; Li, Y.; Knauss, D.M. Poly(phenylene oxide) Copolymer Anion Exchange Membranes. *J. Polym. Sci. Part B Polym. Phys.* **2013**, *51*, 1770–1778. [CrossRef]

22. Hibbs, M.R.; Hickner, M.A.; Alam, T.M.; McIntyre, S.K.; Fujimoto, C.H.; Cornelius, C.J. Transport properties of hydroxide and proton conducting membranes. *Chem. Mater.* **2008**, *20*, 2566–2573. [CrossRef]

23. Poynton, S.D.; Varcoe, J.R. Reduction of the monomer quantities required for the preparation of radiation-grafted alkaline anion-exchange membranes. *Solid State Ion.* **2015**, *277*, 38–43. [CrossRef]

24. Dang, H.S.; Jannasch, P. Exploring Different Cationic Alkyl Side Chain Designs for Enhanced Alkaline Stability and Hydroxide Ion Conductivity of Anion-Exchange Membranes. *Macromolecules* **2015**, *48*, 5742–5751. [CrossRef]

25. Zhu, L.; Pan, Y.; Wang, Y.; Han, J.; Zhuang, L.; Hickner, M.A. Multication Side Chain Anion Exchange Membranes. *Macromolecules* **2016**, *49*, 815–824. [CrossRef]

26. Thomas, O.D.; Soo, K.J.W.Y.; Peckham, T.J.; Kulkarni, M.P.; Holdcroft, S. A Stable Hydroxide-Conducting Polymer. *J. Am. Chem. Soc.* **2012**, *134*, 10753–10756. [CrossRef] [PubMed]

27. Wright, A.G.; Holdcroft, S. Hydroxide-Stable Ionenes. *ACS Macro Lett.* **2014**, *3*, 444–447. [CrossRef]

28. Liu, Y.; Wang, J.; Yang, Y.; Brenner, T.M.; Seifert, S.; Yan, Y.; Liberatore, M.W.; Herring, A.M. Anion Transport in a Chemically Stable, Sterically Bulky α-C Modified Imidazolium Functionalized Anion Exchange Membrane. *J. Phys. Chem. C* **2014**, *118*, 15136–15145. [CrossRef]

29. Lin, X.; Varcoe, J.R.; Poynton, S.D.; Liang, X.; Ong, A.L.; Ran, J.; Li, Y.; Xu, T. Alkaline polymer electrolytes containing pendant dimethylimidazolium groups for alkaline membrane fuel cells. *J. Mater. Chem. A* **2013**, *1*, 7262–7269. [CrossRef]

30. Kim, D.S.; Labouriau, A.; Guiver, M.D.; Kim, Y.S. Guanidinium-Functionalized Anion Exchange Polymer Electrolytes via Activated Fluorophenyl-Amine Reaction. *Chem. Mater.* **2011**, *23*, 3795–3797. [CrossRef]

31. Kim, D.S.; Fujimoto, C.H.; Hibbs, M.R.; Laboriau, A.; Choe, Y.K.; Kim, Y.S. Resonance Stabilized Perfluorinated Ionomers for Alkaline Membrane Fuel Cells. *Macromolecules* **2013**, *46*, 7826–7833. [CrossRef]

32. Maurya, S.; Shin, S.H.; Kim, M.K.; Yun, S.H.; Moon, S.H. Stability of composite anion exchange membranes with various functional groups and their performance for energy conversion. *J. Memb. Sci.* **2013**, *443*, 28–35. [CrossRef]

33. Zha, Y.; Disabb-Miller, M.L.; Johnson, Z.D.; Hickner, M.A.; Tew, G.N. Metal-Cation-Based Anion Exchange Membranes. *J. Am. Chem. Soc.* **2012**, *134*, 4493–4496. [CrossRef] [PubMed]

34. He, Y.; Wu, L.; Pan, J.; Zhu, Y.; Ge, X.; Yang, Z.; Ran, J.; Xu, T. A mechanically robust anion exchange membrane with high hydroxide conductivity. *J. Memb. Sci.* **2016**, *504*, 47–54. [CrossRef]

35. Cheng, J.; Yang, G.; Zhang, K.; He, G.; Jia, J.; Yu, H.; Gai, F.; Li, L.; Hao, C.; Zhang, F. Guanidimidazole-quaternized and cross-linked alkaline polymer electrolyte membrane for fuel cell application. *J. Memb. Sci.* **2016**, *501*, 100–108. [CrossRef]

36. Katzfuss, A.; Gogel, V.; Jörissen, L.; Kerres, J. The application of covalently cross-linked BrPPO as AEM in alkaline DMFC. *J. Memb. Sci.* **2013**, *425–426*, 131–140. [CrossRef]

37. Katzfuss, A.; Poynton, S.; Varcoe, J.; Gogel, V.; Storr, U.; Kerres, J. Methylated polybenzimidazole and its application as a blend component in covalently cross-linked anion-exchange membranes for DMFC. *J. Memb. Sci.* **2014**, *465*, 129–137. [CrossRef]

38. Morandi, C.G.; Peach, R.; Krieg, H.M.; Kerres, J. Novel morpholinium-functionalized anion-exchange PBI-polymer blends. *J. Mater. Chem. A* **2015**, *3*, 1110–1120. [CrossRef]

39. Kerres, J.; Atanasov, V. Cross-linked PBI-based high-temperature membranes: Stability, conductivity and fuel cell performance. *Int. J. Hydrogen Energy* **2015**, *40*, 14723–14735. [CrossRef]

40. Morandi, C.G.; Peach, R.; Krieg, H.M.; Kerres, J. Novel imidazolium-functionalized anion-exchange polymer PBI blend membranes. *J. Memb. Sci.* **2015**, *476*, 256–263. [CrossRef]

41. Pan, J.; Zhu, L.; Han, J.; Hickner, M.A. Mechanically Tough and Chemically Stable Anion Exchange Membranes from Rigid-Flexible Semi-Interpenetrating Networks. *Chem. Mater.* **2015**, *27*, 6689–6698. [CrossRef]

42. Lu, W.; Shao, Z.-G.; Zhang, G.; Zhao, Y.; Yi, B. Crosslinked poly(vinylbenzyl chloride) with a macromolecular crosslinker for anion exchange membrane fuel cells. *J. Power Sources* **2014**, *248*, 905–914. [CrossRef]

43. Aili, D.; Jankova, K.; Li, Q.; Bjerrum, N.J.; Jensen, J.O. The stability of poly(2,2'-(m-phenylene)-5,5'-bibenzimidazole) membranes in aqueous potassium hydroxide. *J. Membr. Sci.* **2015**, *492*, 422–429. [CrossRef]

44. Aili, D.; Jankova, K.; Han, J.; Bjerrum, N.J.; Jensen, J.O.; Li, Q. Understanding ternary poly(potassium benzimidazolide)-based polymer electrolytes. *Polymer* **2016**, *84*, 304–310. [CrossRef]

45. Yang, C.; Wang, S.; Ma, W.; Zhao, S.; Xu, Z.; Sun, G. Highly stable poly(ethylene glycol)-grafted alkaline anion exchange membranes. *J. Mater. Chem. A* **2016**, *4*, 3886–3892. [CrossRef]

46. Ran, J.; Wu, L.; Wei, B.; Chen, Y.; Xu, T. Simultaneous Enhancements of Conductivity and Stability for Anion Exchange Membranes (AEMs) through Precise Structure Design. *Sci. Rep.* **2014**, *4*, 6486–6490. [CrossRef] [PubMed]

47. Kerres, J.; Ullrich, A.; Hein, M.; Gogel, V.; Friedrich, K.A.; Jörissen, L. Cross-Linked Polyaryl Blend Membranes for Polymer Electrolyte Fuel Cells. *Fuel Cells* **2004**, *4*, 105–112. [CrossRef]

48. Danks, T.N.; Slade, R.C.T.; Varcoe, J.R. Comparison of PVDF- and FEP-based radiation-grafted alkaline anion-exchange membranes for use in low temperature portable DMFCs. *J. Mater. Chem.* **2002**, *12*, 3371–3373. [CrossRef]

49. Zhang, Q.; Zhang, Q.; Wang, J.; Zhang, S.; Li, S. Synthesis and alkaline stability of novel cardo poly(aryl ether sulfone)s with pendent quaternary ammonium aliphatic side chains for anion exchange membranes. *Polymer* **2010**, *51*, 5407–5416. [CrossRef]

50. Chromik, A.; Kerres, J. Degradation studies on acid-base blends for both LT and intermediate T fuel cells. *Solid State Ion.* **2013**, *252*, 140–151. [CrossRef]

51. Yang, J.; Xu, Y.; Liu, P.; Gao, L.; Che, Q.; He, R. Epoxides cross-linked hexafluoropropylidene polybenzimidazole membranes for application as high temperature proton exchange membranes. *Electrochim. Acta* **2015**, *160*, 281–287. [CrossRef]

membranes

MDPI

Article

Thin Film Nanocomposite Membrane Filled with Metal-Organic Frameworks UiO-66 and MIL-125 Nanoparticles for Water Desalination

Mohammed Kadhom [1,2], Weiming Hu [3] and Baolin Deng [1,3,*]

[1]   Department of Chemical Engineering, University of Missouri, Columbia, MO 65211, USA;
      makbq6@mail.missouri.edu
[2]   Al-Dour Technical Institute, Northern Technical University, Al-Dour, Saladin, Iraq
[3]   Department of Civil and Environmental Engineering, University of Missouri, Columbia, MO 65211, USA;
      why78@mail.missouri.edu
*   Correspondence: dengb@missouri.edu; Tel.: +1-573-882-0075

Academic Editor: Mohamed Khayet
Received: 10 May 2017; Accepted: 9 June 2017; Published: 14 June 2017

Abstract: Knowing that the world is facing a shortage of fresh water, desalination, in its different forms including reverse osmosis, represents a practical approach to produce potable water from a saline source. In this report, two kinds of Metal-Organic Frameworks (MOFs) nanoparticles (NPs), UiO-66 (~100 nm) and MIL-125 (~100 nm), were embedded separately into thin-film composite membranes in different weight ratios, 0%, 0.05%, 0.1%, 0.15%, 0.2%, and 0.3%. The membranes were synthesized by the interfacial polymerization (IP) of m-phenylenediamine (MPD) in aqueous solution and trimesoyl chloride (TMC) in an organic phase. The as-prepared membranes were characterized by scanning electron microscopy (SEM), transmission electron microscopy (TEM), contact angle measurement, attenuated total reflection Fourier transform infrared (ATR FT-IR) spectroscopy, and salt rejection and water flux assessments. Results showed that both UiO-66 and MIL-125 could improve the membranes' performance and the impacts depended on the NPs loading. At the optimum NPs loadings, 0.15% for UiO-66 and 0.3% for MIL-125, the water flux increased from 62.5 L/m^2 h to 74.9 and 85.0 L/m^2 h, respectively. NaCl rejection was not significantly affected (UiO-66) or slightly improved (MIL-125) by embedding these NPs, always at >98.5% as tested at 2000 ppm salt concentration and 300 psi transmembrane pressure. The results from this study demonstrate that it is promising to apply MOFs NPs to enhance the TFC membrane performance for desalination.

Keywords: metal-organic framework; thin film nanocomposite membrane; reverse osmosis; desalination; nanoparticles

1. Introduction

Reverse osmosis (RO) is now the most employed method of desalination to convert saline or brackish water source to freshwater. The process uses a membrane as a selective barrier through which water molecules, but not the salts or organics, pass under pressure [1]. Desalination by various approaches, especially RO, represents a solution to address the water shortage problem that has become increasingly serious in the recent decades [2]. Development of RO processes and membranes has gone through many stages and forms since the first applicable membrane synthesis in the early 1960s [3]. The state-of-the-art membrane is the thin film composite (TFC) membrane invented by John Cadotte in the early 1980s. The TFC membrane has a polyamide dense layer with a thickness of a few hundred nanometers supported by an ultrafiltration membrane. This polyamide layer is the active layer for salt rejection, formed by the interfacial polymerization of the MPD dissolved in water and TMC dissolved in organic solvent [4].

Numerous researches have been devoted to developing and improving the performance of TFC membranes [5]. One approach is to embed nano-sized materials such as zeolite [6–12], silica [13–18], titanium dioxide [19–22], or carbon nanotubes [23–25] inside the membrane to improve its properties and performance. Another way is to select proper materials and control the manufacturing conditions affecting the membrane's efficiency [6,26–29]. In addition, the engineering process could be optimized to increase the pure water production efficiency and lower the energy cost [30–34].

Metal-Organic Frameworks (MOFs) are a class of materials consisting of an inorganic or metal core surrounded by an organic linker material [35]. MOFs have very high specific surface area, a high number of adsorption sites, different particle structures, distinct pore size and structure, and can be applied for various purposes [35–39]. These include applications for ion exchange [40], light harvesting [41], sensing [42], catalyst [43], drug delivery [44], gas storage [45,46], and separation [47,48]. Such a wide range of applications are possible because MOFs can be easily designed toward desired uses.

MOFs have been widely used for gas separation. Their porous structure can generate exceptionally high pore volume and surface area reaching over 7000 m^2/g [49], which is particularly suitable for gas storage and separation. Beginning with the first MOFs membrane synthesized and used for gas separation [50], different forms of MOFs membranes have been prepared, often on an organic or inorganic support layer. MOFs particles have been incorporated into polyamide layer of thin film nanocomposite (TFN) membrane with improved properties [35,51]. In water treatment, MOFs have been used as adsorbents of heavy metals and as fillers in alumina hollow fiber membranes for water desalination [52,53]. MOFs have also been employed in TFN membrane to treat water in different applications. Lee et al. used A100 (aluminum terephthalate) and C300 (copper benzene-1,3,5-tricarboxylate) as water soluble MOFs, which were embedded and dissolved into the ultrafiltration membrane to increase its porosity [54]. For reverse osmosis, Hu et al. [55] and Gupta et al. [56] reported two simulation studies that used Zeolitic Imidazolate Framework (ZIFs)—MOFs based on zeolite as membranes for water desalination. Xu et al. [57] filled MIL-101(Cr) NPs into the TFN membrane to improve the desalination performance. Findings showed that the water flux increased by 44% comparing to the plain membrane by adding 0.05% MIL-101 NPs to the organic solution, while the salt rejection remained almost the same. Similar outcomes were observed by Ma et al. [58] when embedding UiO-66 in membranes used in forward osmosis. By loading 0.1% NPs to the membrane, the water flux increased from 2.19 to 3.33 LMH/bar. Nevertheless, the use of MOFs membranes in water treatment is still in its infancy when compared with their applications for gas separation.

UiO-66 (MOFs based on zirconium) [59] and MIL-125 (MOFs based on titanium) [60] are water stable, carboxylate ligand–based MOFs [61]. The materials possess many metal atom centers, which give them high structural symmetry and steadiness. Figure 1 shows these structures as reported by Wang et al. [52] for UiO-66 and Devic and Serre for MIL-125 [62]. Properties including their high surface area, chemicals resistance, and functionalization ability made these MOFs good candidates for different applications [63,64].

In this paper, UiO-66 and MIL-125 MOFs nanoparticles were synthesized and then embedded into the TFN membrane. The NPs were incorporated by mixing different amounts of materials (0–0.3 wt %) in the trimesoyl chloride solution during membrane fabrication, and the membranes' performance in reverse osmosis application was tested. MOFs NPs were characterized by SEM, zeta potential measurement, and surface area evaluation, while the membranes' physicochemical properties were examined by SEM, TEM, contact angle, and ATR FT-IR tests. It was found that embedding UiO-66 and MIL-125 NPs to the TFN membrane improved its performance to a level that is considered to be among the best for brackish water desalination.

(a)

Ti
O
C

(b)

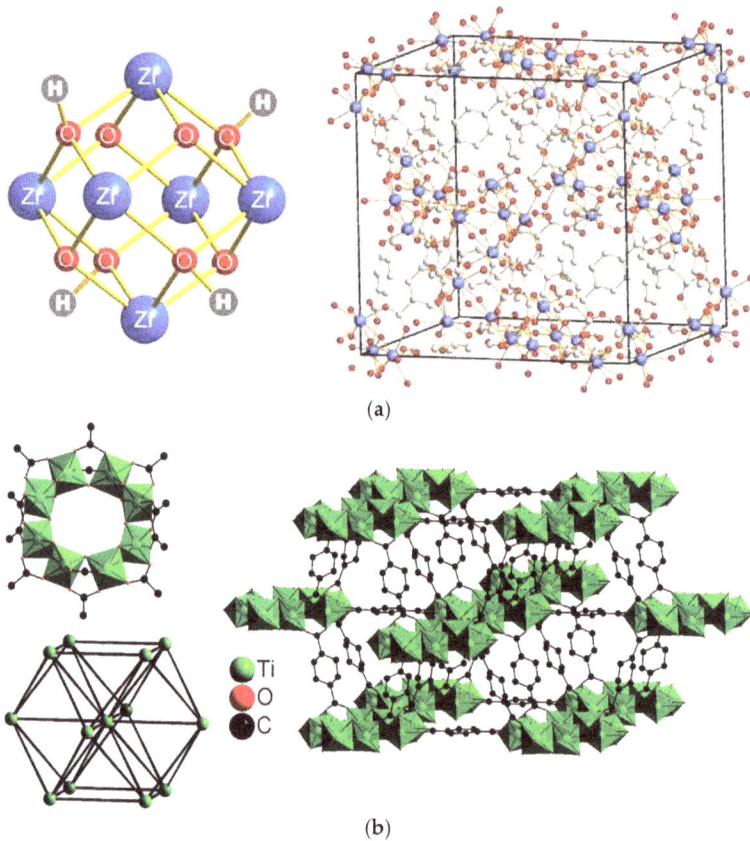

Figure 1. Metal-Organic Frameworks (MOFs) structure: (**a**) UiO66 [52], reproduced with permission from Springer Nature and (**b**) MIL-125 [62], reproduced with permission from Royal Society of Chemistry.

2. Materials and Methods

2.1. Materials

Zirconium (IV) chloride ($ZrCl_4$, 99.5%), titanium (IV) isopropoxide (Ti $[OCH(CH_3)_2]_4$, 97%), 1,4-benzenedicarboxylic acid (BDC, 98%), dimethylformamide (DMF, 99.9%), and methanol (CH_3OH, 99.8%) were obtained from Sigma-Aldrich (St. Louis, MO, USA) for the preparation of MOFs NPs.

The PSU support sheets were prepared by dissolving polysulfone pellets (PSU, M_W = 35,000, Sigma-Aldrich) in DMF solvent. The chemicals for the IP, *m*-phenylenediamine (MPD, $\geq$99%) and trimesoyl chloride (TMC, $\geq$98.5%), were purchased from Fisher Scientific (Pittsburgh, PA, USA) and Sigma Aldrich, respectively. 2,2,4-trimethylpentane (isooctane, 99%) was obtained from Fisher Scientific and used as TMC solvent. Our recent study demonstrated that isooctane could replace commonly used hexane during the IP process as a more environmental-friendly solvent [18]. Millipore DI water (Synergy185, 18.2 MΩ-cm, EMD Millipore Corp., Billerica, MA, USA) was used to prepare MPD solution and for cleaning purposes. Calcium chloride ($CaCl_2$) and sodium chloride (NaCl) were obtained from Fisher Scientific and Sigma Aldrich, respectively. CSA/TEA salt materials, triethylamine (TEA, $\geq$99%) and (1s)-(+)-10-camphorsulfonic acid (CSA, 99%), were purchased from Sigma Aldrich.

2.2. Synthesis and Characterizations of MOFs NPs

UiO-66 and MIL-125 MOFs NPs used in this research were synthesized through the microwave and solvothermal methods, respectively. For UiO-66 preparation, $ZrCl_4$, BDC, and H_2O were dissolved in DMF under stirring in molar ratios of 1:1:1:500, respectively [52]. After a complete mixing, the solution was placed in a GE microwave (Model no. J VM131K 002) and irradiated for 5 min at a power of 220 W. The 5 min irradiation was separated into 4 steps: 2 min, 1 min, 1 min, and 1 min with 1 min relaxing time between each irradiation section to prevent solvent from heating over boiling point. The MOF precipitates appeared during the last 2 steps of irradiation. MIL-125 NPs were prepared in different formula, in which 3 mmol $Ti[OCH(CH_3)_2]_4$, 6 mmol BDC, 50 mL methanol, and 20 mL acetic acid were dissolved in 100 mL DMF [60]. After mixing, the solution was transferred to a 500 mL Teflon liner and put into a metallic digestion bomb at 150 °C for 24 h. Precipitates of both MOFs were washed and centrifuged with DMF and methanol, three times each. The washed samples were dried in an oven overnight at 80 °C and collected for later use.

The NPs size and morphology were tested by SEM (Hitachi S-4700 Field Emission Scanning Electron Microscope, Hitachi, Ltd., Tokyo, Japan). First, the particles were spread on a carbon adhesive disc. After that, they were coated with platinum on a sputter coater (K575x, Emitech, Kent, UK) for 1 min at 20 milliAmps, prior to imaging. Zetasizer (Malvern nano series, Malvern Instruments Ltd., Malvern, UK) was used to measure the NPs zeta potential. 0.1 g of the NPs was placed in 20 mL DI water of pH 5.7 and sonicated for 10 min to avoid aggregation. The specific surface area was determined by N_2-adsorption using the Beckman coulter SA 3100 (Beckman Coulter, Inc., Brea, CA, USA) surface area analyzer, according to the Brunauer-Emmett-Teller (BET) method.

2.3. Preparation of PSU Support Sheets

The casting solution was made by dissolving 15 wt % of PSU grains in DMF. The solution was heated to 60 °C and stirred for at least 5 h, till a colorless solution formed. The solution was removed from the heater and cooled down at room temperature while allowing vent of the organic vapor; then left overnight for degassing.

A solution aliquot was spread over a glass plate and cast to a thickness of 130 μm by a casting knife (EQ-Se-KTQ-150, MTI Corp., Richmond, CA, USA). The glass plate with the PSU solution was directly immersed into a water bath. The colorless solution then turned to a white sheet instantly and detached from the glass in a few seconds, in a process of phase inversion. The support sheets were collected and washed three times in DI water for the removal of any remaining solvent. Finally, the sheets were stored in DI water at 4 °C for at least 24 h before use.

2.4. Preparation of TFN Membranes

The prepared support sheet was placed on a piece of glass with the excess water removed by a squeegee roller. MPD solution was then poured on the PSU sheet for 25 s, then excess solution was removed by the squeegee roller. MPD solution was prepared by dissolving 2 wt % of MPD in DI water. The solution also contained 1% of CAS/TEA salt and 0.01% $CaCl_2$, which were added to improve membrane's hydrophilicity and performance. The membrane with MPD was left for 2 min to dry, then the TMC solution, which was prepared by dissolving 0.15 wt % TMC in isooctane, was added to the sheet for 15 s, and the reaction of TMC and MPD led to the formation of a thin polyamide film by the interfacial polymerization (IP). After disposing the extra TMC solution, the sheets were dried in oven at 80 °C for 6 min for the evaporation of any residual solvent. Finally, the sheets were washed in DI water three times and stored in water at 4 °C for at least 18 h before test. Since the MOFs have higher dispersion affinity in organic solutions than in water due to the relatively hydrophobic organic shells, the MOFs NPs were dispersed in TMC solution in different loading ratios, 0.05, 0.1, 0.15, 0.2, and 0.3 wt %.

2.5. TFN Membrane Characterizations

The prepared membranes were dried at room temperature for around 24 h and stored at 4 °C prior to characterizations. The membrane's morphology was examined by SEM (Hitachi S-4700). The samples were coated with platinum by a sputter coater (Emitech, K575x) for 1 min at 20 milliAmps. After placing the samples inside the SEM device, different voltages were used to achieve various resolutions.

The membrane's cross sectional view was observed using TEM (JEOL JEM-1400, JEOL Ltd., Peabody, MA, USA) device. The samples were prepared by soaking them in a resin (Eponate 12, Ted Pella, Inc., Redding, CA, USA) overnight, then cut by the Reichert–Jung Ultracut E ultramicrotome (Reichert, Depew, NY, USA).

The membrane surface functional groups were assessed by the ATR FT-IR spectroscopy, using the Nicolet 4700 FT-IR with the multi-reflection Smart Performer® ATR accessory (Thermo Electron, Waltham, MA, USA).

The contact angle between DI water drop and membrane's surface was measured to understand the surface hydrophilicity. A sessile drop technique-based contact angle video system, VCA-2500 XE (AST products, Inc., Billerica, MA, USA), was employed for this measurement. Each reported value of contact angle was the average of six measurements tested on different locations.

A cross flow reverse osmosis system was used to measure the water flux and salt rejection as shown in our previous work [18]. The membrane was placed in a filter holder cell (HP Filter Holder, 47 mm, stainless steel, EMD Millipore, Billerica, MA, USA) and tested, under conditions of 300 psi, 25 °C, and 2000 ppm NaCl solution, for eight hours. The permeate water flux was calculated based on its transferred volume per unit time; the LABVIEW software was used to track the water flux, calculated based on Equation (1).

$$J = \frac{V}{A \times t} \tag{1}$$

where J is the water flux (L/m^2 h), V the permeate water volume (L), A the membrane area (m^2), and t the accumulation time (h).

A conductivity meter (HACH Company, Loveland, CO, USA) was employed to measure the total dissolved salts in feed solution and permeate water. NaCl rejection was calculated by Equation (2).

$$R = \left(1 - \frac{C_p}{C_f}\right) \times 100 \tag{2}$$

where R is the salt rejection ratio, C_p the permeate conductivity and C_f the feed conductivity.

The salt solubility maintained the same by controlling the system temperature at 25 °C using external water bath.

3. Results and Discussion

3.1. MOFs NPs Characterizations

UiO-66 and MIL-125 NPs had measured surface areas of 293.6 and 515.5 m^2/g, respectively. Although these numbers seem to indicate the materials are highly porous, they are low compared with what had been reported in the literature [60,65]. The reason for this difference is not clear but the materials' aggregation could have contributed to the difference.

The zeta potential values of UiO-66 and MIL-125 NPs were 7.66 mV and −46.1 mV, respectively, when they were tested at a concentration of 0.1% in DI water. Although both zirconium and titanium could be in the oxidation state of +4, different surface charges, positive and negative, were measured for UiO-66 and MIL-125, respectively. The measured zeta potential of UiO-66 at pH 5.7 was close to the literature-reported value of 0 mV at almost the same pH 5.5 [66]. MIL-125's zeta potential was lower than what was previously reported, around +2 mV at pH of 5.5 [67,68]. This could be due to the use of different titanium raw materials; titanium isopropoxide was used in this research instead of titanium butoxide used in previous studies.

Figure 2a,b presents the SEM morphology of UiO-66 (100–200 nm, cubic shape) and MIL-125 (100 nm, spherical shape) respectively. The sizes of NPs were appropriate to be embedded in TFN membranes, as it would be shown later.

(a) (b)

Figure 2. Scanning electron microscopy (SEM) images for (**a**) UiO-66 and (**b**) MIL-125 MOFs nanoparticles (NPs).

3.2. TFN Membrane Characterizations

ATR FT-IR spectra were collected from the surface of various membrane specimens. The spectra of the PSU support sheet, TFC membrane, and the TFN membranes with UiO-66 and MIL-125 NPs, respectively, were illustrated in Figure 3a,b. The spectral features of the PSU layer and TFC membrane chemical groups are repeated in TFN membranes spectra, in which the peaks corresponding to MOFs were also observed. Generally, by increasing NPs percentage, the reflected spectra had new peaks relevant to the embedded materials. Starting with the PSU spectrum, two featured peaks at 1150 and 1245 cm^{-1} were observed, which could be allocated to the O=S=O symmetric stretching of sulfone group [69] and C–O–C asymmetric stretching of aryl ethyl group, respectively [70]. The peaks at 1298 and 1325 cm^{-1} could indicate the asymmetric stretching of O=S=O sulfone group. At 1488 and 1590 cm^{-1}, the peaks could be attributed to the aromatic C–C stretching [69,70]. By adding polyamide thin film layer, peaks at 1350 and 1610 cm^{-1} appeared due to the N–H deforming (amide III) and C=O (carboxylic), respectively [71]. Furthermore, N–H bending and C–N stretching could be represented by the peak around 1545 cm^{-1} (amide II). The vibration at 1660 cm^{-1} could be assigned to the C=O stretching (amide I) [71,72].

When MOFs NPs were added at very low amounts, new peaks corresponding to these NPs were not apparent. By increasing UiO-66 molar ratio to 0.15 and above, an obvious peak around 1380 cm^{-1} and a tiny one at 1565 cm^{-1} appeared, which could be assigned to the Zr–OH vibration [73–76], as shown in Figure 3a. At 0.3%, the peak disappeared; this might be caused by the particles' aggregation so no proper NP dispersion was obtained. For MIL-125 at 0.3%, a clear peak was observed around 1416 cm^{-1}, which could be assigned to the Ti–O–Ti stretching vibrations [77].

The SEM images of membranes surface with UiO-66 and MIL-125 NPs are illustrated in Figures 4 and 5, respectively. The TFC membrane, for which there was no particles added, showed a leaf-like shape characteristic of normal polyamide layer. Addition of up to 0.05% UiO-66 NPs did not cause much change in leaf-like morphology, as illustrated in Figure 4b. When the loading was increased to 0.1% and 0.15%, the membranes' surface began to change due to NPs filling into the membrane. The nanoparticles covered a wide area and had a good distribution in the membrane structure. At 0.2% and 0.3%, the solid clusters/aggregates were easily observable, which might have an impact on salt rejection. At loading of 0.3%, the NPs aggregated and formed even larger clusters, and covered the surface more sparsely. This is consistent with the FT-IR results and trends, as discussed in the previous section.

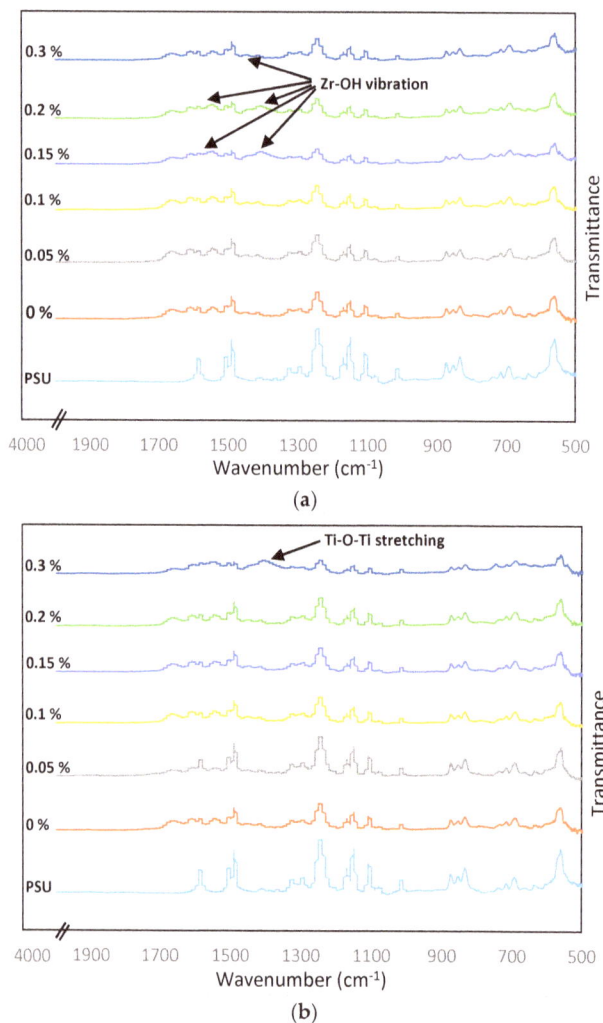

Figure 3. Attenuated total reflection Fourier transform infrared (ATR FT-IR) spectra for (**a**) UiO-66 and (**b**) MIL-125.

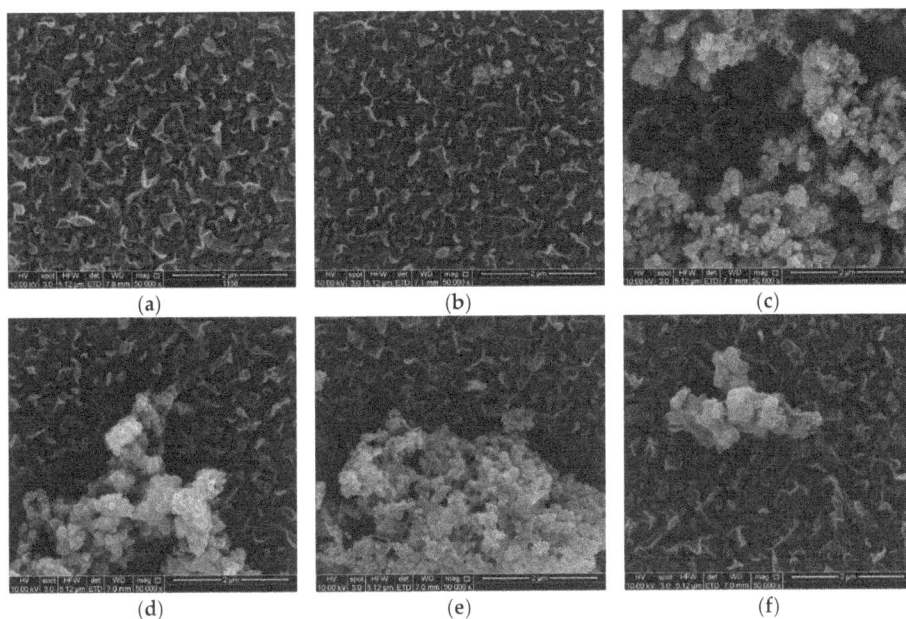

Figure 4. SEM images for membranes surface with UiO-66 NPs injection, (**a**) 0%, (**b**) 0.05%, (**c**) 0.1%, (**d**) 0.15%, (**e**) 0.2%, and (**f**) 0.3%.

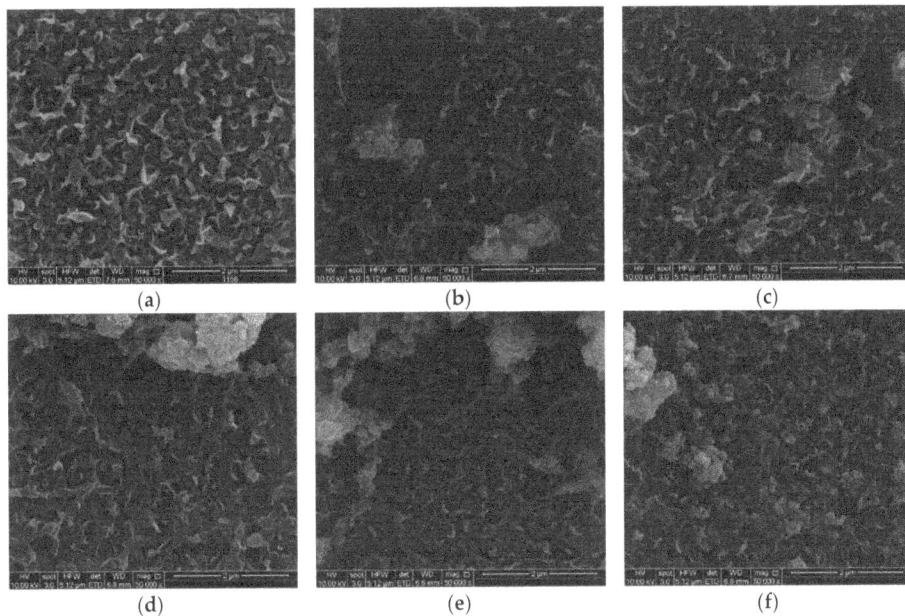

Figure 5. SEM images for membranes surface with MIL-125 NPs injection, (**a**) 0%, (**b**) 0.05%, (**c**) 0.1%, (**d**) 0.15%, (**e**) 0.2%, and (**f**) 0.3%.

In Figure 5, the impact of embedding MIL-125 NPs on the morphology was presented. Unlike UiO-66 filling, MIL-125 appeared to have changed the leaf-shaped morphology, in addition to the aggregation. The leaves are connected to each other, making long leaves. This could be because of MIL-125's quasi-cubic tetragonal structure, or the smaller size of MIL-125 comparing with UiO-66 NPs, which might give better filling into the membrane.

Figure 6 shows the measured contact angles between DI water drop and the membrane's surface. By adding the MOFs NPs, the contact angle did not decrease significantly. The result is different from filling other nanomaterials such as hydrophilic silica [18] and oxidized graphene [78] where significant decreases of contact angle were observed. This may be attributed to the MOFs structure, which contains organic linkers surrounding the metal core. The organic part is slightly hydrophobic, leading to decreased particle hydrophilicity [38]. From the figure, the UiO-66 had more hydrophilic effect than MIL-125. This could be because MIL-125 has more linkers on the surface shielding the metal atom. The observation is also in consistence with the fact that MIL-125 had a higher surface area.

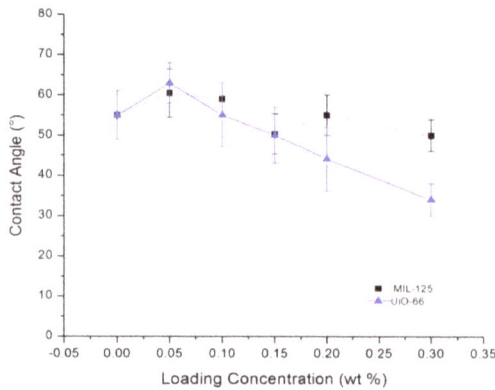

Figure 6. Pure water contact angle with the modified thin film nanocomposite (TFN) membranes surface.

TEM cross section images for the plain membrane and UiO-66 and MIL-125 nanocomposite membranes are illustrated in Figure 7a–c, respectively. Figure 7b is the cross-sectional view of TFN membrane at an optimal UiO-66 NPs loading (0.15%); while Figure 7c is at an optimal MIL-125 NPs loading (0.3%). The TEM examination showed that the particles were filled into the membranes and appeared as black clusters; MIL-125 NPs were highly homogenized with the membrane and affected its texture. From the images, the membrane thickness is found to be 200–400 nm.

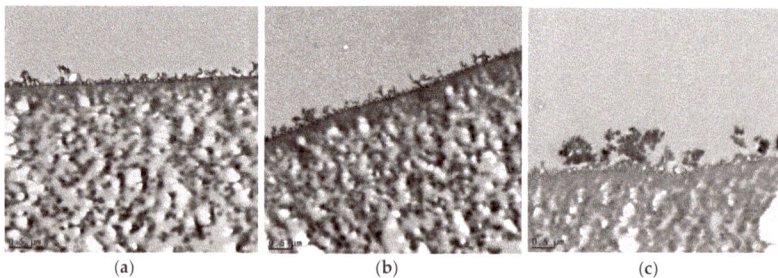

(a) (b) (c)

Figure 7. Transmission electron microscopy (TEM) images of (**a**) thin film composite (TFC) membrane, (**b**) TFN membrane of 0.15% UiO-66, and (**c**) TFN membrane of 0.3% MIL-125.

3.3. TFN Membranes Performance

The salt rejection and water flux for UiO-66 and MIL-125 modified membranes were documented in Figure 8a,b. For UiO-66, the optimal addition was 0.15%, where the permeate water flux and salt rejection increased from 62.5 L/m^2 h and 98.4% to 74.9 L/m^2 h and 98.8%, respectively. From Figure 8a, adding UiO-66 NPs increased NaCl rejection at low loadings, but a further increase of the NPs resulted in a decrease of salt rejection. The rejection increase at low loading could be attributed to UiO-66's pore size (~6.0 Å) [59], which is larger than water molecule (~2.8 Å) but smaller than the hydrated ions (Na$^+$ = 7.16 and Cl$^-$ = 6.64 Å) [79]. Increasing the NPs amount too much could decrease salt rejection, as the particles aggregation may generate cracks in TFN membranes structure. On the other hand, the flux increase by adding NPs reached its maximum at 0.15%. This is likely due to the hydrophilic nature of UiO-66 [59,80]. At percentages higher than 0.15%, the water flux decreased, again likely due to NPs aggregation [13]. Here, both salt rejection and water flux decreased because of particle aggregation. Aggregation may have created micro gaps between particle blocs that allowed the saline water to pass through. The particle aggregates could also block the water transfer through the membrane by the hydrophobic linkers overlapping, thus decreased water flux. Thereby, reduction in both salt rejection and water flux occurred.

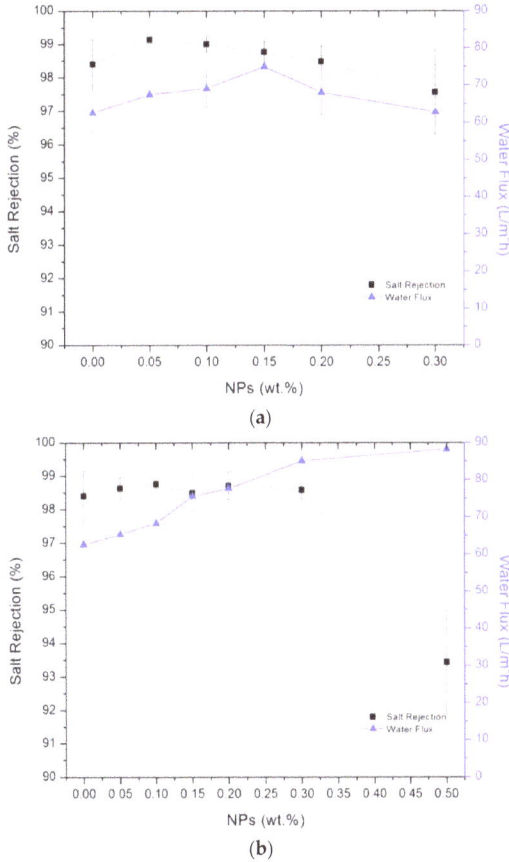

Figure 8. TFN membranes performance with injected (**a**) UiO-66 and (**b**) MIL-125 NPs.

Figure 8b shows the effect of adding MIL-125 NPs to TFN membrane. The water flux increased by increasing NPs loading, for example, with a solid loading of 0.3%, the permeate water flux was increased from 62.5 to 85.0 L/m^2 h, while the salt rejection maintained almost the same (98.4% vs. 98.6%). At filling of 0.5% NPs, a drop in salt rejection happened. The effective pore diameter of MIL-125 is 12.55 Å [60] which is larger than both the water molecule and hydrated ions of Na$^+$ and Cl$^-$. However, the salt rejection remained the same even at a higher loading, the high negative charge of MIL-125 could play a role here as the formed electrical double layer could repel the negatively-charged ions. The reported water flux here at 0.3%, 85.0 L/m^2 h, is considered one of the highest in comparison with the reported values in the literature [81]. The high flux could be due to the NPs hydrophilicity and their large surface area and pore size. Figure 9 shows the permeability of optimal membranes by adding 0.15 wt % of UiO-66 or 0.3 wt % MIL-125 at different feed pressures.

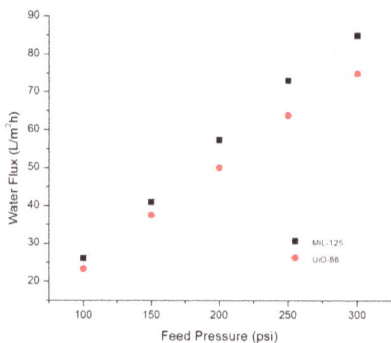

Figure 9. Membranes permeability at an optimal NPs filling.

3.4. Why MOFs for TFN Membranes?

MOFs are valuable materials due to their unique structure and wide applications in various areas [35]. It is anticipated that by the coming decade, mixed matrix membranes based on MOFs could grow, including those for large scale applications [38]. Among various nanocomposite membranes, MOFs-enabled membranes are advantageous because they can overcome the traditional filling problems of low affinity with the membrane, porosity blocking, and segregation [82,83]. The hybrid MOFs structure with organic and inorganic parts is generally compatible with polymeric layers in membrane [84]. This concept is illustrated in Figure 10, where MOFs particles can be more conveniently embedded into the TFN membrane with less gaps than those created by traditional inorganic fillers. Another advantage is the low cost of the polymeric membranes and the vast selection range of linkers toward controlling MOFs shape, morphology, and surface chemistry, making it potentially possible to design membrane for specific applications.

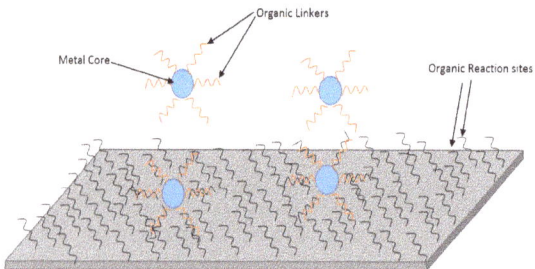

Figure 10. MOFs filling into polymeric substance diagram.

4. Conclusions

Incorporation of UiO-66 and MIL-125 MOFs NPs in reverse osmosis membranes is reported in this study. The results showed that MOFs were embedded well inside the membranes and increased the water flux and salt rejection. Different techniques were used to exam the NPs and TFN membranes' physicochemical properties. The organic linker enhanced the compatibility of MOFs particles with the polyamide thin film. The results indicated that the water flux was increased from 62.5 to 74.9 or 85.0 L/m^2 h at a transmembrane pressure of 300 psi, by addition of 0.15% of UiO-66 or 0.3% of MIL-125, respectively, while salt rejection was slightly increased. These high flux results indicate filling of MOFs NPs in TFN membrane might be advantageous because of the NPs containing an organic part that could link properly to polyamide, and inorganic part that could enhance the membrane performance.

Acknowledgments: We gratefully acknowledge the Higher Committee for Education Development in Iraq (HCED-Iraq) for supporting Mohammed Kadhom. The project was partial supported by the Missouri Water Resources Research Center. We would like to thank Qingsong Yu in the Department of Mechanical and Aerospace Engineering and Maria Fidalgo and Zhiqiang Hu in the Department of Civil and Environmental Engineering at the University of Missouri-Columbia for the access to contact angle video system, BET, and zeta potential device, respectively. We would like to acknowledge Martin Schauflinger from the Electron Microscopy Core at the University of Missouri-Columbia for his help in preparing and characterizing TEM samples.

Author Contributions: Mohammed Kadhom and Baolin Deng conceived and designed the experiments. Weiming Hu prepared the NPs. Mohammed Kadhom performed the experiments, analyzed the date, and wrote the first draft. All authors contributed to put the paper in its final form.

Conflicts of Interest: The authors declare no conflicts of interest.

References

1. Nobel, R.; Stern, S.A. *Membrane Separations Technology Principles and Applications*; Elsevier: Amsterdam, The Netherlands, 1995.
2. Service, R.F. Desalination Freshens Up. *Science* **2006**, *313*, 1088–1090. [CrossRef] [PubMed]
3. Sagle, A.; Freeman, B. Fundamentals of membranes for water treatment. *Tex. Water Dev.* **2004**, *2*, 137–153.
4. Cadotte, J. Interfacially Synthesized Reverse Osmosis Membrane. U.S. Patent 4,277,344, 7 July 1981.
5. Pabby, A.K.; Rizvi, S.S.; Sastre, A.M. *Handbook of Membrane Separations: Chemical, Pharmaceutical, Food, and Biotechnological Applications*; Taylor & Francis Group: Boca Raton, FL, USA, 2015.
6. Dong, H.; Zhao, L.; Zhang, L.; Chen, H.; Gao, C.; Ho, W.S.W. High-flux reverse osmosis membranes incorporated with NaY zeolite nanoparticles for brackish water desalination. *J. Membr. Sci.* **2015**, *476*, 373–383. [CrossRef]
7. Lind, M.L.; Ghosh, A.K.; Jawor, A.; Huang, X.; Hou, W.; Yang, Y.; Hoek, E. Influence of zeolite crystal size on zeolite–polyamide thin film nanocomposite membranes. *Langmuir* **2009**, *25*, 10139–10145. [CrossRef] [PubMed]
8. Kong, C.L.; Shintani, T.; Tsuru, T. Pre-seeding assisted synthesis of a high performance polyamide–zeolite nanocomposite membrane for water purification. *New J. Chem.* **2010**, *34*, 2101–2104. [CrossRef]
9. Huang, H.; Qu, X.; Dong, H.; Zhang, L.; Chen, H. Role of NaA zeolites in the interfacial polymerization process towards a polyamide nanocomposite reverse osmosis membrane. *RSC Adv.* **2013**, *3*, 8203–8207. [CrossRef]
10. Huang, H.; Qu, X.; Ji, X.; Gao, X.; Zhang, L.; Chen, H.; Hou, L. Acid and multivalent ion resistance of thin film nanocomposite RO membranes loaded with silicalite-1 nanozeolites. *J. Mater. Chem. A* **2013**, *1*, 11343–11349. [CrossRef]
11. Dong, H.; Qu, X.Y.; Zhang, L.; Cheng, L.H.; Chen, H.L.; Gao, C.J. Preparation and characterization of surface-modified zeolite–polyamide thin film nanocomposite membranes for desalination. *Desal. Water Treat.* **2011**, *34*, 6–12. [CrossRef]
12. Li, L.; Dong, J.; Nenoff, T.M.; Lee, R. Desalination by reverse osmosis using MFI zeolite membranes. *J. Membr. Sci.* **2004**, *243*, 401–404. [CrossRef]
13. Yin, J.; Kim, E.; Yang, J.; Deng, B. Fabrication of a novel thin-film nanocomposite (TFN) membrane containing MCM-41 silica nanoparticles (NPs) for water purification. *J. Membr. Sci.* **2012**, *423–424*, 238–246. [CrossRef]

14. Wu, H.; Tang, B.; Wu, P. Optimizing polyamide thin film composite membrane covalently bonded with modified mesoporous silica nanoparticles. *J. Membr. Sci.* **2013**, *428*, 341–348. [CrossRef]
15. Ma, X.; Lee, N.; Oh, H.; Hwang, J.; Kim, S. Preparation and characterization of silica/polyamide-imide nanocomposite thin films. *Nanoscale Res. Lett.* **2010**, *5*, 1846–1851. [CrossRef] [PubMed]
16. Tiraferri, A.; Kang, Y.; Giannelis, E.P.; Elimelech, M. Highly hydrophilic thin-film composite forward osmosis membranes functionalized with surface-tailored nanoparticles. *Appl. Mater. Interfaces* **2012**, *4*, 5044–5053. [CrossRef] [PubMed]
17. Jadav, G.L.; Singh, P.S. Synthesis of novel silica-polyamide nanocomposite membrane with enhanced properties. *J. Membr. Sci.* **2009**, *328*, 257–267. [CrossRef]
18. Kadhom, M.; Yin, J.; Deng, B. A thin film nanocomposite membrane with MCM-41 silica nanoparticles for brackish water purification. *Membranes* **2016**, *6*, 1–12. [CrossRef] [PubMed]
19. Emadzadeh, D.; Lau, W.J.; Matsuura, T.; Ismail, A.F.; Rahbari-Sisakht, M. Synthesis and characterization of thin film nanocomposite forward osmosis membrane with hydrophilic nanocomposite support to reduce internal concentration polarization. *J. Membr. Sci.* **2014**, *449*, 74–85. [CrossRef]
20. Yadaa, M.; Inouea, Y.; Akihitoa, G.; Nodab, I.; Torikaia, T.; Wataria, T.; Hotokebuchi, T. Apatite-forming ability of titanium compound nanotube thin films formed on a titanium metal plate in a simulated body fluid. *Colloids Surf. B Biointerfaces* **2010**, *80*, 116–124. [CrossRef] [PubMed]
21. Emadzadeha, D.; Laua, W.J.; Matsuura, T.; Hilal, N.; Ismail, A.F. The potential of thin film nanocomposite membrane in reducing organic fouling in forward osmosis process. *Desalination* **2014**, *348*, 82–88. [CrossRef]
22. Tettey, K.E.; Yee, M.Q.; Lee, D. Photocatalytic and conductive MWCNT/TiO$_2$ nanocomposite thin films. *Appl. Mater. Interfaces* **2010**, *2*, 2646–2652. [CrossRef] [PubMed]
23. Shen, J.; Yu, C.; Ruan, H.; Gao, C.; Bruggen, B.V. Preparation and characterization of thin-film nanocomposite membranes embedded with poly(methyl methacrylate) hydrophobic modified multiwalled carbon nanotubes by interfacial polymerization. *J. Membr. Sci.* **2013**, *442*, 18–26. [CrossRef]
24. Roy, S.; Ntim, S.A.; Mitra, S.; Sirkar, K.K. Facile fabrication of superior nanofiltration membranes from interfacially polymerized CNT-polymer composites. *J. Membr. Sci.* **2011**, *375*, 81–87. [CrossRef]
25. Ma, L.; Dong, X.; Chen, M.; Zhu, L.; Wang, C.; Yang, F.; Dong, Y. Fabrication and Water Treatment Application of Carbon Nanotubes (CNTs)-Based Composite Membranes: A Review. *Membranes* **2017**, *7*, 1–21. [CrossRef] [PubMed]
26. Ghosh, A.K.; Jeong, B.-H.; Huang, X.; Hoek, E.M.V. Impacts of reaction and curing conditions on polyamide composite reverse osmosis membrane properties. *J. Membr. Sci.* **2008**, *311*, 34–45. [CrossRef]
27. Ding, C.; Yin, J.; Deng, B. Effects of polysulfone (PSf) support layer on the performance of thin-film composite (TFC) membranes. *J. Chem. Process Eng.* **2014**, *1*, 1–8. [CrossRef]
28. Kuehne, M.A.; Song, R.Q.; Li, N.N.; Petersen, R.J. Flux enhancement in TFC RO membranes. *Environ. Prog.* **2001**, *20*, 23–26. [CrossRef]
29. Zhao, L.; Ho, W.S.W. Novel reverse osmosis membranes incorporated with a hydrophilic additive for seawater desalination. *J. Membr. Sci.* **2014**, *455*, 44–54. [CrossRef]
30. Gu, H.; Rahardianto, A.; Gao, L.X.; Christof, P.D.; Cohen, Y. Ultrafiltration with self-generated RO concentrate pulse backwash in a novel integrated seawater desalination UF-RO system. *J. Membr. Sci.* **2016**, *520*, 111–119. [CrossRef]
31. Hai, Y.; Zhang, J.; Shi, C.; Zhou, A.; Bian, C.; Li, W. Thin film composite nanofiltration membrane prepared by the interfacial polymerization of 1,2,4,5-benzene tetracarbonyl chloride on the mixed amines cross-linked poly(ether imide) support. *J. Membr. Sci.* **2016**, *520*, 19–28. [CrossRef]
32. Shaffer, D.L.; Yip, N.Y.; Gilron, J.; Elimelech, M. Seawater desalination for agriculture by integrated forward and reverse osmosis: Improved product water quality for potentially less energy. *J. Membr. Sci.* **2012**, *415–416*, 1–8. [CrossRef]
33. Elimelech, M.; Phillip, W.A. The future of sea water desalination: Energy, technology, and the environment. *Science* **2011**, *333*, 712–717. [CrossRef] [PubMed]
34. Fritzmann, C.; Löwenberg, J.; Wintgens, T.; Melin, T. State-of-the-art of reverse osmosis desalination. *Desalination* **2007**, *216*, 1–76. [CrossRef]
35. Li, W.; Zhang, Y.; Li, Q.; Zhang, G. Metal−organic framework composite membranes: Synthesis and separation applications. *Chem. Eng. Sci.* **2015**, *135*, 232–257. [CrossRef]
36. James, S.L. Metal-organic frameworks. *Chem. Soc. Rev.* **2003**, *32*, 276–288. [CrossRef] [PubMed]

37. Li, J.; Sculley, J.; Zhou, H.C. Metal-organic frameworks for separations. *Chem. Rev.* **2012**, *112*, 869–932. [CrossRef] [PubMed]

38. Zornoza, B.; Tellez, C.; Coronas, J.; Gascon, J.; Kapteijn, F. Metal organic framework based mixed matrix membranes: An increasingly important field of research with a large application potential. *Microporous Mesoporous Mater.* **2013**, *166*, 67–78. [CrossRef]

39. Gascon, J.; Aktay, U.; Hernandez-Alonso, M.D.; van Klink, G.P.; Kapteijn, F. Amino-based metal-organic frameworks as stable, highly active basic catalysts. *J. Catal.* **2009**, *261*, 75–87. [CrossRef]

40. An, J.; Rosi, N. Tuning MOF CO_2 adsorption properties via cation exchange. *J. Am. Chem. Soc.* **2010**, *132*, 5578–5579. [CrossRef] [PubMed]

41. Lee, C.Y.; Farha, O.K.; Hong, B.J.; Sarjeant, A.A.; Nguyen, S.T.; Hupp, J.T. Light-Harvesting MetalOrganic Frameworks (MOFs): Efficient Strut-to-Strut Energy Transfer in Bodipy and Porphyrin-Based MOFs. *J. Am. Chem. Soc.* **2011**, *133*, 15858–15861. [CrossRef] [PubMed]

42. Allendorf, M.D.; Bauer, C.A.; Bhakta, R.K.; Houk, R.J.T. Luminescent metal–organic frameworks. *Chem. Soc. Rev.* **2009**, *38*, 1330–1352. [CrossRef] [PubMed]

43. Ma, L.; Abney, C.; Lin, W. Enantioselective catalysis with homochiral metal-organic frameworks. *Chem. Soc. Rev.* **2009**, *38*, 1248–1256. [CrossRef] [PubMed]

44. Rocca, J.D.; Liu, D.; Lin, W. Nanoscale Metal–Organic Frameworks for Biomedical Imaging and Drug Delivery. *Acc. Chem. Res.* **2011**, *44*, 957–968. [CrossRef] [PubMed]

45. Murray, L.J.; Dincă, M.; Long, J.R. Hydrogen storage in metal-organic frameworks. *Chem. Soc. Rev.* **2009**, *38*, 1294–1314. [CrossRef] [PubMed]

46. Hu, Y.H.; Zhang, L. Hydrogen Storage in Metal–Organic Frameworks. *Adv. Mater.* **2010**, *22*, E117–E130. [CrossRef] [PubMed]

47. Li, J.-R.; Kuppler, R.J.; Zhou, H.C. Selective gas adsorption and separation in metal-organic frameworks. *Chem. Soc. Rev.* **2009**, *38*, 1477–1504. [CrossRef] [PubMed]

48. An, J.; Geib, S.J.; Rosi, N.L. High and selective CO_2 uptake in a cobalt adeninate metal-organic framework exhibiting pyrimidine- and amino-decorated pores. *J. Am. Chem. Soc.* **2010**, *132*, 38–39. [CrossRef] [PubMed]

49. Farha, O.K.; Eryazici, I.; Jeong, N.C.; Hauser, B.G.; Wilmer, C.E.; Sarjean, A.A.; Snurr, R.Q.; Nguyen, S.T.; Yazaydın, A.O.; Hupp, J.T. Metal−Organic Framework Materials with Ultrahigh Surface Areas: Is the Sky the Limit? *J. Am. Chem. Soc.* **2012**, *134*, 15016–15021. [CrossRef] [PubMed]

50. Liu, Y. Synthesis of continuous MOF-5 membranes on porous a-alumina substrates. *Microporous Mesoporous Mater.* **2009**, *118*, 296–301. [CrossRef]

51. Shen, J.; Liu, G.; Huang, K.; Li, Q.; Guan, K.; Li, Y.; Jin, W. UiO-66-polyether block amide mixed matrix membranes for CO_2 separation. *J. Membr. Sci.* **2016**, *513*, 155–165. [CrossRef]

52. Wang, C.; Liu, X.; Chen, J.P.; Li, K. Superior removal of arsenic from water with zirconium metal-organic framework UiO-66. *Sci. Rep.* **2015**, *5*, 1–10. [CrossRef] [PubMed]

53. Liu, X.; Demir, N.K.; Wu, Z.; Li, K. Highly water-stable zirconium metal−organic framework UiO-66 membranes supported on alumina hollow fibers for desalination. *J. Am. Chem. Soc.* **2015**, *137*, 6999–7002. [CrossRef] [PubMed]

54. Lee, J.-Y.; Tang, C.Y.; Huo, F. Fabrication of porous matrix membrane (PMM) using metal-organic framework as green template for water treatment. *Sci. Rep.* **2014**, *4*, 3740. [CrossRef] [PubMed]

55. Hu, Z.; Chen, Y.; Jiang, J. Zeolitic imidazolate framework-8 as a reverse osmosis membrane for water desalination: Insight from molecular simulation. *J. Chem. Phys.* **2011**, *134*, 134705. [CrossRef] [PubMed]

56. Gupta, K.; Zhang, K.; Jiang, J. Water desalination through zeolitic imidazolate framework membranes: Significant role of functional groups. *Langmuir* **2015**, *31*, 13230–13237. [CrossRef] [PubMed]

57. Xu, Y.; Gao, X.; Wang, X.; Wang, Q.; Ji, Z.; Wang, X.; Wu, T.; Gao, C. Highly and stably water permeable thin film nanocomposite membranes doped with MIL-101 (Cr) nanoparticles for reverse osmosis application. *Materials* **2016**, *9*, 870. [CrossRef]

58. Ma, D.; Peh, S.B.; Han, G.; Chen, S.B. Thin-film nanocomposite (TFN) membranes incorporated with super-hydrophilic metal−organic framework (MOF) UiO-66: Toward enhancement of water flux and salt rejection. *ACS Appl. Mater. Interfaces* **2017**, *9*, 7523–7534. [CrossRef] [PubMed]

59. Cavka, J.H.; Jakobsen, S.; Olsbye, U.; Guillou, N.; Lamberti, C. A new zirconium inorganic building brick forming metal organic frameworks with exceptional stability. *J. Am. Chem. Soc.* **2008**, *130*, 13850–13851. [CrossRef] [PubMed]

60. Dan-Hardi, M.; Serre, C.; Frot, T.; Rozes, L.; Maurin, G.; Sanchez, C.; Férey, G. A new photoactive crystalline highly porous titanium(IV) dicarboxylate. *J. Am. Chem. Soc.* **2009**, *131*, 10857–10859. [CrossRef] [PubMed]

61. Jasuja, H.; Jiao, Y.; Burtch, N.C.; Huang, Y.-G.; Walton, K.S. Synthesis of cobalt, nickel, copper, and zinc-Based, water-Stable, pillared metal−organic frameworks. *Langmuir* **2014**, *30*, 14300–14307. [CrossRef] [PubMed]

62. Devic, T.; Serre, C. High valence 3p and transition metal based MOFs. *Chem. Soc. Rev.* **2014**, *43*, 6097–6115. [CrossRef] [PubMed]

63. Jeremias, F.; Lozan, V.; Henninger, S.K.; Janiak, C. Programming MOFs for water sorption: Aminofunctionalized MIL-125 and UiO-66 for heat transformation and heat storage applications. *Dalton Trans.* **2013**, *42*, 15967–15973. [CrossRef] [PubMed]

64. Stassen, I.; Styles, M.; van Assche, T.; Campagnol, N.; Fransaer, J.; Denayer, J.; Tan, J.-C.; Falcaro, P.; de Vos, D.; Ameloot, R. Electrochemical film deposition of the zirconium metal−organic framework UiO-66 and application in a miniaturized sorbent trap. *Chem. Mater.* **2015**, *27*, 1801–1807. [CrossRef]

65. Kandiah, M.; Nilsen, M.H.; Usseglio, S.; Jakobsen, S.; Tilset, M.; Larabi, C.; Quadrelli, E.A.; Bonino, F.; Lillerud, K.P. Synthesis and stability of tagged UiO-66 Zr-MOFs. *Chem. Mater.* **2010**, *22*, 6632–6640. [CrossRef]

66. Seo, Y.S.; Khan, N.A.; Jhung, S.H. Adsorptive removal of methylchlorophenoxypropionic acid from water with a metal-organic framework. *Chem. Eng. J.* **2015**, *270*, 22–27. [CrossRef]

67. Wang, H.; Yuan, X.; Wu, Y.; Zeng, G.; Dong, H.; Chen, X.; Leng, L.; Wu, Z.; Peng, L. In situ synthesis of In2S3@MIL-125(Ti) core-shell microparticle for the removal of tetracycline from wastewater by integrated adsorption and visible-light-driven photocatalysis. *Appl. Catal. B Environ.* **2016**, *186*, 19–29. [CrossRef]

68. Wang, H.; Yuan, X.; Wu, Y.; Zeng, G.; Chen, X.; Leng, L.; Wu, Z.; Jiang, L.; Li, H. Facile synthesis of amino-functionalized titanium metal-organic frameworks and their superior visible-light photocatalytic activity for Cr(VI) reduction. *J. Hazard. Mater.* **2015**, *286*, 187–194. [CrossRef] [PubMed]

69. Deng, H.; Xu, Y.; Chen, Q.; Wei, X.; Zhu, B. High flux positively charged nanofiltration membranes prepared by UV-initiated graft polymerization of methacrylatoethyl trimethyl ammonium chloride (DMC) onto polysulfone membranes. *J. Membr. Sci.* **2011**, *366*, 363–372. [CrossRef]

70. Tarboush, B.A.; Rana, D.; Matsuuraa, T.; Arafat, H.; Narbaitz, R. Preparation of thin-film-composite polyamide membranes for desalination using novel hydrophilic surface modifying macromolecules. *J. Membr. Sci.* **2008**, *325*, 166–175. [CrossRef]

71. Lee, H.; Im, S.; Kim, J.; Kim, H.; Kim, J.; Min, B. Polyamide thin-film nanofiltration membranes containing TiO2 nanoparticles. *Desalination* **2008**, *219*, 48–56. [CrossRef]

72. Shawky, H.; Chae, S.; Lin, S.; Wiesner, M. Synthesis and characterization of a carbon nanotube/polymer nanocomposite membrane for water treatment. *Desalination* **2011**, *272*, 46–50. [CrossRef]

73. Simmons, C. Corrosion of heavy-metal fluoride glasses. In *Corrosion of Glass, Ceramics and Ceramic Superconductors: Principles, Testing, Characterization and Applications*; Noyes Publications: Park Ridge, NJ, USA, 1992.

74. Toullec, M.L.; Simmons, C.J.; Simmons, J.H. Infrared spectroscopic studies of the hydrolysis reaction during leaching of heavy-metal fluoride glasses. *J. Am. Chem. Soc.* **1988**, *71*, 219–224. [CrossRef]

75. Sarkar, D.; Mohapatra, D.; Ray, S.; Bhattacharyya, S.; Adak, S.; Mitra, N. Synthesis and characterization of sol–gel derived ZrO2 doped Al2O3 nanopowder. *Ceram. Int.* **2007**, *33*, 1275–1282. [CrossRef]

76. Liu, H.; Sun, X.; Yin, C.; Hu, C. Removal of phosphate by mesoporous ZrO2. *J. Hazard. Mater.* **2008**, *151*, 616–622. [CrossRef] [PubMed]

77. Vetrivel, V.; Rajendran, K.; Kalaiselvi, V. Synthesis and characterization of Pure Titanium dioxide nanoparticles by Sol- gel method. *Int. J. ChemTech Res.* **2015**, *7*, 1090–1097.

78. Perreault, F.; Tousley, M.E.; Elimelech, M. Thin-film composite polyamide membranes functionalized with biocidal graphene oxide nanosheets. *Environ. Sci. Technol. Lett.* **2014**, *1*, 71–76. [CrossRef]

79. Nightingale, E.R.J. Phenomenological theory of ion solvation. Effective radii of hydrated ions. *J. Phys. Chem.* **1959**, *63*, 1381–1387. [CrossRef]

80. Furukawa, H.; Gándara, F.; Zhang, Y.-B.; Jiang, J.; Queen, W.L.; Hudson, M.R.; Yaghi, O.M. Water adsorption in porous metal-organic frameworks and related materials. *J. Am. Chem. Soc.* **2014**, *136*, 4369–4381. [CrossRef] [PubMed]

81. Zhao, L.; Chang, P.C.-Y.; Ho, W. High-flux reverse osmosis membranes incorporated with hydrophilic additives for brackish water desalination. *Desalination* **2013**, *308*, 225–232. [CrossRef]

82. Nik, O.G.; Chen, X.Y.; Kaliaguine, S. Amine-functionalized zeolite FAU/EMT-polyimide mixed matrix membranes for CO_2/CH_4 separation. *J. Membr. Sci.* **2011**, *379*, 468–478. [CrossRef]

83. Bae, T.-H.; Liu, J.; Lee, J.S.; Koros, W.J.; Jones, C.W.; Nair, S. Facile high-yield solvothermal deposition of inorganic nanostructures on zeolite crystals for mixed matrix membrane fabrication. *J. Am. Chem. Soc.* **2009**, *131*, 14662–14663. [CrossRef] [PubMed]

84. Centrone, A.; Yang, Y.; Speakman, S.; Bromberg, L.; Rutledge, G.C.; Hatton, T.A. Growth of metal−organic frameworks on polymer surfaces. *J. Am. Chem. Soc.* **2010**, *132*, 15687–15691. [CrossRef] [PubMed]

Article

Membranes of Polymers of Intrinsic Microporosity (PIM-1) Modified by Poly(ethylene glycol)

Gisela Bengtson, Silvio Neumann and Volkan Filiz *

Helmholtz-Zentrum Geesthacht, Institute of Polymer Research, Max-Planck-Strasse 1, 21502 Geesthacht,
Germany; gisela.bengtson@hzg.de (G.B.); silvio.neumann@hzg.de (S.N.)
* Correspondence: volkan.filiz@hzg.de; Tel.: +49-4152-87-2425; Fax: +49-4152-87-2499

Academic Editor: Klaus Rätzke
Received: 27 April 2017; Accepted: 31 May 2017; Published: 5 June 2017

Abstract: Until now, the leading polymer of intrinsic microporosity PIM-1 has become quite famous
for its high membrane permeability for many gases in gas separation, linked, however, to a rather
moderate selectivity. The combination with the hydrophilic and low permeable poly(ethylene glycol)
(PEG) and poly(ethylene oxides) (PEO) should on the one hand reduce permeability, while on the
other hand enhance selectivity, especially for the polar gas CO_2 by improving the hydrophilicity
of the membranes. Four different paths to combine PIM-1 with PEG or poly(ethylene oxide) and
poly(propylene oxide) (PPO) were studied: physically blending, quenching of polycondensation,
synthesis of multiblock copolymers and synthesis of copolymers with PEO/PPO side chain.
Blends and new, chemically linked polymers were successfully formed into free standing dense
membranes and measured in single gas permeation of N_2, O_2, CO_2 and CH_4 by time lag method.
As expected, permeability was lowered by any substantial addition of PEG/PEO/PPO regardless the
manufacturing process and proportionally to the added amount. About 6 to 7 wt % of PEG/PEO/PPO
added to PIM-1 halved permeability compared to PIM-1 membrane prepared under similar conditions.
Consequently, selectivity from single gas measurements increased up to values of about 30 for
CO_2/N_2 gas pair, a maximum of 18 for CO_2/CH_4 and 3.5 for O_2/N_2.

Keywords: polymers of intrinsic microporosity (PIM-1); poly(ethylene glycol); copolymers;
gas separation; membranes

---

1. Introduction

Polymers of intrinsic microporosity have been and continue to be the objects of many research
papers (see, e.g., citations in [1]) since Budd and McKeown launched their first publication about
PIM-1 in 2004 [2].

PIM-1 is a hydrophobic polymer formed as a molecular ladder; the polymer chains cannot pack
tightly because of their contorted molecular structure, provided by spiro fused rings in the polymer
chain. Thus, intrinsic nanopores are formed resulting in a large inner surface [3]. Membranes prepared
from this polymer show a large fractional free volume and therefore high permeability for many
gases [4], but especially for CO_2 (4000 to 10,000 Barrer) because of an also very high solubility [5].
However, the trade-off between permeability and selectivity, as it was evaluated by Robeson [6],
consequently results for PIM-1 in a moderate permselectivity of CO_2 over N_2 of about 16 to 20 and this
is not at all comparable to the selectivity values obtained with membranes made from polymers such
as PEBAX® (Arkema, Colombes, France) or Polyactive™ (Polyvation, Groningen, The Netherlands),
ranging from 30 to 40 [7]. Both latter polymers are block copolymers, containing poly(ethylene (or
higher equivalents) oxide) (PEO) chains for soft blocks, thus performing the transport of CO_2 [8]. As a
drawback, they are of low overall permeability for gases (e.g., 100 to 300 Barrer CO_2). The hard blocks
(polyamides) necessary for mechanical stability do not contribute to permeability at all and PEOs in

general show low permeability [9]. The PEO blocks in PEBAX® and Polyactive™ derive of ethylene or butylene glycol units and they are rather short to avoid crystallization of PEO because crystallites are almost impermeable to any gases and would additionally reduce the overall permeability [10].

In combining PIM-1 and low molecular PEG- or PEO-blocks, it is expected to get a reduced permeability but an enhanced selectivity for CO_2 due to its preferred solubility in PEGs and also due to a reduction of permeability of non-polar gases like N_2. In recently published research Wu et al. [11] followed this idea by blending PIM-1 and PEGs of different molecular weights (2 up to 20 kg/mol) in concentrations up to 5 wt % PEGs to enhance the selectivity of CO_2 over N_2 as well as over CH_4, clearly without crucifying permeability too much. They observed a reduction to about half of the permeability of pure PIM-1 blended with 2.5 wt % PEG. With increasing molecular weight of PEG the permeability increased somewhat again coupled to a slight decrease in selectivity. Therefore, they chose PEG with 20 kg/mol for further experiments regarding PEG content and feed pressure. At 3.5 wt % PEG (20 kg/mol), a permeability of almost 2000 Barrer CO_2 and a selectivity for CO_2/CH_4 of 39 was achieved. Selectivity of CO_2 over N_2 was highest with 2.5 wt % PEG (2 kg/mol), with a value of 23.

In our study, we followed four different paths to combine PIM-1 and PEG or PEO. First, we also blended PIM-1 with PEGs of different molecular weights to evaluate the optimum amount of PEG and to check on the influence of PEG chain length. Long PEO chains are favorable for CO_2 solubility [10] but tend to crystallize and separate from PIM-1 phase. Therefore, we favored PEGs with smaller chain lengths of ≤ 5 kg/mol. The permeability of blended membranes should not be reduced too much; the influence of the added PEGs on solubility of CO_2 and other gases is going to be estimated. A similar approach was followed in [12,13] by blending PEBAX® (Arkema, France) with PEGs in different amounts, although in these cases it was thought as an attempt to increase the rather low pristine permeability of the block copolymer.

Along the following Paths 2–4, we strived for a chemical bondage between PIM-1 and PEO by different substitution reactions. In Path 2, we added PEG-monomethylethers, bound by esterification to trihydroxybenzoic acid (THBA), as quenching compounds to terminate the polycondensation reaction forming PIM-1. This resulted in (di)block copolymers with rather large PIM-1 blocks fenced by small PEO end group(s). The procedure was performed with PEGs of different chain lengths.

Within Path 3, we combined small PIM-1 blocks (at least smaller blocks than in Paths 1 and 2) with different PEO-containing co-monomers (PEG di-substituted by esterification with dihydroxybenzoic acid (DHBA)) to form multiblock copolymers (PIM1-*b*-PEO). This approach increased the influence of PEO further by avoiding large parts/blocks of pure PIM-1.

Within the fourth path, PEO were built in as side chains to the PIM-1 main chain by copolymerization with PEO substituted anthracene based maleimides. The synthesis of these maleimides followed the procedure published in [14]. The roof shaped tetrahydroxy-anthracene monomers were employed in polycondensation reaction together with the PIM-1 antecessors.

The temperature is surely an important factor for permeability of polymeric structures, with permeability being the product of diffusivity and solubility of gases in polymers [15]. While diffusion increases with rising temperature because of a higher mobility of the polymer chains, solubility usually decreases. However, PEG/PEO chains tend to crystallize with increasing molecular weight and crystallites do not contribute to permeability. At elevated temperature, these crystallites will melt, influencing diffusivity and solubility favorably [10]. Therefore, we also include measurements of gas permeability at elevated temperatures to inquire the influence of melting PEO chains.

2. Experimental

Methods

(1)   For descriptions of Equipment used in NMR, FT-IR, SEC, TGA and DSC, see our former paper [14].

(2)   Size Exclusion Chromatography (SEC) was performed in $CHCl_3$ solution, detection by Refractive Index (RI) with polystyrene standards and/or Multi Angle Light Scattering (MALS) using dn/dc increment measured at different concentrations by RI detection.

(3)   Details of the synthesis of monomers and polymers are given in Supplementary Materials S1.

(4)   Details of the gas separation calculation are given in Supplementary Materials S2.

3. Results and Discussion

The aim of this study in polymeric membrane research is the combination of PIM-1, the rigid hydrophobic high flux polymer, with hydrophilic poly(ethylene glycol) (PEGs) of various polymer weights to form stable membranes of improved gas separation properties, especially for CO_2. Low molecular PEGs are hydrophilic viscous liquids that are (in principle) quite permeable to gases because of their very good solubility, e.g., for CO_2, but obviously they lack the required mechanical stability necessary for a useful membrane. Higher molecular PEGs (>1000 g/mol) are waxy solids, but their crystalline structure diminishes their permeability considerably [10].

In the following (Sections 3.1–3.4), four different approaches have been evaluated to combine PIM-1 with PEGs and PEO/PPO, described schematically in Table 1. In Section 4, a summary of the different paths is given.

Table 1. Description and schematic (not true to scale) presentation of the four different approaches to combine PEG and PEO (ellipsoids) and PIM-1 (rectangles), physically mixed or chemically bound. Anthracene based copolymer (Section 3.4, Path 4) is depicted as triangle to distinguish from PIM-1.

Path 1 (Section 3.1)	Path 2 (Section 3.2)	Path 3 (Section 3.3)	Path 4 (Section 3.4)
Blend membranes from PIM-1 and different low molecular PEGs	Quenching to form PEO end groups of PIM-1 chains	PIM1(X)-*b*-PEO(Y)—multiblock copolymers	PEO/PPO in side chain of PIM1 copolymer

3.1. Blend Membranes (Path 1)

Our blending of PEGs in PIM-1 membranes is a similar approach that was already followed in [11] with a PEG of 20 kg/mol. In our study, however, we used PEGs of low molecular weight, 200, 1000 and 2000 g/mol; PEGs were added in 5 and 10 wt % relative to polymer mass by dissolving in PIM-1 casting solutions (solvent $CHCl_3$). The solvent was evaporated slowly in a nitrogen stream during 24 h to allow a sufficient packing of polymer chains. Dense membranes of 70–100 μm thickness were formed. They are flexible and mostly transparent with exception of the blends with PEG-2000, these are slightly turbid, caused supposedly by some partial internal phase separation. However, by blending with even longer chained PEG of 20 kg/mol phase, separation was already observed at 2.5 wt % [11].

The prepared membranes were measured with single gases (N_2, O_2, CO_2, and CH_4) in a time lag apparatus [16,17] at pressure differences below 1 bar. In the course of Path 1 we decided to compare to the measured values of a freshly prepared PIM-1 membrane originating from the identical PIM-1 batch (M_w ca. 300 kg/mol, preparation at 150 °C as described in [18] procedure III) we used for all experiments of blending. To confirm that the blend membranes are (thermally) stable, TGA was conducted before and after a time lag measurement cycle at different temperatures ranging from 20 to 80 °C. Especially low molecular PEG-200 is sufficiently volatile to be squeezed out of the membrane by repeated differences in pressure. However, less than 1% of the membranes weight was lost during the first time lag cycle measured with 4 different gases and temperatures up to 80 °C. These small losses could be easily explained by removal of some residual solvent. No further losses were observed during repetitions of time lag cycle. These results are in stark contrast to the report on blends of

PEBAX® and PEGs of low molecular weight (<500 g/mol) losing about 30 wt % PEG during time lag measurement [13].

Additionally, we used TGA measurements up to 500 °C to estimate the actual blend composition by the observed specific decompositions/weight losses of the PIM-1-PEG blends. While PIM-1 in Argon atmosphere is stable up to at least 430 °C, decomposition of the blending PEGs starts already below 400 °C. Therefore, the weight loss up to 400 °C was taken for a measure of the actual PEG content. Deviations from calculated compositions (5 and 10 wt %) were probably caused by the water content of the "as received" PEG samples in use for blending or by solvent residues within the PIM-1 polymer.

Setup and details of gas separation measurements are given in Supplementary Materials S2. Gas separation measurements at 30 °C are presented in Table 2, together with the measured values for the pristine PIM-1 we used for blending.

Table 2. Time lag measurements at 30 °C of four gases using membranes formed by blending PIM-1 with different PEGs.

	PEG-Content wt % [d]	Permeability (Barrer [b])				Selectivity [c]		
		N_2	O_2	CO_2	CH_4	O_2/N_2	CO_2/N_2	CO_2/CH_4
PIM-1 [a]	0	360	1030	7160	590	2.8	19.9	12.1
+ PEG-200	6.2	115	310	2335	199	2.7	20.2	11.7
+ PEG-1000	5.3	120	361	2776	211	3.0	23.0	13.1
+ PEG-2000	5.0	100	314	2513	177	3.2	25.5	14.2
+ PEG-200	10.0	40	122	1097	78	3.0	26.8	14.1
+ PEG-1000	12.2	28	95	806	46	3.3	28.4	17.6
+ PEG-2000	11.0	20	74	660	37	3.5	30.9	18.0

[a] M_w ca. 300 kg/mol (SEC/MALS detector); [b] 1×10^{-10} (cm³(STP)cm cm^{-2} s^{-1}cm Hg^{-1}); [c] Calculated from single gas measurements; [d] Content of PEG as calculated from TGA.

As expected from the very different basic permeability of PIM-1 and PEG, the PEG-blended membranes showed a considerable reduction of permeability compared to the measured values of pristine PIM-1 membrane. The reduction depends directly on the amount of PEG added. With ca. 5 wt % of PEG added, the permeability was reduced to 30–40%, with ca. 10 wt % PEG the permeability dropped to 10–20% of the original PIM-1 value. From these results we concluded that addition of more than 10 wt % PEG is inappropriate. Permeability reduction of CO_2 and O_2, however, was less severe compared to N_2 and CH_4 and therefore improved selectivity for CO_2/N_2 and for O_2/N_2 gas pairs were calculated by division of single gas measurements (see Table 2).

In Figure 1 are shown diffusivity (Figure 1a) and solubility coefficients (Figure 1b) of three blended membranes (ca. 10 wt % of the three different PEGs) calculated from time lag experiments (see Supplementary Materials S2) and set in relation to the pristine PIM-1 (=100%). As expected, the solubility of CO_2 is reduced less compared to O_2, N_2, and CH_4 (Figure 1b). It is known that enhanced solubility is the reason for the high flux of CO_2 in PIM-1 and other polymers [19], and also in PEGs [13] while N_2 has a low solubility in PIM-1 as well as in PEGs. Therefore, selectivity of CO_2 over N_2 increased up to a value of 30 at 10 wt % PEG content. The chain length of the PEGs affected solubility only slightly but with PEG-1000 in favor.

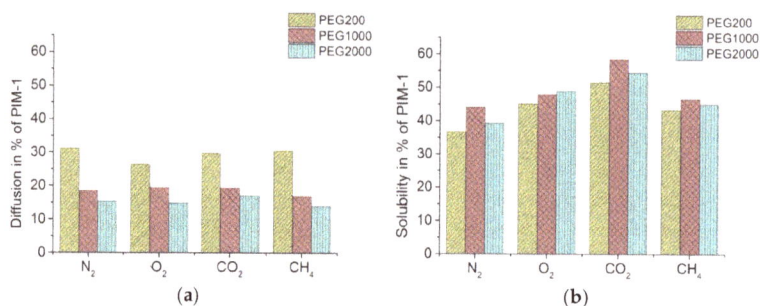

Figure 1. Calculated diffusivity (**a**) and solubility (**b**) of PIM-1 + PEG-blends (ca. 10 wt% PEGs) in relation to plain PIM-1 (=100%); results of time-lag experiments with single gases at 30 °C.

Within PIM-1, membrane diffusion of O_2 is highest (2.41 × 10^{-6} cm^2/s) followed by CO_2 (0.95) and N_2 (0.86), while CH_4 is slowest (0.35), and introduction of PEG does not change this order. All measured gases are affected quite evenly (Figure 1a). Diffusivity is higher in blends with PEG-200 than in PEG-1000 and PEG-2000, caused probably by the absence of any crystallinity within the originally liquid PEG-200, the higher amount of the small PEG molecules and the hence better distribution of the small PEG-200 chains into the intrinsic microporosity of the PIM-1 ladders. The larger PEG molecules represent parts within the PIM-1 matrix that are of low permeability and consequently slow down diffusion of all gases.

Blends of PIM-1/PEG-2000 were measured at different temperatures from 30 to 80 °C to inquire for the influence of PEG crystallites. While PEG-200 is liquid, PEG-1000 and PEG-2000 are solids and show distinctive melting points (40 and 54 °C, respectively). PEG-2000 has the highest M_w, a melting point of 53.8 °C and a crystallinity of ca. 70% was calculated from DSC (crystallinity calculated after [20]). The blend of PIM-1 + 11 wt % PEG-2000 in DSC measurement showed a (slightly) lowered melting peak at 49 °C and a very low crystallinity (<1%). It was concluded that crystallization of PEG-2000 was successfully inhibited within the PIM-1 matrix although the membrane was not fully transparent. Consequently, it was not expected to observe a deviation in thermal dependency of permeability or diffusion when compared to PIM-1. In Figure 2, permeability of four gases at temperatures from 20 to 80 °C is depicted and the prospected melting region of PEG-2000 is marked by a grey zone. Permeability of N_2, O_2 and CH_4 increase linearly with temperature, while permeability of CO_2 decreases also linearly with T; the different behaviors caused by the considerable decreasing solubility of CO_2 within the polymer with rising temperature. As expected from measured crystallinity, the permeability showed no inconsistency in the curves depending on temperature.

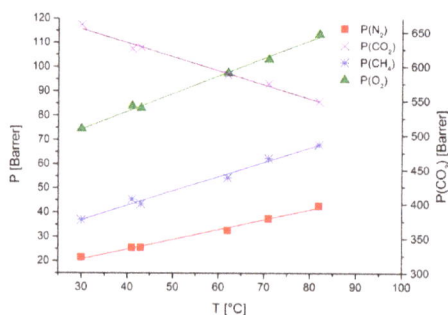

Figure 2. PIM-1 + 11 wt% PEG-2000: Permeability of four different gases at temperatures up to 80 °C. Colored in grey is the melting region prospected for PEG-2000.

Because of the clearly different behavior of CO_2 from the other measured gases, in Figure 3 diffusion and solubility of CO_2 in pristine PIM-1 and PIM-1 blended by 11% PEG-2000 versus temperature are compared. Solubility of CO_2 (Figure 3, left side, triangles) in blended PIM-1 membrane (pale triangles) decreased exponentially from a high level and very similar compared to PIM-1 (dark triangles), at 80 °C only 20% of the solubility coefficient at 30 °C is left. According to the low degree of crystallinity, diffusivity of CO_2 could be described by exponential curves too (Figure 3, right side, squares). The increase in diffusivity from 30 to 80 °C is about 2.5 times for PIM-1 (dark squares) and 4 times for the blended membrane (pale squares) involving the pronounced miscibility of PIM-1 and PEG-2000 at higher temperature. The other three gases, measured but not depicted, showed a similar behavior in diffusion as well as a decrease in solubility but on a much lower level.

Gas permeation is clearly dominated by PIM-1, and the selective slow down by PEG results in better selectivity for the very soluble CO_2 compared to N_2 and CH_4 (see Table 2).

At higher temperature, however, this effect is more than neutralized by the decreasing solubility of CO_2 (Figure 3).

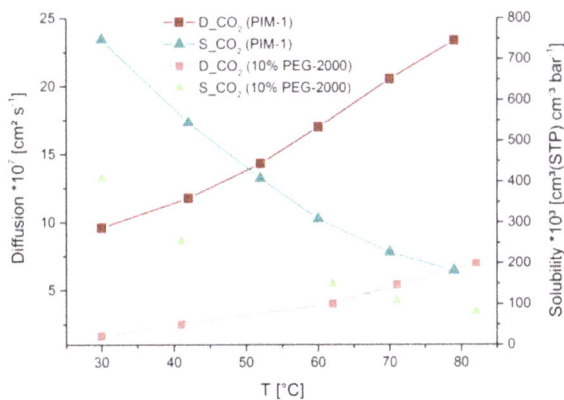

Figure 3. Time lag measurement of CO_2 with PIM-1 (dark color) and PIM-1 blended with 11 wt % PEG-2000 (light color): Diffusion (left, squares) and solubility (right, triangles) of CO_2 depending on temperature.

From these orienting results derived from blending PIM-1 and PEG, it was concluded that the amount of PEG should stay below 10 wt %, or even lower, to rescue a better part of the high permeability of PIM-1. Introduction of PEG, even one as crystalline as PEG-2000, into PIM-1 membrane does not change the temperature dependency of permeability and diffusion. Melting of PEG chains seems to have no recognizable influence on permeation behavior. Consequently, further evaluation of higher temperatures was skipped for the following paths; all following measurements were performed at 30 °C.

3.2. Quenching to Form PEO End Groups of PIM-1 Chains (Path 2)

Path 2 to introduce PEG and PEO into PIM-1 polymer employed the use of PEG-substituted trihydroxy-benzoic compounds to terminate the PIM-1 polycondensation reaction. Details of synthesis are given in Supplementary Materials S1. PIM-1 was prepared as described in [1] by low temperature polycondensation reaction of equimolar amounts of **1** and **2** catalyzed by potassium carbonate (see upper part of Scheme 1). The reaction runs smoothly and reproducibly in DMF at 55 °C. To achieve a film and membrane forming polymer an average M_w of about 50 kg/mol is necessary and this particular molecular weight is obtained after ca. 3 h of reaction time. Usually, the polycondensation reaction would be terminated by precipitation in water or methanol restoring the hydroxy end groups.

By adding a compound, e.g., 1,2-dihydroxybenzene, as quencher, providing two aromatic hydroxyl groups, the chains' growth would be stopped at the fluoro substituted ends. The fluoro end groups in this case are quenched preferentially by the small and therefore more easily accessible molecules of the quencher instead of PIM-1 chains. To introduce PEG structures by this method, we prepared esters of different poly(ethylene glycol) monomethyl ethers and trihydroxybenzoic acid (**3a–c**) (Scheme 1) by standard synthetic procedures and used them for quenching. THBA provides the necessary phenolic OH-groups to react as phenolate in alkaline solution with the PIM-1 fluoro ends. Within these experiments, THBA was chosen over dihydroxybenzoic acid (DHBA) to improve the availability of hydroxy groups for quenching. Poly(ethylene glycol) monomethyl ethers with molecular weights of 350 (**3a**), 750 (**3b**) and 5000 (**3c**) g/mol were employed. The latter one was included to improve the amount of PEG in comparison to a PIM-1 polymer of 50 kg/mol.

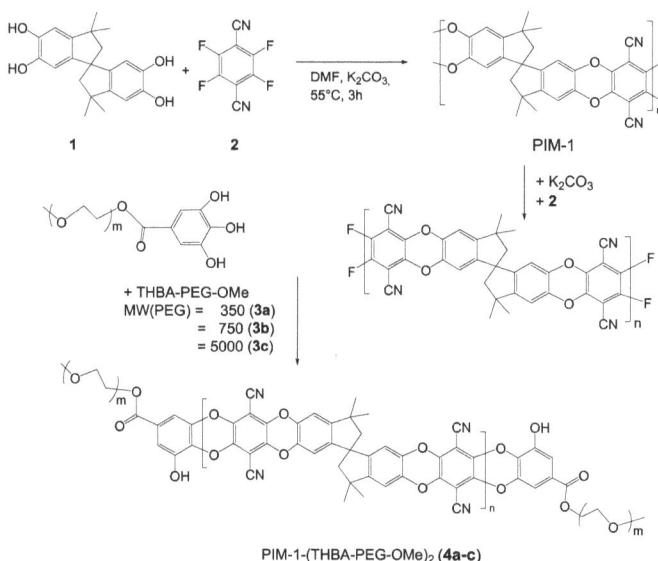

Scheme 1. Preparation of PIM-1-(THBA-PEG-OMe)$_2$ polymers (**4a–c**) by termination of PIM-1 polycondensation with THBA-PEG-OMe (**3a–c**).

A growing PIM-1 polymer chain has fluoro as well as hydroxy end groups; therefore, a simple quenching would lead to positioning of the quenching molecule at one end of each polymer chain. To improve the amount of end groups to be quenched by THBA-PEG-OMe (**3a–c**) most end groups favorably should be fluoro substituted. Unfortunately it is not feasible to run the reaction with a considerable excess of **2** from the start, while this leads to rather short chains of low molecular weight polymer (8000 to 10,000 g/mol) without any film forming properties (compare to Section 3.3, Figure 5). Therefore, to achieve entire fluoro substitution on both ends of the polymer chains an additional small amount of **2** was added after 3 h of reaction time to react for another 15 min with the already present phenolates. The respective trihydroxy-benzoic acid-PEG-monoester **3** was added to the reaction solution, stirred for another 0.5 h and the polymer was worked up as usual by precipitation in water. The isolated polymers PIM1-(THBA-PEG-OMe)$_2$ (**4a–c**) are film forming and readily soluble in CHCl$_3$. They were characterized by ^{1}H-NMR, FT-IR, SEC with RI/MALS (multi angle light scattering) detector, TGA and DSC (see *Polymerization example 3* in Supplementary Materials S1).

In ^{1}H-NMR the signals of the PEG, –OCH$_2$ groups at 3.6 ppm are clearly visible and the carboxylic group of the respective glycol esters is found at 1700 cm^{-1} in FT-IR. The molecular weights measured by SEC are quite similar to PIM-1 prepared under identical conditions and within the usually observed deviations (M_w = 48–55 kg/mol). Regarding the low molecular weights of PEG-350 and PEG-750, it was clear that the overall PEG content of polymers **4a–b** is rather low, a considerable amount was achieved only by adding PEG-5000 in polymer **4c**. In TGA measurements, the decomposition of the PEO-containing polymers **4a–c** was observed to identify the PEO-part of polymers **4a–c**. While PIM-1 is stable up to 430 °C, decomposition of PEO-containing polymers starts already below 350 °C. Therefore, it seemed reasonable to use the weight loss up to 400 °C (splitting off of PEG-esters) for a measure of the PEO containing part. Measured mass changes of the first decomposition step, re-calculated to the solely PEO content, are 0.6 (**4a**), 0.9 (**4b**) and 3.1 (**4c**) wt %. Calculation of PEG-contents via ^{1}H-NMR integrals of the polymers gave similar values for **4a** of 0.5 and for **4b** of 0.9 wt %. **4c** was difficult to evaluate from ^{1}H-NMR spectrum because of enclosed amounts of the quencher **3c**. After a series of washing steps, finally a content of 2.6 wt % was estimated. A probable reason for deviation from originally aspired higher content could be an only partly successful quenching; especially for the long chained PEG-5000-OMe with its restricted mobility the reaction time was probably insufficient to quench all possible fluoro ends of PIM-1.

The polymers **4a–c** were successfully formed into dense membranes by solution casting from CHCl$_3$. Gas permeability of the polymers **4a–c** with four different gases at 30 °C were measured in the time lag apparatus and presented in Table 3 in comparison to a freshly prepared PIM-1 membrane of comparable molecular weight (50 kg/mol). As expected by the low contents, the influence of the PEO end groups on gas permeation performance is quite small, especially for the low molecular PEGs added in **4a** and **4b**. The gas permeability and selectivity for **4a** and **4b** changed only slightly compared to a PIM-1 membrane prepared similarly.

Table 3. Time lag results at 30 °C for PIM-PEG-polymers **4a–c** synthesized by quenching with PEG-esters of trihydroxybenzoic acid (**3a–c**) and compared to a freshly prepared PIM-1 membrane synthesized along the first part of Scheme 1.

	PEG-Content wt % [a,b]	N$_2$	O$_2$	CO$_2$	CH$_4$	O$_2$/N$_2$	CO$_2$/N$_2$	CO$_2$/CH$_4$
PIM-1 (50 kg/mol)	0	530	1480	10,000	900	2.8	18.8	10.9
Polymer (**4a**)	0.6 (0.5)	490	1360	9270	830	2.8	18.9	11.2
Polymer (**4b**)	0.9 (0.9)	450	1265	8700	780	2.8	19.2	11.2
Polymer (**4c**)	3.1 (2.6)	140	370	2210	270	2.7	15.8	8.2

[a] Because of TGA measurements of the 1st decomposition step; [b] In brackets: Calculated by ^{1}H-NMR integrals.

By differentiating time lag measurements in diffusivity and solubility, it was found that **4a** and **4b** had a slightly reduced diffusivity compared to PIM-1 but similar or even slightly enhanced (in case of CH$_4$) solubility (Figure 4). These results explain quite well the observed permeability. Polymer **4c**, however, showed a serious reduction in permeability, and selectivity of CO$_2$ over N$_2$ was lower than with PIM-1 membrane. Despite its low content of PEO, it lost more than half of diffusivity and one third of the solubility compared to PIM-1 values (Figure 4). However, diffusivity of N$_2$ is affected less than the other gases and this caused the lowered selectivity of CO$_2$ and O$_2$ on N$_2$ compared to PIM-1. The long chain of PEO-5000 has an original crystallinity of more than 95%; the crystallinity of quenching compound **3c** is still at ca. 67%. From DSC measurements, a crystallinity of polymer **4c** of ca. 1% was calculated (after [21]).

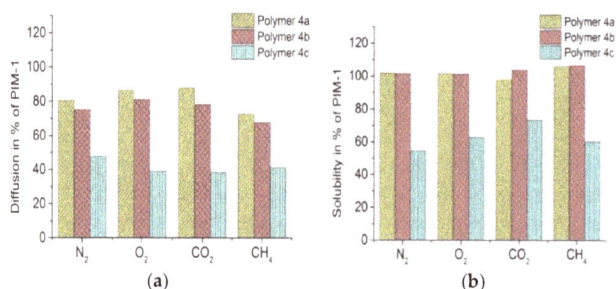

Figure 4. Diffusivity (**a**); and solubility (**b**) coefficients of polymers **4a–c**, calculated from time lag measurements at 30 °C and set in relation to PIM-1 membrane (see Table 3).

To Summarize, the effect of the PEG-mono-esters (**3a–c**) as quenching compounds in polycondensation and on PIM-1 performance is quite limited. The low molecular PEG-compounds **3a** and **3b** react (quench) better because of their enhanced mobility in solution. Their achieved permeability and selectivity are quite comparable to PIM-1. PEG-5000-ester **3c** with its 100 (O–CH$_2$–CH$_2$) repetition units is too large to gain a similar degree of substitution to PEG-350 and PEG-750. It reduces permeability, diffusion and solubility considerably, but not in the desired direction.

3.3. Formation of PIM-1-b-PEO Multiblock Copolymers (Path 3)

Within the polymers described in Path 2 (Section 3.2), the PIM-1 fraction is relatively large compared to the PEG part, e.g., 50 kg/mol versus 0.2–5 kg/mol. While the PIM-1 fraction is already a film forming polymer, the prepared PEO containing polymers represented a PIM-1 intercepted by more or less small inclusions of PEO. Unfortunately, PEGs of high molecular weight, e.g., 50 kg/mol, are not useful because of their high crystallinity and consequently very low permeability [10]. Therefore, on Path 3, we decided to reduce the chain length of PIM-1 to approximate the level of the used PEGs. Multiblock copolymers were formed by two consequent steps of polycondensation combining low molecular PIM-1 treated to end in fluoro groups and di-esters of dihydroxybenzoic acid formed with PEGs of different chain length (see Scheme 2).

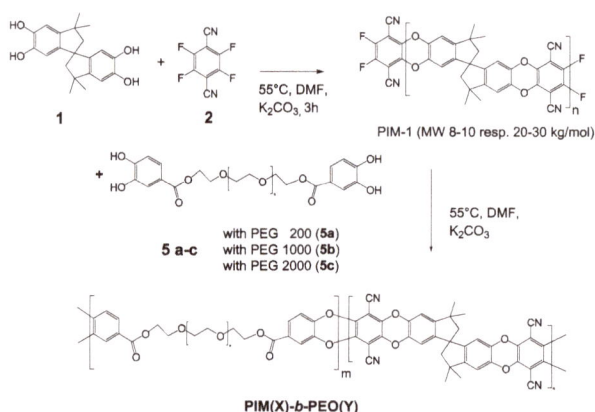

Scheme 2. Preparation of **PIM(X)-b-PEO(Y)** multiblock copolymers in two reaction steps. Step 1: polycondensation with excess of **2** over **1** to achieve low molecular PIM-1 (1.1:1 = 8–10 kg/mol, and 1.025:1 = 20–30 kg/mol); and Step 2: polycondensation with DHBA-di-esters **5a–c**.

The influence of excess of tetrafluoro-1,4-dicyanobenzene (**2**) on the molecular weight of the formed PIM-1 was examined in a series of polycondensations that ran exactly 3 h of reaction time at 55 °C under identical conditions. The resulting PIM-1 polymers were dissolved in CHCl$_3$ and measured in SEC; achieved molecular weights were determined by two different detection modes: Refractive index (RI) and multi angle laser light scattering (MALS). The achieved molecular weights are depicted in Figure 5 as a function of the ratio of **2** versus **1** (detection: RI (squares) and MALS (triangles)). At an equimolar relation of hydroxy (OH) and fluoro (F)-groups and a moderate excess of K$_2$CO$_3$ (2.04 molar) the achieved molecular weight is in the range of 45 to 55 kg/mol with PDI of ca. 3, comparable to the results of Guiver et al.: M_w of 65 kg/mol and PDI of 2.2 at 60 °C within 3 h [22]. An increase of the relation F:OH up to 1.025:1 (2.5% excess of **2** above **1**) came out to about 20–30 kg/mol with PDI of 2.2 to 3.2. A relation of F:OH = 1.1:1 (10% excess of **2**) resulted in approximately 8–10 kg/mol, with PDI of 2. These results are fairly reproducible, by using the identical conditions, the respective molecular weight of the prepared PIM-1 can be directed by the ratio of F:OH monomers.

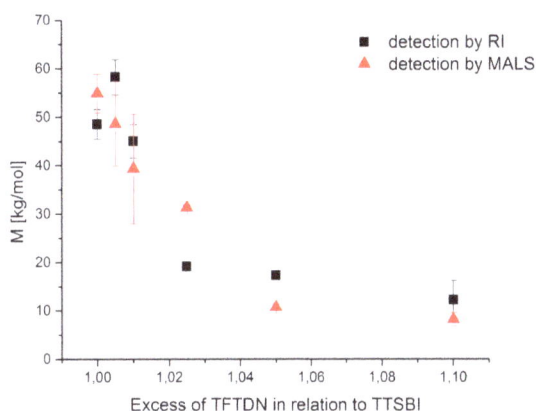

Figure 5. Molecular weight of PIM-1 prepared by polycondensation with excess of **2** above **1**. Reaction conditions: K$_2$CO$_3$ (2.04 molar towards **1**), 3 h at 55 °C in DMF. SEC measurements detected by RI (squares) and MALS (triangles).

Subsequently, low molecular PIM-1 of defined molecular weight reacted in a successive second polycondensation step at 55 °C, with one of di-esters **5a–c** (Scheme 2) and catalyzed by an additional amount of K$_2$CO$_3$. Monomers **5a–c** were prepared from DHBA and poly(ethylene glycol) (200, 1000, and 2000 kg/mol) by standard method identical to monoesters in Section 3.2 (see Supplementary Materials S1. *Monomer synthesis example 2*). The slowly formed multiblock copolymers were named **PIM(X)-*b*-PEO(Y)**: X gives the approximately molecular weight of PIM-1, Y the molecular weight of the PEG as given by the manufacturer, both numbers representing kg/mol units. Progress of the reaction was followed by sampling and subsequent analysis in SEC (see below). The reactivity of the PEG- di-DHBA-esters **5a–c** as tetrahydroxy component in polycondensation reaction is definitely lower compared to **1** and therefore reaction times needed to obtain film forming polymers (M_w of at least 40 kg/mol) turned out to be longer than with PIM-1(1–14 days). Final work up of polycondensation reactions was by precipitation in water, filtering and drying at 60 °C (see Supplementary Materials S1. *Polymerization example 4*).

3.3.1. SEC of PIM(25)-*b*-PEO(2)

In Figure 6 are depicted typical SEC measurements of samples taken during an extended poly-condensation resulting finally in **PIM(25)-*b*-PEO(2)** with rather high molecular weight. These samples were prepared by precipitation in water and worked up either by extraction with $CHCl_3$ (samples from the start) or by direct filtration. All samples were measured in SEC (liquid phase $CHCl_3$). Black lines show the eluted volumes, differentiated by reaction time and detected by RI, the red lines were results of MALS detection. The black line at the bottom of Figure 6 (PIM-1 reactant) represents the starting PIM-1 sample with F-end groups and approximately 20–25 kg/mol with a broad peak at 16–17 mL eluted volume. Monomer PEG-2000-di-DHBA-ester (**5c**) was added and after 60 min the next black line was measured. During progress of reaction the initial PIM-1 peak broadened towards lower elution volume (=higher molecular weights). Additional samples were drawn after 10 h, 45 h and 75 h of reaction time. Finally, the product is represented by the black line on top of Figure 6. Distribution of molecular weights is very broad. The final M_w measured by RI detection was calculated by universal calibration (polystyrene (PS)) standards to average 380 kg/mol with quite large deviations. However, a basic problem with RI detection is the universal calibration by polystyrene standards. The contorted structure of PIM-1 (and the similar polymers in our study), introduced by the spiro-formation of **1**, is the reason for the apparent intrinsic microporosity by hindering the polymer chains to pack efficiently and thus forming molecular nano-sized pores in between the chains. As the product of a polycondensation reaction, PIM-1 shows a rather broad mixture of different chain lengths (PDI >> 1), therefore many deviations towards higher and also lower molecular weights. Average molecular weight of polymers is usually estimated by SEC on calibrated separation columns and in most SEC labs the universal calibration by commercial available PS standards and detection by refraction index (RI) and/or UV signals is the acknowledged standard method. Separation on SEC columns also depends on the hydrodynamic volume of the polymers and PIM-1 with its contorted structure is quite different from PS. Therefore, this standard method provides not always reliable molecular weight values. However, from SEC measurements at different concentrations and detection by RI the determination of the refractive index (RI) increment dn/dc is available comparing the deviation from respective PS standards. Using the d_n/d_c value measured for the employed polymers, it is possible to detect also by Multi Angle Light Scattering (MALS) (method see for example [23]) and to achieve more reliable molecular weight values for PIM-1 and the structural variations that are described in this study.

The red lines in Figure 6 display the detection of the identical samples by MALS, the final product (on top) was calculated to 130 kg/mol/PDI 1.1 by MALS using the measured value of $d_n/d_c = 0.174$. The results for M_w and M_n obtained by MALS detection are in most cases more reasonable than those by RI detection. Although, during these experiments it was also observed that detection by MALS is of rather low sensitivity to molecular weights below ca. 15 kg/mol. The first two red lines in a comparable experiment to produce **PIM(10)-*b*-PEG(1)** (sample of low molecular PIM-1 and reaction after 60 min) show even lower intensities above 18 mL of elution volume compared to RI signals. The restricted sensitivity of the laser light for small polymer molecules explains also the rather low values of PDI (1.1 to 2.1) calculated from M_w and M_n estimated by MALS in comparison to PDI values of 3 to >10 calculated from RI detection. Therefore, it was decided to follow polycondensation reactions by a combination of RI and MALS to get the complete information.

All prepared PIM(X)-*b*-PEG(Y) multiblock copolymers were characterized by ^{1}H-NMR, FT-IR, SEC with RI/MALS detector, TGA and DSC (see Supplementary Materials S1). In ^{1}H-NMR the signal of PEO at 3.6 ppm is clearly visible and because of the relatively high PEO-diester content, compared to Path 2 in Section 3.2, also the aromatic protons of DHBA (>7 ppm) as well as the ester $COOCH_2$ group at 3.8 ppm are detected. The integrals of the PIM-1 block (two singlets at 6.8 and 6.4 ppm representing the aromatic hydrogen of spirobisindane **1**) and of the PEO block (OCH_2 at 3.6 ppm) are sufficiently separated to allow calculation of the approximate content of PEO within the polymers (see Table 4). All synthesized polymers are eventually film forming; SEC analysis (MALS) proved M_w to be in the range of 45 up to 70 kg/mol with PDI of 1.1–2.1 (Table 4).

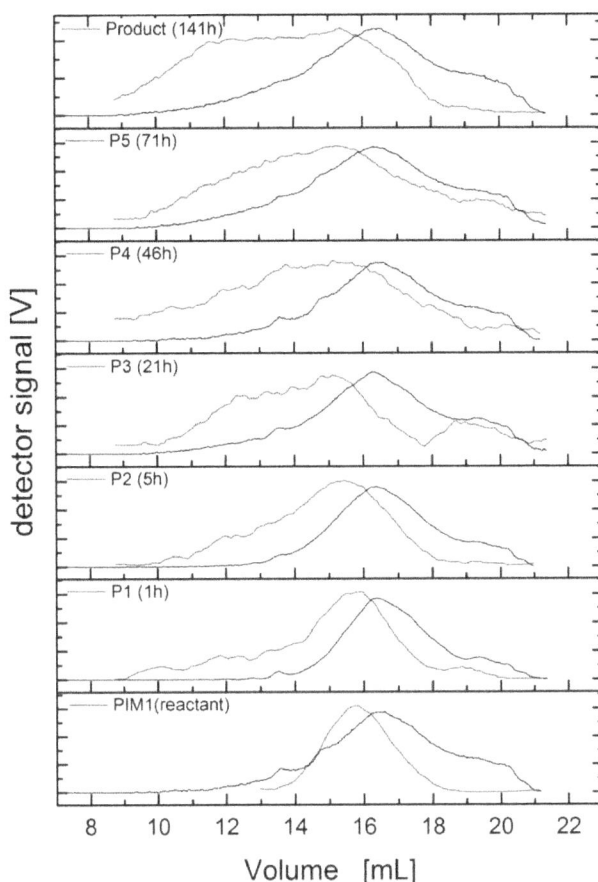

Figure 6. SEC during polycondensation of **PIM(10)-*b*-PEO(1)**; detection by RI (black lines) and by MALS (red lines).

Table 4. PEO content by ^{1}H-NMR and TGA, molecular weight of film forming polymers by SEC and MALS detection, and measured crystallinity of multiblock copolymers PIM(X)-*b*-PEG(Y) from DSC.

Multiblock Copolymer [a,b]	Intended PEO Content wt %	PEO Content (^{1}H-NMR) wt %	PEO Content (TGA) wt %	M_w (MALS) kg/mol	PDI	Calc. Crystallinity (DSC) %
PIM(10)-*b*-PEO(0.2)	2	2.5	2.5	n.m.		0.06
PIM(10)-*b*-PEO(1)	9	10.8	9.3	67	1.3	0.7
PIM(10)-*b*-PEO(2)	16	n.m.	18	41	2.1	n.m.
PIM(25)-*b*-PEO(0.2)	1	1.2	1.6	57	1.1	0.08
PIM(25)-*b*-PEO(1)	4	3.6	4.1	43	1.3	0.11
PIM(25)-*b*-PEO(2)	7	7	7.5	44	1.1	0.14

[a] M_w of PIM-1 was roughly estimated by reaction time and excess of TFTDN (see Figure 5); [b] M_w of PEG as given by deliverer.

The approximate PEO content of the multiblock copolymers was calculated also from TGA measurements: A 1st (minor) weight loss step up to 200 °C was caused by the loss of some adsorbed water and residual solvent, while the PEO-containing DHBA-diester block disintegrates in the 2nd step between 250 °C and 400 °C. Depending on the PEG chain included, the 2nd weight loss step was converted to the solely PEO content, e.g., the polymer block with monomer **5a** (PEG-200) contains

42% PEO, **5b** (PEG-1000) 78% and **5c** (PEG-2000) 88%. A 3rd step above 450 °C marked the final decomposition of the PIM-1block.

From DSC measurements the crystallinity was calculated (Table 4). Multiblock copolymers with PEO-200 showed a very low crystallinity below 0.1%, with PEO-1000 and PEO-2000 the crystallinity was still below 1%, which proves the quite even distribution of the blocks within the respective polymers.

All prepared multiblock-copolymers are sufficiently soluble in $CHCl_3$ to allow the formation of dense membranes by solution casting. However, the resulting membranes are rather brittle and not as flexible as PIM-1. They required membrane thicknesses of 70 to 100 μm to withstand the pressure difference in time lag apparatus. Permeability and selectivity results of PIM-*b*-PEG multiblock copolymer membranes at 30 °C are presented in Table 5. A direct comparison to a respective PIM-1 membrane is not possible because neither PIM-1 of 10 nor 25 kg/mol is film forming. **PIM(10)-*b*-PEO(2)** was prepared solely to check the feasibility of the polycondensation reaction, the PEO content of 16–18 wt % is definitely too high for tests in gas separation and these membranes are therefore not included in Table 5.

Table 5. Permeability and selected selectivity of PIM-*b*-PEG multiblock copolymers of different PEO content from time lag measurements with four gases at 30 °C.

Multiblock Copolymers	PEG wt %	Permeability (Barrer)				Selectivity		
		N_2	O_2	CO_2	CH_4	CO_2/N_2	O_2/N_2	CO_2/CH_4
PIM(25)-*b*-PEO(0.2)	1.6	260	700	5240	480	20.1	2.7	10.9
PIM(10)-*b*-PEO(0.2)	2.5	250	650	4660	420	18.6	2.6	11.1
PIM(25)-*b*-PEO(1)	4.1	200	520	3830	330	19.1	2.6	11.6
PIM(25)-*b*-PEO(2)	7.5	150	380	2690	250	18.0	2.5	10.8
PIM(10)-*b*-PEO(1)	9.3	140	370	3000	270	22.3	2.8	11.1

In Figure 7, permeability of the usual set of four gases is depicted versus the approximate PEO content and turned out to be in good linear dependency. A graphical extension to 0% of PEO leads to quite reasonable permeability of 300 (N_2), 800 (O_2), 5800 (CO_2) and 500 (CH_4) Barrer for the hypothetical PIM-1 block and matches approximately the values of PIM-1 membrane mentioned in Table 2.

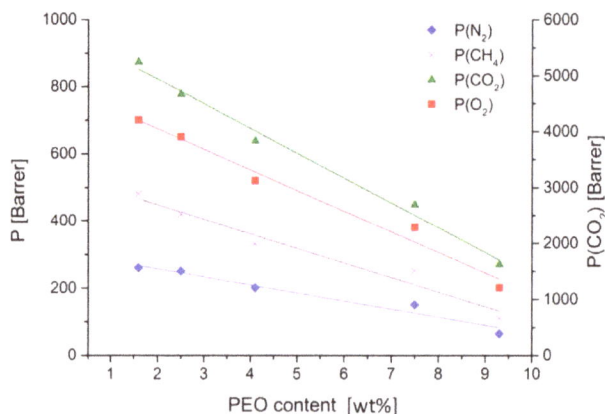

Figure 7. Permeability of 4 different gases depicted versus PEO content (values from TGA, see Table 4) in PIM-*b*-PEO multiblock copolymers, measured in time lag apparatus at 30 °C. Left: Permeability of N_2, O_2, CH_4; right: CO_2.

PIM(25)-*b*-PEO(0.2) represents the lowest PEO content (ca. 1 wt %) and showed rather high permeabilities, e.g., more than 5000 Barrer of CO_2. The separation behavior is also similar to PIM-1 with separation factors of 20 for CO_2 over N_2 and 10.9 for CO_2 over CH_4. With rising content of PEO up to 7–9 wt % permeability decreased to about a half of the 1 wt % PEO value but selectivity at higher PEO-contents increases only slightly (see Table 5, right side). Differentiating the time lag measurements into diffusion and solubility gives the already expected result that diffusion is linearly reduced with rising PEO content for all four measured gases. Solubility coefficients, however, stayed almost the same with O_2 and N_2 (ca. 0.02), while the solubility of CH_4 (0.1–0.12) as well as solubility of CO_2 (0.50–0.60 $cm^3/(cm^3$ bar)) increased slightly.

3.4. Path 4: PEG as Side Chain in PIM-1-Copolymer

To introduce a PEG side chain into a PIM-1-copolymer we used the monomer tetrahydroxy-9,10-dibutylanthracene (**6**) as vehicle and added maleimides to **6** in a similar chemistry as published recently by Khan et al. [14]. **6** is easy to prepare, although in low yield (ca. 25%), from 1,2-dimethoxybenzene and pentanal following the synthesis described in [23] using pentanal instead of acetaldehyde. It shows good solubility in many common solvents, caused presumably by the butyl substitution in 9,10-position. Methyl substitution in the related tetrahydroxy-9,10-dimethylanthracene resulted in a hardly soluble component [14].

By Diels–Alder-reaction at elevated temperature anthracene **6** offers the possibility to add a dienophilic substituted double bond (see Scheme 3). Maleimide-(PEO)$_{44}$ **7a** (Specific Polymers, Castries, France) is commercially available and has a molecular weight of 2032 g/mol providing a PEO chain of 1936 g/mol. The maleic double bond adds successfully to tetrahydroxy-monomer **6** at 140 °C in 24 h and with a good yield (90%). The product of cycloaddition, monomer **8a**, was characterized by ^{1}H-NMR, IR, TGA and DSC (see Supplementary Materials S1); it shows a distinctive melting point of 46 °C determined by the long PEO side chain.

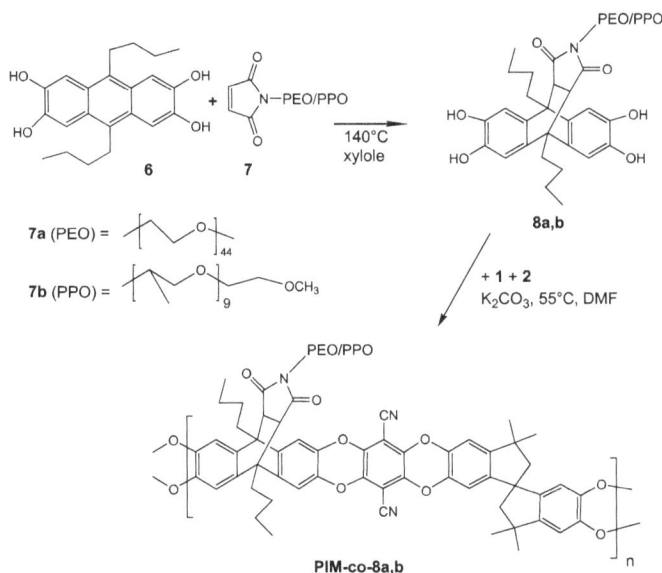

Scheme 3. Preparation of tetrahydroxy-monomers **8a** and **8b**, and in the following polycondensation to **PIM1-co-8a** and **PIM1-co-8b** (55 °C in DMF/3–5 days).

Tetrahydroxy-monomer **8b** was prepared analogously by Diels–Alder addition of maleimide-methoxypolypropylene glycol (**7b**), with a PPO chain of ca. 600 g/mol, to **6** (see Scheme 3). **7b** was synthesized by a low temperature method [24] from commercially available Jeffamine®-M600 (Huntsman, Germany) and maleic anhydride. It has a molecular weight of 677 g/mol and adds to **6** at similar conditions as mentioned above. **8b** was obtained in 70% yield as highly viscous oil and was also fully characterized by ^{1}H-NMR, FT-IR, TGA and DSC (see Supplementary Materials S1).

Both components **8a** and **8b** were successfully employed as comonomers in PIM-1 polycondensation in various relations to **1**. Reaction conditions are similar to PIM-1 polycondensation but with prolonged reaction times of at least three days at 55 °C to achieve a film forming polymer (see Supplementary Materials S1. *Polymerization example 5*). Introduction of the long hydrophilic PEO chain changes the quality of the (co)polymer decisively, e.g., copolymers with content of **8a** higher than 40 wt % became in fact water soluble and too soft for membrane formation. However, the experience of the Paths 1–3 suggested a desirable PEG content below 10 wt % to avoid exceeding reduction of gas permeability. Therefore it required only 2.5 mol-% of **8a** for tetrahydroxy component together with **1** in polycondensation with **2** to obtain a calculated overall PEO content of 9.4 wt %. The melting point of the long PEO chain of **8a** was also observed in DSC measurement of the formed copolymer **PIM1-co-8a-1**.

Within copolymer **PIM1-co-8a-1** averagely every 40th molecule of **1** is substituted by **8a**. Nevertheless, in FT-IR the O-C stretching at 1100 cm^{-1} can be seen perfectly, filling in a gap within the PIM-1 spectrum. ^{1}H-NMR spectrum shows the prominent OCH$_2$ singlet at 3.64 ppm, but because of the low molar quantity of **8a** only OCH$_2$ is visible, while the aromatic and aliphatic hydrogen peaks of **8a** are overlaid by PIM-1. For reasons of comparison and without any impact of **8a** the singlet of PIM-1 at 6.42 ppm was suited best for calculation of composition, integration estimates the PEO content to 2.7 mol %. From the 1st step of decomposition in TGA at an onset of 377 °C, about 9.1 wt% weight loss was identified that represents the Retro-Diels–Alder reaction splitting off the maleimide-PEO **7a**, corrected to PEO a content of 9 wt % was estimated. SEC with MALS detection ($d_n/d_c = 0.2611$) revealed molecular weight of ca. 40 kg/mol and PDI of 1.4. Transparent membranes were formed from CHCl$_3$ solutions of this copolymer; they are quite stable and were measured in time lag apparatus with the usual set of four gases (results in Table 6).

Table 6. Time lag measurements of PIM-copolymers **8a,b** with PEO and PPO in side chain at 30 °C.

Polymer	PEO/PPO-Content (TGA) wt%	Permeability (Barrer)				Selectivity		
		N$_2$	O$_2$	CO$_2$	CH$_4$	O$_2$/N$_2$	CO$_2$/N$_2$	CO$_2$/CH$_4$
PIM-co-8b-1	6.0	110	400	2650	160	3.6	24.1	16.5
PIM-co-8b-2	8.6	73	230	1735	123	3,2	23.8	14.1
PIM-co-8a-1	9.0	64	190	1570	107	2.9	24.5	14.7

Monomer **8b** contains a shorter chain of about 600 g/mol that additionally introduce a branching originating from its polypropylene glycol (PPO) basis. It should be less hydrophilic than poly(ethylene glycol). It was added in 5 and 10 mol % of tetrahydroxy component in PIM-polycondensation, resulting in copolymers of 6 and 11 wt% calculated PPO content (**PIM1-co-8b-1** and **PIM1-co-8b-2**), respectively. In ^{1}H-NMR spectrum the multiplet between 3.36 and 3.6 ppm covers OCH$_2$, CH as well as OCH$_3$ signals within the PPO-chain. Compared to the singlet of the PIM-1unit at 6.42 ppm, the PPO content was calculated to 4.5 mol % (**PIM1-co-8b-1**) and 8.8 mol % (**PIM1-co-8b-2**). Furthermore, TGA measurements include the Retro-Diels–Alder reaction as the loss of maleimide **7b** (observed as a first weight loss step below 400 °C), re-calculation to PPO content yielded 6.0 wt % PPO in **PIM1-co-8b-1** and 8.6 wt % PPO in **PIM1-co-8b-2**, respectively. These latter values are included in Table 6. Both new copolymers dissolve easily in CHCl$_3$. SEC was measured with MALS detection to 46 kg/mol and 42 kg/mol, with PDI = 1.2 for both polymers. By casting from CHCl$_3$ solution, stable membranes could be formed.

Gas permeability of PIM-copolymers with PEO and PPO in side chain was measured at 30 °C in time lag apparatus; results are presented in Table 6. Caused by the quite high PEO/PPO content of 6–9 wt %, permeability is low and selectivity is enhanced compared to a pristine PIM-1 membrane, as presented in Table 2. The difference between PEO and PPO is not very pronounced when considering that a PPO chain of 600 g/mol in **8b** is rather short compared to the PEO chain of ca. 2000 g/mol in **8a**. The low permeability is caused primarily by a reduction to 30–40% diffusivity (see Figure 8a, directly compared to PIM-1 value given in Table 2), as was already observed within Path 1. Solubility is affected less, especially for CO_2 and CH_4 (Figure 8b).

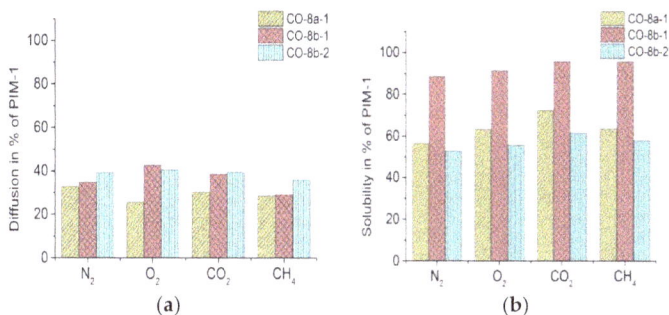

Figure 8. Diffusivity (**a**); and solubility coefficients (**b**) of polymers **PIM-CO-8**, calculated from time lag measurements at 30 °C and in relation to pristine PIM-1 membrane (in Table 2).

4. Summary of All Approaches

Permeability of PIM-1 membrane in gas separation depends on several variables such as process of preparation and casting solvent [22], as well as pre-/treatment [25]. Values, e.g., for CO_2 permeability at 20–30 °C have been presented in the literature, ranging from 2300 Barrer [4] to 12,600 Barrer [5]. We also observed the phenomenon that was already mentioned by Guiver et al. [22] that PIM-1 membranes prepared from $CHCl_3$ or CH_2Cl_2 have definitely higher permeability than those made from tetrahydrofurane or dichlorobenzene under otherwise identical conditions. To exclude differences on this account, all membranes in our study were cast from the same solvent and, fortunately, we could use $CHCl_3$ for all membranes. As with most high flux glassy polymers, a severe ageing/compaction over time takes place reducing permeability significantly within a rather short measure of time [26]. Actually, we measured permeability of N_2 and O_2 of several membrane examples from our study for at least 100 h but observed a quite similar ageing behavior like PIM-1. However, ageing of PIM was not the task of our study and would have gone beyond the scope of this article.

Apparently, it is all together no easy task to define a base line to rank newly developed membranes on PIM-1 basis. In this study, we referred to PIM-1 mostly to directly measure equivalent values (Path 1) or chose a fresh batch of PIM-1 prepared under identical conditions (Paths 2 and 4). In the case of Path 3, we had no reference value; therefore, we referred to the lowest concentration of PEO and to the graphical reduction of measured permeability to 0% of PEO. All membranes were measured as soon as possible after casting to avoid a noticeable influence of ageing effects.

The outstanding of the four gases examined in this study with new combinations of PIM-1 and PEG and PEO/PPO is CO_2, displaying the highest permeability because of a high solubility in PIM-1 as well as in PEG. The other measured gases, N_2, O_2, and CH_4, follow the behavior of CO_2 on regard of diffusivity on a similar level, but they are much less soluble.

In Figure 9, a summary of all paths is presented, comparing CO_2 permeability of PEG/PEO/PPO containing membranes by means of the actual PEG/PEO/PPO content. Addition of trend lines leads the eyes and allows consequently comparison of all approaches to combine PIM-1 and PEG/PEO/PPO. Two data points for pristine PIM-1 permeability are included (from Tables 2 and 3), these both already displaying the variety of measured CO_2 permeability of PIM-1 membranes under research. In Figure 9, it is obvious that every substantial addition of PEG or PEO to PIM-1 leads to a linear reduction in membrane permeability, caused by the low permeability of PEG(s) in comparison to the superior properties of PIM-1. Differences within the paths are diminishing with increase of PEO content towards and above 10 wt %. An amount of PEO above 10 wt % reduces permeability of CO_2 clearly below 1000 Barrer and consequently the other gases in our study were reduced below 100 Barrer. Nevertheless, the effect of an increased hydrophilicity of produced membranes is noticeable in all four paths.

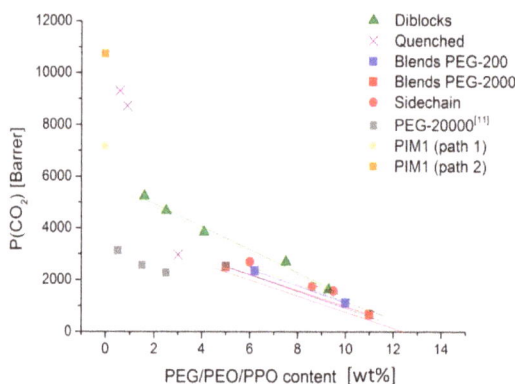

Figure 9. Graphical summary of all approaches compares measured CO_2 permeability (time lag, single gases, 30 °C) versus PEG and PEO/PPO-content from this work: blends with PEG-200 and PEG-2000 (dark squares), quenched polymers (crosses), multiblock-copolymers (triangles), and side chain-copolymers (circles). Data from [11] are also included: Blends of PIM-1 with PEG 20 kg/mol (pale squares).

The very low addition (<<1 wt %) in case of Path 2 (Figure 9, crosses) seems not substantial enough to change the permeability of PIM-1 noticeably. Without the quenching by dihydroxybenzoic esters **3a–b** but simply precipitating in H_2O membranes made of this PIM-1 batch showed about 10,000 Barrer of CO_2 and a selectivity of ca. 19 for CO_2/N_2 and that did not change by quenching with **3a–b** (Table 3). The high values of permeability of these membranes are achieved by a short reaction time in polycondensation and a rather low molecular PIM-1 of ca. 50 kg/mol. By adding a large PEO chain like PEG 5000 (**3c**) resulting in polymer **4c** the influence of PEO on permeability becomes definitely visible by a severe decrease to less than 3000 Barrer CO_2. The low selectivity of polymer **4c** for CO_2/N_2 of ca.16, however, points to a similar phase separation by extended regions of PEG as in [11] when a blend of PIM-1 with 2.5 wt % PEG-20000 (pale squares) has a still quite high permeability of CO_2 of ca. 2300 Barrer but also a rather high permeability of N_2 of 140 Barrer. Although, the membranes prepared from polymer **4c** were optically rather transparent.

Preparation of blends of PIM-1 and different PEGs (Path 1, Figure 9, dark squares) is surely the simplest method to combine PIM-1 and PEG. Membranes prepared from the physical mixtures within Path 1 behave quite similar: With increasing PEG content, permeability of all measured gases decreased and selectivity of CO_2/N_2 increased slightly (from 18 up to 30). Influence of PEG chain length of the employed PEGs between 200 and 2000 g/mol in our study is not very pronounced and can be distinguished merely in high resolution of the graphs. Wu et al. [11] presented only one blend membrane with a content as high as 5 wt % but prepared with PEG-20,000 (Figure 9, pale squares). Permeability of CO_2 of this membrane (2550 Barrer) is comparable to our own blend with 5 wt % PEG-2000 (2500 Barrer, Figure 9, dark square) but permeability of N_2 (200 Barrer) is higher with PEG-20000, resulting in a selectivity of $CO_2/N_2 = 12$, compared to 25 with PEG-2000 (N_2 permeability 100 Barrer) in our study. Their experiments in consequence concentrated on low amounts of PEG-20000 to save a better part of PIM-1 permeability. It should be mentioned that they used a PIM-1 of lower initial permeability, which explains the shift to lower permeability in Figure 9 (pale squares) compared to our results. Nevertheless, they achieved lower selectivity of CO_2 with respect to N_2 in the range of 16–12, obviously reducing N_2 permeability less than CO_2 by addition of high molecular PEG. This could be caused by the formation of separated phases within the membranes forming nanochannels that discriminate only larger molecules like CH_4. The remarkable found of their work indeed was an increase of selectivity for CO_2 over CH_4 up to 39, compared to a selectivity of 10–12 of pristine PIM-1. The low molecular PEGs in our work allow a continuous membrane phase in most cases with a similar reduction of all measured gases in diffusion, compensated by an only moderate reduction of solubility of the highly soluble CO_2. Therefore, selectivity of CO_2/CH_4 is influenced only moderately up to a value of 14, except the blend with PEG-2000 that reached a value of 18.

Graphical prolongation of the trend lines of our experiments with PEG-200 and PEG-2000 to 0% of PEG leads to 4000 to 4300 Barrer of CO_2, while the original PIM-1 batch had shown a permeability of more than 7000 Barrer. The dependency of P on addition of PEG and PEO seems to be non-linear at low amounts as the experience of Path 2 already pointed to.

Multiblock copolymers **PIM(x)-*b*-PEO(Y)** contain PIM-1 parts of quite low molecular weight (10 and 25 kg/mol) and blocks of even smaller PEO (0.2, and 1.2 kg/mol), but these blocks are divided evenly among the polymer chain without a chance of phase formation. Pure PIM-1 polymers of low molecular weight like that are not film forming therefore no comparable membranes could be provided. In Figure 9 multiblock copolymers (triangles) show the highest permeability of all paths on regard of the PEO content. At ca. 5 wt % of PEO a multiblock copolymer membrane would show 3500 Barrer of CO_2, about 1000 Barrer more than a membrane made from PIM-1-PEG blend of similar PEG chain length. A graphical prolongation to 0% PEO achieved about 5600 Barrer CO_2 of the remaining PIM-1 polymer, a value definitely higher than that calculated from the PEG blended membranes in Path 1.

PEO and PPO introduced as a side chain was described in Path 4, using the different co-monomers **8a** and **8b** to partly substitute tetrahydroxyspirobisindane (**1**) in PIM-1 polycondensation. The roof shaped bridged anthracene introduced an additional bend angle to the already contorted PIM-1 chain as described in [27]. PEO and PPO side chains fixed to the ethane bridge adding up to 6 to 10 wt% PEO and PPO within the copolymers **PIM1-co-8a** and **PIM1-co-8b**, respectively. Obtained permeability is located within the values measured on Path 1 and Path 3. PEO and PPO chains are of limited mobility as in Path 3 but pure PIM-1 parts within are quite large and therefore these polymers are more comparable to Path 1.

In Figure 10 is depicted exemplarily a section from Robeson's plot (the "upper bound" of 2008 [6] represented by the black line) for selectivity of CO_2/N_2 versus permeability of CO_2 at 30 °C. Included are only PIM-1 values from our study, one of them is above the upper bound, one below. Permeability of our PIM-1–PEO combinations is reduced compared to the pristine PIM-1 and therefore, following Robeson's trade-off, an increase in selectivity for CO_2/N_2 distinctively above 20 is achieved in most cases, however, by relinquishing a considerable part of permeability (Figure 9). Diagrams made up for CO_2/CH_4 (increase in selectivity from 10 up to 14–18), or O_2/N_2 (increase in selectivity from 2.8 up to 3.5) (neither shown) provide identical information but on a much lower (as well permeability as selectivity) level.

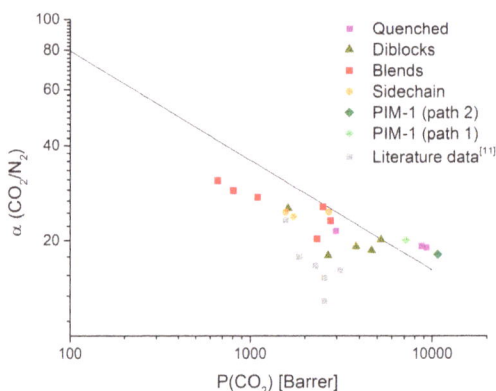

Figure 10. Selectivity of CO_2/N_2 versus permeability of CO_2 at 30 °C: blends with PEG-200/-2000 (dark squares), quenched polymers (crosses), multiblock-copolymers (triangles), side chain-copolymers (circles), and, for comparison data, from [11]: Blends of PIM-1 with PEG 20 kg/mol (pale squares). The depicted section of Robeson's upper bound for CO_2/N_2 was extracted from [6].

5. Conclusions

In our study, we tested four different paths to combine PIM-1 with up to 10 wt % PEG and PEO/PPO to enhance the hydrophilicity of the mixed polymer considerably. Physical mixtures and new, chemically linked polymers were successfully formed into free standing dense membranes and measured in single gas separation of N_2, O_2, CO_2 and CH_4 in time lag method. Melting of long PEG chains (2–5 kg/mol) at elevated temperatures up to 80 °C had no observable influence on permeability; further measurements were performed at 30 °C.

Permeability was lowered by any substantial addition of PEG/PEO/PPO, almost regardless the manufacturing process and quite proportional to the added amount. It can be stated that roughly about 6 to 7 wt % of PEG/PEO/PPO in PIM-1 combination halved the permeability of the original PIM-1 membrane. Selectivity calculated from single gas measurements increased slightly up to values of about 30 for CO_2/N_2, a maximum of 18 for CO_2/CH_4 and 3.5 for O_2/N_2. In all combinations, PIM-1 is the highly permeable part and obtained gas fluxes are defined by the amount and the distribution of PEG and PEO. Blending of PIM-1 and PEGs of different chain length led to phase separation at higher concentrations of PEGs and an even lower selectivity than with PIM-1. An even distribution within the multiblock copolymers **PIM(X)-*b*-PEO(Y)** resulted in slightly higher permeability at similar concentration than a PEO part with a long chain, as in **4c** or **PIM1-co-8a**.

By differentiating permeability in diffusivity and solubility, it was shown that addition of PEG/PEO/PPO affected diffusivity of all measured gases quite uniformly. Solubility of CO_2, O_2, and CH_4, however, was reduced less than solubility of N_2 resulting in increased selectivity for the former gases. Concentration optimum for the combination of PIM-1 and PEO between gaining solubility for selected gases and losing not too much permeability for all gases under consideration is in the range of 2–5 wt % PEO.

For perspective, these modified PIM-1 membranes of enhanced hydrophilicity should be useful in other membrane techniques such as nanofiltration and pervaporation, by combining the very good chemical resistance of PIM-1 with a selective permeance because of a higher hydrophilicity. Addition and even distribution of nanoparticles or other more polar fillers should also be facilitated in presence of PEG chains.

Supplementary Materials: The following are available online at http://www.mdpi.com/2077-0375/7/2/28/s1, S1: Synthesis of Monomers and Polymers, S2: Gas Separation.

Acknowledgments: The authors want to thank Silke Dargel for her valuable work in lab, and Petra Merten and Maren Brinkmann for performing many SEC measurements. Muntazim Munir Khan is thanked for provision of tetrahydroxy-9,10-dibutylanthracene (**6**).

Author Contributions: Gisela Bengtson and Volkan Filiz conceived and designed the experiments; Silvio Neumann performed the experiments, Gisela Bengtson and Silvio Neumann analyzed the data; Gisela Bengtson and Volkan Filiz wrote the paper.

Conflicts of Interest: The authors declare no conflict of interest.

References

1. Carta, M.; Clarizia, G.; Jansen, J.C.; McKeown, N.B. Gas Permeability of Hexaphenylbenzene Based Polymers of Intrinsic Microporosity. *Macromolecules* **2014**, *47*, 8320–8327. [CrossRef]
2. Budd, P.M.; Makhseed, S.; McKeown, N.B.; Msayib, K.J.; Tattershall, C.E. Polymers of intrinsic microporosity (PIMs): Robust, solution-processable, organic nanoporous materials. *Chem. Commun.* **2004**, 230–231. [CrossRef] [PubMed]
3. Budd, P.M.; Elabas, E.S.; Ghanem, B.S.; Makhseed, S.; McKeown, N.B.; Msayib, K.J.; Tattershall, C.E.; Wang, D. Solution processed, organophilic membrane derived from a polymer of intrinsic microporosity. *Adv. Mater.* **2004**, *16*, 456–459. [CrossRef]
4. Budd, P.M.; Tattershall, C.E.; Ghanem, B.S.; Reynolds, K.J.; McKeown, N.B.; Fritsch, D. Gas separation membranes from polymers of intrinsic microporosity. *J. Membr. Sci.* **2005**, *251*, 263–269. [CrossRef]
5. Budd, P.M.; Ghanem, B.S.; Msayib, K.J.; Fritsch, D.; Starannikova, L.; Belov, N.; Sanfirova, O.; Yampolskii, Y.; Shantarovich, V. Gas permeation parameters and other physicochemical properties of a polymer of intrinsic microporosity: Polybenzodioxane PIM-1. *J. Membr. Sci.* **2008**, *325*, 851–860. [CrossRef]
6. Robeson, L.M. The upper bound revisited. *J. Membr. Sci.* **2008**, *320*, 390–400. [CrossRef]
7. Brinkmann, T.; Pohlmann, J.; Wind, J.; Wolff, T.; Esche, E.; Müller, D.; Wozny, G.; Hoting, B. Pilot scale investigations of the removal of carbondioxide from hydrocarbon gas streams using poly(ethyleneoxide)–poly(butylene terephthalate)PolyActiveTM) thin film composite membranes. *J. Membr. Sci.* **2015**, *489*, 237–247. [CrossRef]
8. Freeman, B. (Ed.) *Membrane Gas Separation*; Wiley: Chichester, UK, 2011; p. 265.
9. Car, A.; Yave, W.; Peinemann, K.-V. PEG modified poly(amide-b-ethylene oxide) membranes for CO_2 separation. *J. Membr. Sci.* **2008**, *307*, 88–95. [CrossRef]
10. Freeman, B.D.; Lin, H. Gas solubility, diffusivity and permeability in poly(ethylene oxide). *J. Membr. Sci.* **2004**, *239*, 105–117.
11. Wu, X.M.; Peng, J.L.; Qu, Y.; Zhu, A.M.; Liu, Q.L. Towards enhanced CO_2 selectivity of the PIM-1 membrane by blending with polyethyleneglycol. *J. Membr. Sci.* **2015**, *493*, 147–155.
12. Metz, J.; Wessling, M. Gas-Permeation Properties of Poly(ethylene oxide) Poly(butyleneterephthalate) Block Copolymers. *Macromolecules* **2004**, *37*, 4590–4597. [CrossRef]
13. Lillepärg, J.; Shishatskiy, S. Stability of blended polymeric materials for CO_2 separation. *J. Membr. Sci.* **2014**, *467*, 269–278. [CrossRef]
14. Khan, M.M.; Neumann, S.; Rahman, M.M.; Abetz, V.; Filiz, V. Synthesis, characterization and gas permeation properties of anthracene maleimide-based polymers of intrinsic microporosity. *RSC Adv.* **2014**, *4*, 32148–32160. [CrossRef]
15. Wijmans, J.G.; Baker, R.W. The solution-diffusion model: A review. *J. Membr. Sci.* **1995**, *107*, 1–21. [CrossRef]
16. Al-Ismaily, M.; Kruczek, B. A shortcut method for faster determination of permeability coefficient from time lag experiments. *J. Membr. Sci.* **2012**, *423–424*, 165–179. [CrossRef]
17. Mason, C.R.; Al-Harbi, N.M.; Budd, P.M.; Bernardo, P.; Bazzarelli, F.; Clarizia, G.; Jansen, J.C. Polymer of Intrinsic Microporosity Incorporating Thioamide Functionality: Preparation and Gas Transport Properties. *Macromolecules* **2011**, *44*, 6471–6479. [CrossRef]
18. Fritsch, D.; Carta, M.; McKeown, N.B. Synthesis and Gas Permeation Properties of Spirobischromane-Based Polymers of Intrinsic Microporosity. *Marcomol. Chem. Phys.* **2011**, *212*, 1137–1146. [CrossRef]

19. Vopička, O.; Du, N.; Li, N.; Guiver, M.D.; Sarti, G.C. Mixed gas sorption in glassy polymeric membranes: II. CO_2/CH_4 mixtures in a polymer of intrinsic microporosity (PIM-1). *J. Membr. Sci.* **2014**, *459*, 264–276. [CrossRef]

20. Li, Y.; Huang, C.; Liu, G. Crystallization of poly(ethylene glycol) in poly(methyl methacrylate) networks. *Mater. Sci.* **2013**, *19*, 147–151. [CrossRef]

21. Fan, W.; Tian, W.; Zhu, X.; Zhang, W. Differential analysis on precise determination of molecular weight of triblock copolymer using SEC/MALS and MALDI-TOF MS. *Polym. Test.* **2014**, *40*, 116–123. [CrossRef]

22. Song, J.; Dai, Y.; Robertson, G.P.; Guiver, M.D.; Thomas, S.; Pinnau, I. Linear High Molecular Weight Ladder Polymers by Optimized Polycondensation of Tetrahydroxytetramethylspirobisindane and 1,4-Dicyanotetrafluorobenzene. *Macromolecules* **2008**, *41*, 7411–7417. [CrossRef]

23. Balaban, T.S.; Krische, M.J.; Lehn, J.-M. Hierarchic Supramolecular Interactions within Assemblies in Solution and in the Crystal of 2,3,6,7-Tetrasubstituted 5,5′-(Anthracene-9,10-diyl)bis[pyrimidin-2-amines]. *Helv. Chim. Acta* **2006**, *89*, 333–351. [CrossRef]

24. Mehta, N.B.; Lui, F.F.; Brooks, R.E. Maleamic and Citraconamic Acids, Methyl Esters, and Imides. *J. Org. Chem.* **1960**, *25*, 1012–1015. [CrossRef]

25. Jue, M.L.; McCool, B.A.; Finn, M.G.; Lively, R.P. Effect of Nonsolvent Treatments on the Microstructure of PIM-1. *Macromolecules* **2015**, *48*, 5780–5790. [CrossRef]

26. Swaidan, R.; Litwiller, E.; Pinnau, I. Physical Aging, Plasticization and Their Effects on Gas Permeation in "Rigid" Polymers of Intrinsic Microporosity. *Macromolecules* **2015**, *48*, 6553–6563. [CrossRef]

27. Emmler, T.; Fritsch, D.; Budd, P.M.; Chaukura, N.; Ehlers, D.; Raetzke, K.; Faupel, F. Free Volume Investigation of Polymers of Intrinsic Microporosity (PIMs): PIM-1 and PIM1 Copolymers Incorporating Ethanoanthracene Units. *Macromolecules* **2010**, *43*, 6075–6084. [CrossRef]

membranes

MDPI

Article

Fabrication of Defect-Free Cellulose Acetate Hollow Fibers by Optimization of Spinning Parameters

Xuezhong He

Department of Chemical Engineering, Faculty of Natural Sciences, Norwegian University of Science and Technology, NO-7491 Trondheim, Norway; xuezhong.he@ntnu.no; Tel.: +47-7359-3942; Fax: +47-7359-4080

Academic Editor: Klaus Rätzke
Received: 15 April 2017; Accepted: 31 May 2017; Published: 5 June 2017

Abstract: Spinning of cellulose acetate (CA) with the additive polyvinylpyrrolidone (PVP) in *N*-methyl-2-pyrrolidone (NMP) solvent under different conditions was investigated. The spinning parameters of air gap, bore fluid composition, flow rate of bore fluid, and quench bath temperature were optimized based on the orthogonal experiment design (OED) method and multivariate analysis. FTIR and scanning electron microscopy were used to characterize the membrane structure and morphology. Based on the conjoint analysis in Statistical Product and Service Solutions (SPSS) software, the importance of these parameters was identified as: air gap > bore fluid composition > flow rate of bore fluid > quench bath temperature. The optimal spinning condition with the bore fluid (water + NMP (85%)), air gap (25 mm), flow rate of bore fluid (40% of dope rate), and temperature of quench bath (50 °C) was identified to make high PVP content, symmetric cross-section and highly cross-linked CA hollow fibers. The results can be used to guide the spinning of defect-free CA hollow fiber membranes with desired structures and properties as carbon membrane precursors.

Keywords: spinning; cellulose acetate; hollow fiber membrane; orthogonal experiment design; conjoint analysis

---

1. Introduction

Carbon membranes have been studied in the last few years as a promising candidate for energy-efficient gas separation technology due to their improved permselectivity, thermal and mechanical stability, and chemical stability compared to those that are already used [1–4]. Hollow fiber carbon membranes are usually prepared by the carbonization of hollow fiber precursors. How to prepare cheaper, defect-free hollow fiber precursors becomes a key issue in the fabrication of carbon membranes. Many studies have reported that polymer membranes such as polyimide, polyacrylonitrile (PAN), and cellulose were used as the precursor for carbon membranes [2,5–7]. Cellulose is the most abundant biorenewable material with many important commercial applications. However, the potential of cellulosic materials has not been fully exploited for use as the precursors for carbon membranes because cellulose cannot be dissolved in conventional solvents due to strong inter- and intra-molecular hydrogen bonding [8]. The recently developed Lyocell process uses *N*-methylmorpholine-*N*-oxide (NMMO) to dissolve cellulose directly from biomass, and was reported to have higher efficiency compared to other processes [9], but this process embodies significant engineering challenges with regard to solvent stability, safety, and recovery. Ionic liquids are green solvents that have recently been reported to dissolve the cellulose and spin cellulose fibers [8,10,11]. However, ionic liquids are very expensive, and many of them are still not commercially available.

Fortunately, the regeneration of spun cellulose acetate (CA) hollow fiber membranes provided a potential solution to make cellulosic hollow fibers. Some studies have reported the spinning of cellulose acetate fibers [12–18], but mainly used in reverse osmosis (RO) and ultrafiltration (UF), and only a few of them were used as precursors for carbon membranes. Defect-free precursors are crucial for the preparation of high-performance carbon membranes. Thus, in this work, we will investigate and optimize the spinning parameters to obtain defect-free CA hollow fiber membranes.

In the present work, the well-known dry-jet wet spinning technology was used to fabricate thin and defect-free CA hollow fiber membranes. This process consists of the formation of nascent membrane, followed by the interfacial phase separation within the air gap. After that, the nascent membrane was immersed in a non-solvent (water) quench bath at a certain temperature where the phase separation occurred throughout the rest of the membranes. Many parameters, such as air gap, bore fluid composition, flow rate of bore fluid, and temperature of quench bath, etc. can affect the final fiber structure and morphology. Qin [13] and Chung [19] reported that air gap length during the spinning greatly affected the performance of membranes, and an increase in air gap resulted in a significant decrease in membrane permeation. The orthogonal experimental design (OED) method is well used in the multi-factor optimization field, and can consider the effects of all investigated parameters while significantly reducing the experimental runs. This study aims to introduce the OED method to optimize the spinning parameters. The defect-free CA hollow fiber membranes were spun under the optimal spinning conditions. The cellulosic-based membranes can be regenerated from the spun CA membranes by deacetylation treatment, and further used as precursors for carbon membranes preparation.

2. Materials and Methods

2.1. Materials

CA (M_W 100,000, average acetyl content: 39.8%) was purchased from the ACROS (Pittsburgh, PA, USA). Polyvinylpyrrolidone (PVP K30, M_W 10,000) was purchased from Sigma (Darmstadt, Germany). The solvent, N-methyl-2-pyrrolidone (NMP > 99.5%) was purchased from Merck (Darmstadt, Germany).

2.2. Spinning of CA Fibers

CA hollow fiber membranes were spun using the well-known dry-jet wet spinning process [13,20]. The dope solution consisted of CA, NMP, and the additive PVP (used to increase the porosity of the carbon membrane). A schematic diagram for the spinning process is shown in Figure 1. The extrusion rate for dope and bore fluid were controlled by two gear pumps, respectively. A double spinneret (ID/OD, 0.5/0.7 mm) was used in this study with the aim of spinning defect-free hollow fibers by controlling the spinning parameters. In order to systematically investigate the effects of spinning parameters and reduce the experimental times, an orthogonal experimental design (OED) method together with multivariate analysis was introduced to optimize the spinning conditions. The factors and levels for the OED are given in Table 1, and Statistical Product and Service Solutions (SPSS) software was used to generate the experimental plan.

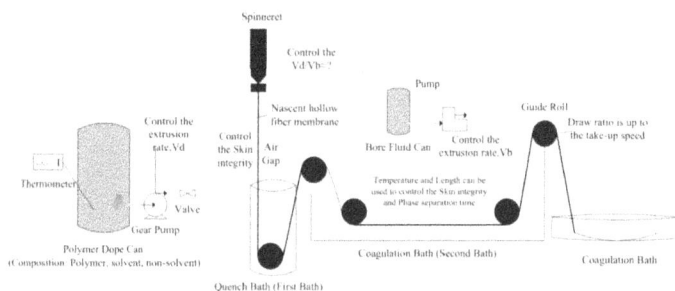

Figure 1. Schematic diagram of membrane spinning process.

Table 1. Factors and levels for orthogonal experimental design (OED) of spinning conditions.

Level	Bore Fluid Composition	Air Gap	Flow Rate of Bore Fluid *	Quench Bath Temp.
1	H_2O	15 mm	20%	20 °C
2	H_2O + NMP (85%)	25 mm	40%	50 °C
3		35 mm	60%	

* 100% means the same flow rate as the dope flow rate.

2.3. Measurement and Characterization

Fourier transform infrared spectroscopy (FTIR) spectra of the samples were obtained from the Bruker Tensor 27 FTIR instrument (Billerica, MA, USA), which was used to determine the acetyl content, PVP content, and cross-linking degree between CA and PVP in the spun hollow fibers. The morphology of spun CA hollow fibers were characterized by a scanning electron microscope (SEM) (Zeiss SUPRA 55VP, Oberkochen, Germany).

3. Results and Discussion

3.1. Experimental Results

The FTIR spectra of the pure CA, the pure PVP, the physical mixture of CA and PVP, and spun hollow fiber membrane were shown in Figure 2. The characteristic adsorption peaks of 1030 cm^{-1}, 1230 cm^{-1}, and 1740 cm^{-1} are attributed to the ether group (ν_{C-O-C}), acetyl ester group ($\nu_{CH_3-C=O}$), and carbonyl group ($\nu_{C=O}$) of CA, respectively. The peak at 1665 cm^{-1} is attributed to the carbonyl group in the PVP. Moreover, the assumption that an additional hydrogen bond may form between the CA and tertiary amide group of PVP is possible because the IR spectrum displays a strong absorption band in the region of 2250–2700 cm^{-1} that is characteristic of hydrogen bond for tertiary amide [21], as illustrated in Figure 3.

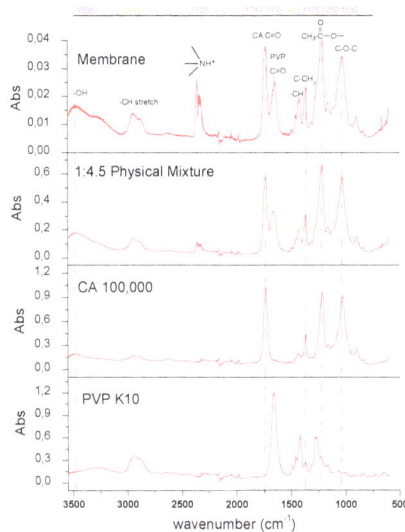

Figure 2. Fourier transform infrared spectroscopy (FTIR) spectra of cellulose acetate (CA) and polyvinylpyrrolidone (PVP), 1:4.5 Physical Mixture (CA and PVP power mixed) and Membrane (spun CA/PVP hollow fibers).

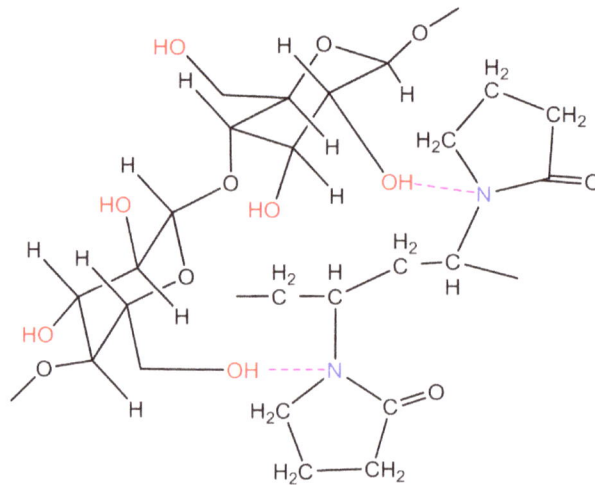

Figure 3. Structure for the formation of hydrogen bond between CA and PVP.

The FTIR spectra for spun membranes of OED are shown in Figures 4 and 5. The absorption ratios of A2320 cm^{-1}/A1030 cm^{-1} and A1665 cm^{-1}/A1030 cm^{-1} were used to characterize the cross-linking degree and the PVP content in the membrane, respectively. The cross sections of membrane morphology were characterized by SEM. Table 2 gives the OED results for different spinning conditions.

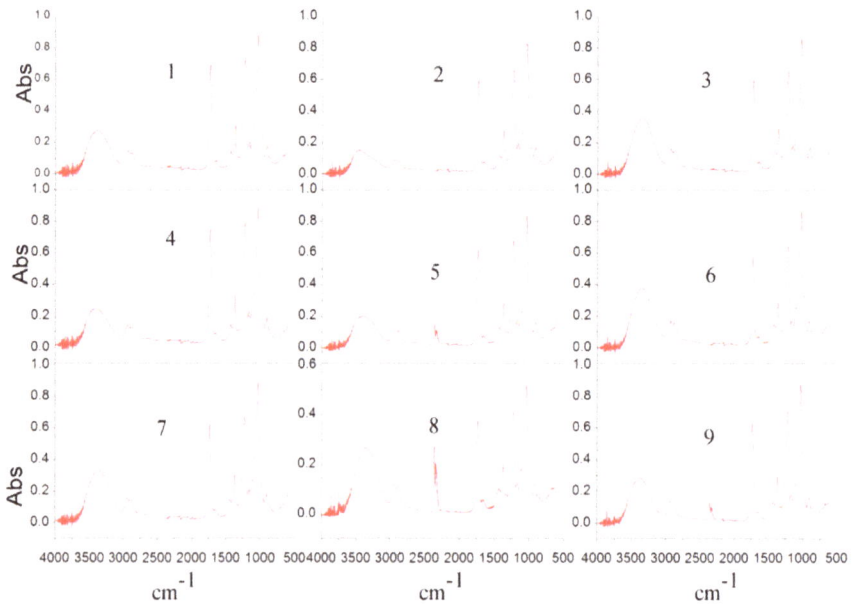

Figure 4. FTIR spectra of the spun hollow fiber membranes for the experiment plan, which were shown in Table 2.

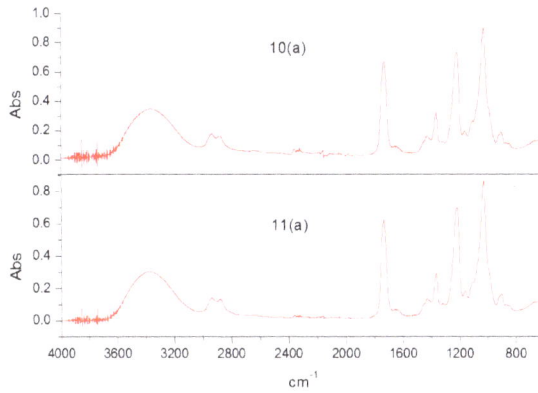

Figure 5. FTIR spectra of spun membranes for holdout experiments, which were shown in Table 2.

Table 2. The OED results for the optimization of spinning parameters.

No.	Bore Fluid	Air Gap (mm)	Flow Rate of Bore Fluid (%) *	Quench Bath Temperature (°C)	Cross-Linking Degree [c]	PVP Content (%)	Membrane Morphology
1	water	15	40	20	5.19	9.41	
2	water	35	20	50	3.14	9.01	
3	water + NMP (85%)	15	60	50	3.68	9.11	
4	water	25	60	20	5.63	10.08	
5	water + NMP (85%)	35	40	20	8.61	8.59	
6	water + NMP (85%)	25	20	20	3.03	8.16	
7	water	25	40	50	4.07	9.41	
8	water	15	20	20	21.47	10.83	
9	water	35	60	20	7.34	7.75	
10(a)	water	15	60	20	2.79	7.72	

Table 2. *Cont.*

No.	Bore Fluid	Air Gap (mm)	Flow Rate of Bore Fluid (%) *	Quench Bath Temperature (°C)	Cross-Linking Degree [c]	PVP Content (%)	Membrane Morphology
11(a)	water + NMP (85%)	25	60	20	7.17	10.37	
12(b)	water	25	40	20			
13(b)	water + NMP (85%)	25	40	20			

3.2. Conjoint Analysis

The conjoint analysis in the SPSS package was used to analyze the results of the orthogonal experimental design [22]. The utilities (part-worth) reflect the importance of each factor level. The range (highest minus lowest) of the utility values for each factor provides a measurement of how important the factor was to overall preference [23]. Factors with larger utility ranges play a more significant role compared to those with smaller ranges. The importance score (IMP) of factor i (%) is calculated as:

$$IMP_i = 100 \frac{Range_i}{\sum\limits_{i=1}^{p} Range_i} \tag{1}$$

where p = factor number. If several subjects are used in the analysis, the importance of each factor is separately calculated for each subject, and then averaged. For the prediction, the probability of each simulation (p_i) can be estimated according to the three different methods: (1) The maximum utility model determines the probability as the number of respondents predicted to choose the case divided by the total number of respondents; (2) The BTL (Bradley–Terry–Luce) model determines the probability as the ratio of one case utility to that for all simulation cases; and (3) The logit model is similar to BTL, but uses the natural log of the utilities instead of the utilities. The conjoint analysis method reported in our previous work [23] was used to estimate the part worth of the contribution from each factor's level in this work, and the response is the combination of three subjects (the PVP content, cross-linking degree, and membrane morphology; these parameters will accordingly influence the microporosity, mechanical strength, and structure and morphology of carbon membrane). The importance of each factor was calculated separately for each subject, and then averaged. The correlations of Pearson's R and Kendall's τ were 0.964 and 0.957, respectively, which indicated a good consistency between the estimated results from the model and the experimental data. Table 3 shows the utilities (part-worth) of each factor level, and averaged importance score of each factor. Higher utility values indicate greater preference in the selection of spinning condition.

Table 3. Utilities and averaged importance scores for different factors.

Factors and Levels		Utility	Averaged Importance Score (%)
Bore fluid composition	water	−0.917	28.731
	water + NMP(85%)	0.917	
Air gap	15 mm	−1.111	29.467
	25 mm	0.889	
	35 mm	0.222	
Flow rate of bore fluid	20%	−0.778	27.860
	40%	0.889	
	60%	−0.111	
Quench bath Temp.	20 °C	−0.750	13.942
	50 °C	0.750	
(Constant)		5.556	

The range of the utility values (averaged importance score) for each factor provides a measure of the importance of each factor to the overall performance. Factors with greater averaged importance score play a more significant role than those with smaller values. From Table 3, one could find that the importance of these four factors was sorted as follows:

- air gap > bore fluid composition > flow rate of bore fluid > quench bath temperature.

It can be clearly seen that air gap was the most important parameter of the spinning process, which kept a good consistency with the previous results [13,19]. Therefore, the length of air gap needs to be well controlled during the spinning process to prepare defect-free CA hollow fiber membranes with desired structure and property. Moreover, all the utilities are expressed in a common unit, and can be summed to give the total utility of any combination. Table 4 shows the comparison between an arbitrary selected combination of factor level and the optimal spinning condition. It can be found that utility of the optimal condition (Case 2) is much higher compared to Case 1.

Table 4. An example for combination of different spinning conditions.

Case	Utility					Total Utility	Remarks
	Bore Fluid Composition	Air Gap	Flow Rate of Bore Fluid	Quench Bath Temp	Constant		
1	water (−0.917)	35 mm (0.222)	60% (−0.111)	20 °C (−0.75)	5.556	4	
2	water + NMP (85%) (0.917)	25 mm (0.889)	40% (0.889)	50 °C (0.75)	5.556	9.001	Optimal

3.3. Predictions

The real power of conjoint analysis is to predict the performance (structure and property) of the spun hollow fibers that have not been investigated in the experiments—the simulation cases (see No. 12 and 13 in Table 2). The simulation results are given in Table 5. It was found that the utility of Case 2 was larger than that of Case 1 across the three response variables (PVP content, cross-linking degree, and membrane morphology) in this study. All three models—maximum utility, Bradley–Terry–Luce (BTL), and logit—indicated that simulation Case 2 would be preferred. In order to validate this simulation result, CA hollow fiber membranes were spun under these two conditions and characterized by FTIR, as shown in Figure 6. The PVP content (A1665 cm^{-1}/A1030 cm^{-1}) was higher and cross-linking degree was weaker for Case 2, which indicated that the membrane spun from Case 2 was better for use as a precursor for carbonization. The predicted score of Case 2 was also higher compared to Case 1. Thus, the generated model from the conjoint analysis can be used to guide the spinning of hollow fibers with the desired structure and properties.

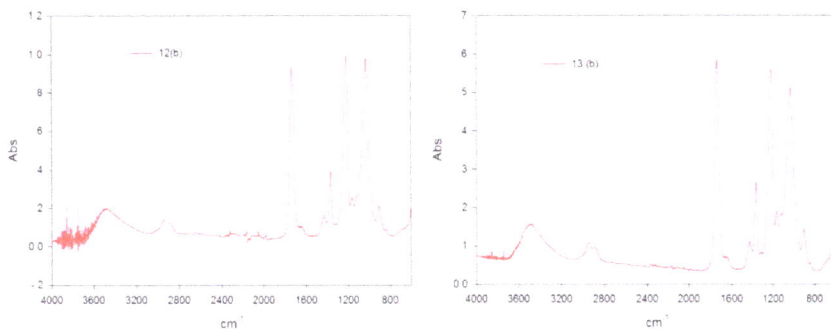

Figure 6. FTIR spectra for simulation cases, which has been mentioned in Table 2.

Table 5. Simulation results by conjoint analysis. BTL: Bradley–Terry–Luce.

Card Number	Score	Maximum Utility(a)	BTL	Logit
1	5.667	33.3%	43.7%	26.9%
2	7.500	66.7%	56.3%	73.1%

4. Conclusions

CA hollow fiber membranes were spun from a dope solution containing (CA + PVP)/NMP at different conditions using a dry-jet wet spinning process. The orthogonal experimental design method was firstly used for the optimization of spinning conditions. The experimental results showed that the importance of these four factors was sorted as:

air gap > bore fluid composition > flow rate of bore fluid > quench bath temperature.

The spinning parameter *air gap* was identified as the most important factor during the spinning process, which kept the consistency with the results reported in the literature. Moreover, the optimal spinning condition with a bore fluid composition (water + NMP (85%)), an air gap (25 mm), a flow rate of bore fluid (40% of dope flow rate), and a temperature of quench bath (50 °C) was obtained for making cellulose acetate hollow fibers with high PVP content, symmetrical cross-section, and high cross-linking degree. The proposed OED method can be well used for the optimization of spinning conditions, and the simulation results based on conjoint analysis can be applied to guide the spinning of hollow fibers with desired structure and properties.

Acknowledgments: The author would like to acknowledge the Research Council of Norway for the funding of this work in the CO_2Hing project (PETROMAKS2 #267615).

Conflicts of Interest: The author declares no conflict of interest.

References

1. Barsema, J.N. Carbon Membranes Precursor, Preparation And Functionalization. Ph.D. Thesis, University of Twente, Enschede, The Netherlands, 2007.
2. Favvas, E.P.; Kapantaidakis, G.C.; Nolan, J.W.; Mitropoulos, A.C.; Kanellopoulos, N.K. Preparation, characterization and gas permeation properties of carbon hollow fiber membranes based on Matrimid(R) 5218 precursor. *J. Mater. Process. Technol.* **2007**, *186*, 102–110. [CrossRef]
3. Lie, J.A. Synthesis, Performance and Regeneration of Carbon Membranes for Biogas Upgrading—A Future Energy Carrier. Ph.D. Thesis, Norwegian University of Science and Technology, Trondheim, Norway, 2005.
4. Saufi, S.M.; Ismail, A.F. Development and characterization of polyacrylonitrile (PAN) based carbon hollow fiber membrane. *Songklanakarin J. Sci. Technol.* **2002**, *24*, 843–854.
5. Vu, D.Q.; Koros, W.J.; Miller, S.J. High Pressure CO_2/CH_4 Separation Using Carbon Molecular Sieve Hollow Fiber Membranes. *Ind. Eng. Chem. Res.* **2002**, *41*, 367–380. [CrossRef]
6. David, L.I.B.; Ismail, A.F. Influence of the thermastabilization process and soak time during pyrolysis process on the polyacrylonitrile carbon membranes for O_2/N_2 separation. *J. Membr. Sci.* **2003**, *213*, 285–291. [CrossRef]
7. Soffer, A.; Koresh, J.; Saggy, S. Separation Device. U.S. Patent 4,685,940, 11 August 1987.
8. Sun, N.; Swatloski, R.P.; Maxim, M.L.; Rahman, M.; Harland, A.G.; Haque, A.; Spear, S.K.; Daly, D.T.; Rogers, R.D. Magnetite-embedded cellulose fibers prepared from ionic liquid. *J. Mater. Chem.* **2008**, *18*, 1–9. [CrossRef]
9. Rosenau, T.; Potthast, A.; Sixta, H.; Kosma, P. The chemistry of side reactions and byproduct formation in the system NMMO/cellulose (Lyocell process). *Prog. Polym. Sci.* **2001**, *26*, 1763–1837. [CrossRef]
10. Swatloski, R.P.; Spear, S.K.; Holbrey, J.D.; Rogers, R.D. Dissolution of Cellose with Ionic Liquids. *J. Am. Chem. Soc.* **2002**, *124*, 4974–4975. [CrossRef] [PubMed]
11. Zhang, H.; Wu, J.; Zhang, J.; He, J. 1-Allyl-3-methylimidazolium Chloride Room Temperature Ionic Liquid: A New and Powerful Nonderivatizing Solvent for Cellulose. *Macromolecules* **2005**, *38*, 8272–8277. [CrossRef]

12. Jie, X.; Cao, Y.; Qin, J.-J.; Liu, J.; Yuan, Q. Influence of drying method on morphology and properties of asymmetric cellulose hollow fiber membrane. *J. Membr. Sci.* **2005**, *246*, 157–165. [CrossRef]

13. Qin, J.-J.; Li, Y.; Lee, L.-S.; Lee, H. Cellulose acetate hollow fiber ultrafiltration membranes made from CA/PVP 360 K/NMP/water. *J. Membr. Sci.* **2003**, *218*, 173–183. [CrossRef]

14. Liu, H.; Hsieh, Y.-L. Ultrafine fibrous cellulose membranes from electrospinning of cellulose acetate. *J. Polym. Sci. Part B Polym. Phys.* **2002**, *40*, 2119–2129. [CrossRef]

15. Son, W.K.; Youk, J.H.; Lee, T.S.; Park, W.H. Electrospinning of Ultrafine Cellulose Acetate Fibers: Studies of a New Solvent System and Deacetylation of Ultrafine Cellulose Acetate Fibers. *J. Polym. Sci. Part B Polym. Phys.* **2004**, *42*, 5–11. [CrossRef]

16. Hao, J.-H.; Dai, H.-P.; Yang, P.-C.; Wei, J.-M.; Wang, Z. Cellulose acetate hollow fiber performance for ultra-low pressure reverse osmosis. *Desalination* **1996**, *107*, 217–221. [CrossRef]

17. He, X.; Lie, J.A.; Sheridan, E.; Hägg, M.-B. Preparation and Characterization of Hollow Fiber Carbon Membranes from Cellulose Acetate Precursors. *Ind. Eng. Chem. Res.* **2011**, *50*, 2080–2087. [CrossRef]

18. He, X.; Hägg, M.-B. Structural, kinetic and performance characterization of hollow fiber carbon membranes. *J. Membr. Sci.* **2012**, *390–391*, 23–31. [CrossRef]

19. Chung, T.-S.; Hu, X. Effect of air-gap distance on the morphology and thermal properties of polyethersulfone hollow fibers. *J. Appl. Polym. Sci.* **1997**, *66*, 1067–1077. [CrossRef]

20. Qin, J.-J.; Wong, F.-S.; Li, Y.; Liu, Y.-T. A high flux ultrafiltration membrane spun from PSU/PVP (K90)/DMF/1,2-propanediol. *J. Membr. Sci.* **2003**, *211*, 139–147. [CrossRef]

21. Tang, H.T. *Spectroscopic Identification for Organic Compounds*; Beijing University Press: Beijing, China, 1992.

22. SPSS Conjoint™ 17.0. Available online: http://www.sussex.ac.uk/its/pdfs/SPSS_Conjoint_17.0.pdf (accessed on 15 May 2017).

23. He, X.; Hagg, M.-B. Optimization of Carbonization Process for Preparation of High Performance Hollow Fiber Carbon Membranes. *Ind. Eng. Chem. Res.* **2011**, *50*, 8065–8072. [CrossRef]

Review

Short Review on Predicting Fouling in RO Desalination

Alejandro Ruiz-García [1,*], Noemi Melián-Martel [2] and Ignacio Nuez [3]

[1] Department of Mechanical Engineering, University of Las Palmas de Gran Canaria,
     Las Palmas de Gran Canaria 35017, Spain
[2] Department of Process Engineering, University of Las Palmas de Gran Canaria,
     Las Palmas de Gran Canaria 35017, Spain; noemi.melian@ulpgc.es
[3] Department of Electronic and Automatic Engineering, University of Las Palmas de Gran Canaria,
     Las Palmas de Gran Canaria 35017, Spain; ignacio.nuez@ulpgc.es
*   Correspondence: alejandro.ruiz@ulpgc.es; Tel.: +34-928-451-888

Received: 8 June 2017; Accepted: 20 October 2017; Published: 24 October 2017

Abstract: Reverse Osmosis (RO) membrane fouling is one of the main challenges that membrane manufactures, the scientific community and industry professionals have to deal with. The consequences of this inevitable phenomenon have a negative effect on the performance of the desalination system. Predicting fouling in RO systems is key to evaluating the long-term operating conditions and costs. Much research has been done on fouling indices, methods, techniques and prediction models to estimate the influence of fouling on the performance of RO systems. This paper offers a short review evaluating the state of industry knowledge in the development of fouling indices and models in membrane systems for desalination in terms of use and applicability. Despite major efforts in this field, there are gaps in terms of effective methods and models for the estimation of fouling in full-scale RO desalination plants. In existing models applied to full-scale RO desalination plants, neither the spacer geometry of membranes, nor the efficiency and frequency of chemical cleanings are considered.

Keywords: reverse osmosis; membrane fouling; fouling indices; predicting models

---

1. Introduction

Despite improvements and advances in our knowledge of water desalination, one of the main challenges of membrane technology, particularly in Reverse Osmosis (RO) technology, has been how to deal with membrane fouling [1–3].

Membrane fouling results from the accumulation of undesirable materials on, in or near the membrane and involves one or more of the following types [3,4]: (a) particulate and colloidal matter deposition on the membrane surface [5]; (b) organic fouling [6]; (c) scaling and inorganic fouling [7]; and (d) biofouling due to adhesion and bacterial growth on the surface of the membrane generating a layer of gel [8].

The consequences of this inevitable phenomenon have a negative effect on the performance of the desalination system (decline in water production over time for constant pressure operations or an increase in required feed) that requires costly pretreatment, higher operating pressures and frequent chemical cleanings, which can damage membranes, degrade permeate quality, and hasten membrane replacement. This additionally increases water cost and energy consumption [9,10]. Therefore, one of the most important challenges is to understand the factors involved in membrane fouling and the subsequent reduction of permeate flux or applied pressure increase that is inevitably associated with membrane processes.

A great deal of research has been carried out to this field in the last 30 years, and although desalination technology is being extensively studied, much remains to be done and researched in the field of membrane fouling. Research that has been undertaken focuses on six key areas: (1) characterization of foulant agents by autopsy studies of membrane elements; (2) understanding of fouling mechanisms; (3) indices for predicting fouling; (4) modeling for full-scale systems; (5) optimization of pre-treatment and chemical cleaning; and (6) optimizing the membrane material and enhanced module design. The first four areas attempt to address directly how fouling occurs and how to predict it, while the others focus more on the mitigation and prevention of fouling, as for example through the use of antifouling membranes [11–15].

Focusing on attempts to address directly and predict model membrane fouling, several fouling prediction tools and techniques have been developed to describe membrane fouling [16–20]. The traditional and most widely-applied fouling indices in RO systems are the Silt Density Index (SDI) and the Modified Fouling Index (MFI). However, these indices have a limitation in predicting RO fouling rate such as lack of precision with small (<0.45 μm) foulant agents [21–23].

Some recent research has focused on modifying these methods and using membranes more similar to those of RO in order to evaluate fouling potential [17,22,24,25], while another research focus is the proposal of prediction models based on experience in full-scale RO desalination plants [26–29].

This paper provides a critical review evaluating the state of industry knowledge in the development of fouling indices and models in membrane systems for desalination in terms of use and applicability.

2. Membrane Fouling Indices

The Silt Density Index (SDI) and Modified Fouling Index (MFI) are common parameters or indices to determine the fouling potential (mainly colloidal) of feedwaters in RO systems. Microfiltration (MF) membranes with a pore size of 0.45 μm, which is larger by several orders of magnitude than the pore size of the RO membranes, are used to calculate these indices. Although these indices were developed to evaluate RO membrane fouling, they can also serve as a reference in the evaluation of fouling in porous membranes like MF and Ultrafiltration (UF).

These indices are based on conventional and dead-end filtrations, while commercial applications are performed in cross-flow filtration. This implies that the flow conditions in the module are not taken into account, though this is a crucial parameter in the optimization of the process. However, the experimental determination of these data is very simple and frequently used.

2.1. Silt Density Index

SDI is used to predict the colloidal fouling potential of feedwaters in RO systems and the efficiency of pre-treatments. SDI measurement is performed using the standard ASTM D4189 [30]. The feedwater is filtrated in dead-end mode by an MF membrane with a diameter of 47 mm and a pore size of 0.45 μm at a constant pressure of 207 kPa (30 psi). The two time intervals measured at the beginning of filtration are the initial (t_i) and final (t_f) time to collect 500 mL of permeate, respectively. The third time interval (t) can be 5, 10 or 15 min, which is the period between t_i and t_f. SDI is calculated by the following Equation (1):

$$SDI = \frac{1 - \frac{t_i}{t_f}}{t} \cdot 100 \tag{1}$$

Generally, membrane manufactures suggest a value below three for SDI, but four or five are also acceptable values. Most pre-treatment studies are based on $SDI_{15} < 3$. Standard ASTM D4189 [30] specifies that the membranes must have a mean pore size of 0.45 ± 0.2 μm, and the values of SDI obtained with membranes of different suppliers, which present differences in their morphology (porosity, for example), may differ.

SDI has its limitations, and a lack of reliability has been demonstrated in several studies [31–33]. SDI is a static measurement of resistance assuming lineal permeate flux decline. This allows good results to be obtained when the water has a high quality, as the initial and final fluxes would be similar. However, the use of SDI may not be appropriate when the water has a high fouling potential, since SDI has no linear relation with the colloidal content. In this case, derivation of this index is very empirical and is not based on any mechanisms of fouling [31,34]. For these reasons, SDI should not be used as input in the mathematical model to predict fouling rates [35]. To overcome the limitations of SDI, J.C. Schippers and J. Verdouw [31] proposed a different parameter: the Membrane Fouling Index (MFI).

2.2. Modified Fouling Index

MFI (also called $MFI_{0.45}$) is a parameter based on the filtration mechanism of layer deposition or cake formation and takes into account the mechanism of the reduction of flow that takes place in membrane systems. Therefore, it better represents the operating conditions of the membranes than SDI and can be used to measure water with a high and low fouling potential.

MFI [36] is determined using similar equipment and procedures as SDI, except that the volume of permeate water is measured in 30-s intervals over 15 min of filtration. In this period, the data of permeate volume and t are collected. A better understanding of the experimental data that are obtained is achieved by using Equation (2) as proposed by J.C. Schippers and J. Verdouw [31]. Equation (2) is based on the resistances-in-series model and considers that fouling resistance is due to cake formation on membrane surfaces. Equation (2) shows a lineal relation between t/V (s/L) and V (L). The slope of this equation is the value of MFI (Equation (3)).

$$\frac{t}{V} = \frac{\mu \cdot R_{\mathrm{m}}}{\Delta p \cdot A} + \frac{\mu \cdot \alpha \cdot C_{\mathrm{b}}}{2 \cdot \Delta p \cdot A^2} \cdot V \tag{2}$$

$$MFI = \frac{\mu \cdot \alpha \cdot C_{\mathrm{b}}}{2 \cdot \Delta p \cdot A^2} \tag{3}$$

where Δp (Pa) is the transmembrane pressure, μ (Pa s) is the water viscosity, R_{m} (m^{-1}) is the hydraulic resistance of the membranes, α (m/kg) is the specific resistance of the cake, A (m^2) is the membrane surface, V (L) is the volume and C_{b} (kg/m^3) is the concentration of particles in feedwater.

MFI is determined in the second region of the curve t/V vs. V (Figure 1). It can be divided into three stages: blocking filtration, cake filtration (lineal) and cake filtration with clogging and/or cake compressure. In case of a high concentration of colloids, the graph t/V vs. V has non-linear behavior throughout the entire period, so MFI is calculated from the first region [37].

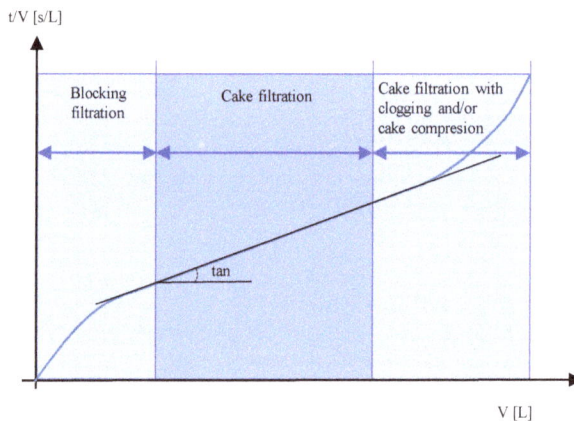

Figure 1. Ratio of filtration time and filtrate volume (t/V) as a function of filtrate volume (V) [38]. Copyright Elsevier, 2012.

Membrane manufacturers suggest using $MFI < 1 \, \text{s}/\text{L}^2$ and a maximum value of $4 \, \text{s}/\text{L}^2$ to control membrane fouling. Most studies have been based on a target value less than $1 \, \text{s}/\text{L}^2$. In practice, the calculation of MFI is complex, so in most cases, SDI is calculated. Some recent research has focused on modifying these methods in order to study the applicability of multiple $MFIs$ to evaluate the fouling potential of feed water in a full-scale RO plant [22].

The term $\alpha \cdot C_b$ is usually called fouling index I. If α and C_b are known, I can be calculated using Equation (4) [6,39–41]:

$$I = \alpha \cdot C_b \tag{4}$$

Following the theory of cake deposition or formation, when there is no compaction, the value of cake resistance is R_c. It can be rewritten as Equation (5) [31,39]:

$$R_c = \frac{I \cdot V}{A} = \frac{\alpha \cdot C_b \cdot V}{A} \tag{5}$$

The index I is related to MFI [31] with the parameters α and C_b (Equation (6)):

$$MFI = \frac{\mu \cdot I}{2 \cdot \Delta p \cdot A^2} \tag{6}$$

MFI depends on the operating conditions of the filtration, Δp and A according to Equation (6). A normalization in the same condition as SDI is required. Otherwise, I does not depend on operating conditions, so the parameter α does not vary as a result of the effect of cake compressibility. It can be considered that I is already a normalized value of MFI, which depends on pressure and membrane surface (Equation (6)). However, values of MFI under different conditions of filtration with the same water sample are not the same as for I [42]. Equation (5) is rewritten as follows:

$$I = \frac{R_c}{V/A} \tag{7}$$

The fouling index can be interpreted as a fouling parameter referring to the increase in cake resistance (R_c) divided by the specific permeate volume (V/A) (by cake formation as the only type of fouling mechanism).

The value of R_c of the deposited foulants on the membrane surface can be calculated knowing I and C_b (Equation (6)). However, the specific resistance of the cake (or permeability of the cake) is affected by the pressure applied, and that effect can be represented (as a first approximation) by an empirical expression in the form of Equation (8) [43].

$$\alpha = \alpha_0 \cdot \Delta p^n \tag{8}$$

where α_0 is the cake-specific resistance at reference pressure and Δp is the pressure gradient working with the reference pressure. n is the compressibility coefficient. The effects of pressure and compressibility on the characteristics of the cake and colloidal dispersion is a complicated topic that is still under investigation.

Index I is defined by Equation (4), and its value is calculated by the experimental determination of MFI (Equation (6)). The parameter I is related to the fouling potential of feedwater, which is defined by multiplication of two characteristics: its specific resistance α and concentration C_b.

2.3. Indices Derived from SDI

A. Alhadini et al. [44] proposed a normalized SDI (SDI^+). This index takes into consideration the temperature (T), Δp, R_m and different fouling mechanisms by using a line chart assuming cake filtration and 100% particle retention. In the same work, they proposed the volume-based SDI (SDI_v). This fouling index compares the initial flow rate with the flow rate after the filtration of the standard volume. SDI_v has a linear relationship with the particle concentration if complete blocking is the dominant fouling mechanism in the test. They concluded that SDI_v is a better index to estimate the fouling potential of feedwater in RO than SDI.

2.4. Indices Derived from MFI

J.C. Schippers and J. Verdouw [31] showed that MFI depends on the membrane molecular weight cut-off. Few authors have developed procedures to calculate MFI using membranes with smaller pore size. Table 1 shows a summary of procedures for calculating MFI, as well as the indices, parameters and methods used to measure the fouling potential. The advantages and disadvantages of each procedure have been commented on in works referenced in Table 1 and others [38,45].

S.F.E. Boerlage et al. [35] showed that the MF membrane (0.45 μm) used for MFI was not suitable for fouling of small size colloids. This fouling can happen in RO membranes if the pre-treatment does not separate these particles. The same authors [35] developed $MFI - UF$ at constant pressure ($MFI - UF_{\text{const. pressure}}$). This procedure uses a UF membrane instead of an MF membrane to separate more particles, but it can take more than 20 h.

The aforementioned fouling indices have been measured at constant pressure, whereas most membrane systems works at constant flux. S.F.E. Boerlage et al. [46] further developed the $MFI - UF_{\text{const. pressure}}$ in order to adapt it to constant flux conditions. This resulted in a noticeable difference in the duration of the test compared to $MFI - UF$ at constant pressure; $MFI - UF$ at constant flux ($MFI - UF_{\text{cont. flux}}$) could be obtained in 2 h.

Recently, S. Khirani et al. [42] proposed $NF - MFI$ using a Nanofiltration (NF) membrane to measure MFI. As is shown in Table 1, the $NF - MFI$ is measured at constant pressure. Khirani et al. [42] showed that fouling potential could be measured by the $NF - MFI$, even for small organic particles. Although this method is a step towards obtaining more realistic fouling indices, the mode of operation was still at constant pressure and dead-end flow.

Modified methods for measuring MFI have the disadvantage that they require a long measuring time with more complex systems than SDI or MFI itself. The filtration mode is dead-end flow, so it is not close to real conditions in terms of hydrodynamic flux in RO process. Cross-flow hydrodynamic conditions influence the selective deposition of smaller particles or colloids, which are the most likely to be deposited on membranes, as illustrated in Figure 2.

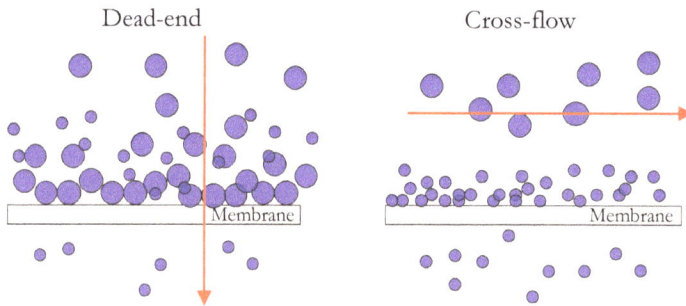

Figure 2. Filtration modes: dead-end and cross-flow.

Due to the balance between the convection flow and the backscattering of particles, the larger particles with higher backscattering speeds tend to move away from the surface of the membrane, whereas the smaller particles are preferably deposited as soiling agents. These cross-flow hydrodynamic conditions lead to a different composition and structure of the cake when compared to the final blind filtration [47].

Figure 2 shows the cross-flow filtration. All foulants in the feedwater are deposited or passed through the membrane, as in the case of the measurements of SDI, MFI and $MFI - UF$, while in cross-flow filtration, foulants are fractionated by selective deposition. These hydrodynamic effects could lead to inaccuracies in the extension of SDI and MFI that is performed in dead-end flow.

To take into consideration, the effect of small particles or colloids in MFI, S.S. Adham and A.G. Fane [48] proposed the use of a selective MF membrane to be operated in cross-flow mode. They called this index the Cross-Flow Sampler-MFI ($CFS - MFI$). After MF membrane filtration (colloid matter passes though this membrane), MFI/SDI is measured as shown in Table 1. Although this method is a better approach, the cross-flow MF is separated from the measurement device in dead-end flow, so $CFS - MFI$ is determined in discontinued mode.

M.A. Jaaved et al. [49] calculated $CFS - MFI$ in continuous operation mode with MFI (in dead-end flow) directly connected to CFS. Recently, L.N. Sim et al. [50] applied CFS to $MFI - UF_{cont.}$ to simulate selective colloidal deposition in real RO systems. The proposed index is known as $CFS - MFI_{UF}$ and uses a UF membrane for MFI. The particles that pass across CFS and that are deposited on the UF membrane will foul the RO membranes. In a later work, L.N. Sim et al. [51] combined $CFS - MFI_{UF}$ with a Cake-Enhanced Osmotic Pressure (CEOP) [52] model to predict the cross-flow RO fouling profile under constant flux filtration. They concluded that incorporating the CEOP effect was a very promising method in predicting colloidal fouling in RO.

J. Choi et al. [53] proposed procedures for measuring MFI with different types of membranes (Table 1). The test was called the Combination Fouling Index-MFI ($CFI - MFI$). It takes into account various foulant agents separated by different membranes. However, the proposed approach is not simple since several types of membranes are required. Although the different measurement systems of MFI improve the prediction of fouling in RO membranes, they are complex and require long times to be determined.

An evaluation of membrane fouling potential by Multiple Membrane Array System (MMAS) was proposed by Y. Yu et al. [54]. MF, UF and NF membranes were connected in series to calculate three MFI values, particle-MFI, colloid-MFI and organic-MFI. MMAS allows the simultaneous separation of target foulants from the feed water and the evaluation of fouling potential of the feed water focusing on the target foulants. The authors suggested that the MMAS could give valuable information about the best candidates for pretreatment and the fouling influence on full-scale RO desalination plants.

M. W. Naceur [55] determined the Dimensionless Fouling Index (DFI). It can be interpreted as the ratio of the membrane resistance to that of the cake due to the concentration of feedwater. The authors mentioned that this index requires performing experiments under different operating conditions to be properly validated.

Table 1. Summary of various methods, indices and parameters used in fouling evaluation (adapted from [38,45,56]). *SDI*, Silt Density Index; *MF*, Microfiltration; *MFI*, Modified Fouling Index; *UF*, Ultrafiltration; *CFS*, Cross-Flow Sampler; *RO*, Reverse Osmosis; *MMAS*, Multiple Membrane Array System; *DFI*, Dimensionless Fouling Index; *CEOP*, Cake-Enhanced Osmotic Pressure.

Methods, Indices and Parameters	Characteristics	Equation	Comments
SDI (1995, [30])	- Membrane: MF 0.45 µm (flat sheet) - Foulant: particulate matter - Operation mode: dead-end and constant pressure - Fouling mechanisms: none - Test: time vs. volume	$SDI = \frac{1-\frac{t_i}{t_f}}{t}\cdot 100$	Disadvantages: *SDI* is a standardized method (ASTM D4189), but empirical, and it is not based on fouling mechanisms. It is not related to foulant concentration in feedwater. It does not take into account the temperature or variation in membrane resistance.
MFI (J.C. Schippers and J. Verdouw, 1980 [31])	- Membrane: MF of 0.45 µm (flat sheet) - Foulant: particulate matter - Operating mode: dead-end and constant pressure - Fouling mechanisms: cake filtration - Test: *t/V* vs. *V* (i.e., each 30 s)	$MFI = \frac{\mu\cdot I}{2\cdot\Delta p\cdot A^2}$	Characteristics: $MFI_{0.45}$ is an improved version of *SDI* and is related to cake filtration theory. The fouling index *I* is obtained from the slope of the lineal region of the graph *t/V* vs. *V* (filtrated volume). Disadvantages: It is not very accurate as foulant agents with a diameter less than 0.45 µm pass across the membrane.
SDI⁺ (A. Alhadini et al., 2011 [44])	- Membrane: MF 0.45 µm (flat sheet) - Foulant: particulate matter - Operation mode: dead-end and constant pressure - Fouling mechanisms: none - Test: time vs. volume	—	Characteristics: SDI^+ is a normalization of *SDI* taking into consideration the variation of temperature, pressure and membrane resistance. Different fouling mechanisms could be assumed based on line charts and parameters' calculation. Disadvantages: It is not very accurate as foulant agents with a diameter less than 0.45 µm pass across the membrane.
SDI_b (A. Alhadini et al., 2011 [44])	- Membrane: MF 0.45 µm (flat sheet) - Foulant: particulate matter - Operation mode: dead-end and constant pressure - Fouling mechanisms: none - Test: time vs. volume	$SDI_b = \frac{100\cdot A_{300}}{V_{f0}}\left(1-\frac{t_1}{t_2}\right)$	Characteristics: SDI_b showed a more linear relationship with foulant concentration in feedwater than standard *SDI*. Besides, it is independent of testing parameters, such as temperature and pressure, and less sensitive to membrane resistance. Disadvantages: It is not very accurate as foulant agents with a diameter less than 0.45 µm pass across the membrane.

Table 1. Cont.

Methods, Indices and Parameters	Characteristics	Equation	Comments
$MFI - UF_{const.\ pressure}$ (S.F.E. Boerlage et al., 1997 [57])	- Membrane: UF (hollow fiber, 13 kDa) - Foulant: particulate matter - Operating mode: dead-end and constant pressure. - Fouling mechanisms: cake filtration - Test: t/V vs. V or $\Delta t/\Delta V$ vs. V (i.e., each 10 s)	$$MFI - UF = \frac{\mu \cdot I}{2 \cdot \Delta p \cdot A^2}$$ $$MFI - UF = \frac{\mu \cdot \alpha_0 \cdot C_b \cdot \Delta p^\omega}{2 \cdot \Delta p \cdot A^2}$$	Characteristics: UF membrane is used instead of MF, so colloidal fouling can be detected. α_0 is a constant, ω the compressibility factor of the cake and C_b the concentration of particles in the feedwater. Disadvantages: $MFI - UF_{const.\ pressure}$ is not able to show fouling behavior in constant flow precesses. Twenty hours are required to obtain a measurement, and the method to obtain the deposition factor is tedious. Although the UF membrane used in the tests is capable of retaining particles and colloidal matter, it is not efficient enough to retain organic matter.
$MFI - UF_{const.\ flux}$ (S.F.E. Boerlage et al., 2004 [46])	- Membrane: UF (flat sheet, 10–200 kDa) - Foulant: colloids - Operating mode: dead-end and constant flux. - Fouling mechanisms: cake filtration - Test: Δp vs. t or $\Delta t/\Delta V$ vs. V	$$MFI - UF_{const.\ flux} = \frac{\mu \cdot g \cdot c \cdot I}{2 \cdot \Delta p_0 \cdot A_0^2}$$ $$MFI - UF = \frac{\mu \cdot \alpha_0 \cdot C_b \cdot \Delta p^\omega}{2 \cdot \Delta p \cdot A^2}$$	Characteristics: The operating mode is constant flow as happens in the majority of actual RO processes. The fouling index I is obtained from the slope of the graph NDP (Net Driven Pressure) vs. filtration time. Δp_0 is the standard pressure (2 bar). Disadvantages: The test is performed under conditions of accelerated flow that do not allow representation of the behavior of fouling to flows of 20–30 L/m²·h. As with $MFI - UF_{const.\ pressure}$, the deposition of particles is considered through a deposition factor, and although through the UF it is possible to retain particulate matter and colloids, it is not enough to retain the organic matter present in the feed. Despite the improvements of $MFI - UF_{const.\ flux}$, the measurement cannot be simulated in cross-flow.
$NF - MFI$ (S. Khirani et al., 2006 [42])	- Membrane: NF - Foulant: organic matter - Operating mode: dead-end and constant pressure. - Fouling mechanisms: cake filtration - Test: $I/(V/A)$ vs. V/A	$$MFI - NF = \frac{\mu \cdot I}{2 \cdot \Delta p \cdot A^2}$$	Characteristics: The test tries to take into consideration the organic matter in the feedwater. Disadvantages: The test is carried out under constant pressure, and the deposition factor of particles in cross-flow is not considered. The total retention of organic matter is not achieved in this procedure.

Table 1. Cont.

Methods, Indices and Parameters	Characteristics	Equation	Comments
CFS – MFI (S.S. Adham and A.G. Fane, 2008 [48])	- Membrane: MF - Foulant: particulate matter - Operating mode: cross-flow and dead-end (separated)/constant pressure. - Fouling mechanisms: cake filtration - Test: t/V vs. V	$CFS - MFI = \dfrac{\mu \cdot a \cdot C_b}{2 \cdot \Delta p \cdot A^2} = \dfrac{\mu \cdot I}{2 \cdot \Delta p \cdot A^2}$	Characteristics: This index incorporates the hydrodynamic behavior of the cross-flow in the measurement of the fouling index. CFS allow small particle to pass across the MF membrane to be deposited on the MF membrane located in MFI in dead-end flow. Disadvantages: Discontinued operating mode.
CFS – MFI (M.A. Javeed et al., 2009 [49])	- Membrane: MF - Foulant: particulate matter - Operating mode: cross-flow and dead-end / constant pressure - Fouling mechanisms: cake filtration - Test: t/V vs. V	$CFS - MFI = \dfrac{\eta_{20^\circ C} \cdot a \cdot C_b}{2 \cdot \Delta p \cdot A^2} = \dfrac{\eta_{20^\circ C} \cdot I}{2 \cdot \Delta p \cdot A^2}$	Characteristics: $CFS - MFI$ is measured in continuous mode. Disadvantages: It uses the same MF membrane as in MFI, and the operating mode is at constant pressure.
CFS – MFI$_{UF}$ (L.N. Sim et al., 2011 [58])	- Membrane: MF and UF - Foulant: colloids - Operating mode: cross-flow and dead-end, constant flow - Fouling mechanisms: cake filtration - Test: Δp vs. t	$CFS - MFI_{UF} = \dfrac{\mu \cdot I'}{2 \cdot \Delta p \cdot A^2}$	Characteristics: This index takes into account the hydrodynamic effect of cross-flow and the deposition factor. I' is the modified resistivity of the cake. $CFS - MFI_{UF}$ can be a more precise method to determine the effect of fouling agents on the RO process. The method is easy due to its short time of filtration.
CFI (J. Choi et al., 2009 [53])	- Membrane: MF and NF - Foulant - Operating mode: constant pressure - Fouling mechanisms: - Test: t/V vs. V	$CFI = \dfrac{\mu \cdot a \cdot C_b}{2 \cdot \Delta p \cdot A^2} = \dfrac{\mu \cdot I}{2 \cdot \Delta p \cdot A^2}$ $CFI = w_1 \cdot M_1 + w_2 \cdot M_2 + w_3 \cdot M_3 + w_4$	Characteristics: It is a combination of various indices, denoted as $MFI - HL$ (using a Hydrophilic MF membrane), $MFI - HP$ (using a MF Hydrophobic membrane) and $MFI - UF$ (using a hydrophilic UF membrane). This test tries to take into consideration all types of foulant agents using different membranes. M_1 is the value of $MFI - HL$, M_2 is the value of $MFI - HP$, and M_3 is the value of $MFI - UF$. The weighting factors w_1, w_2, w_3 and w_4 depend on the characteristics of the membrane. Disadvantages: The method is difficult since it requires different types of membranes, and the procedure to obtain CFI is very tedious. In addition, the fouling index is still measured under constant pressure conditions.

Table 1. Cont.

Methods, Indices and Parameters	Characteristics	Equation	Comments
MMAS (Y. Yu et al., 2010 [54])	- Membrane: MF, UF and NF - Foulant: particulate, colloids and organic matter - Operating mode: dead-end flow and constant pressure - Fouling mechanisms : - Test: t/V vs. V	—	Characteristics: MF, UF and NF membranes are connected in series for simultaneous separation of target foulants. This index was shown to be precise and selective in the prediction of the fouling potential of different feedwaters. Disadvantages: The method is not simple since it requires different types of membranes to determine the particle-MFI, colloid-MFI and organic-MFI. Furthermore, the fouling indices are still measured under constant pressure conditions.
DFI (M. W. Naceur, 2014 [55])	- Membrane: MF of 0.45 μm (flat sheet) - Foulant: particulate matter - Operating mode: dead-end and constant pressure - Fouling mechanisms: cake filtration - Test: t/V vs. V	$DFI = \frac{R_m^2}{2rC}$	Characteristics: The experimental procedure is similar to MFI. By introducing the equation of Ruth in the model, the authors obtained a dimensionless fouling index, which is a simple linear equation. Disadvantages: Experimental work to validate DFI was not carried out, so the accuracy of this index has not been validated.
"Normalized Fouling Rate" (NFR) (H.R. Rabie et al. 2001 [21])	- Membrane: - Foulant: - Operating mode: - Fouling mechanisms: - Test: t/V_s vs. V_s	—	Characteristics: This method is used to analyze data from a pilot plant in a large-scale facility. NFR is the curve of the graph t/V_s vs. V_s, where V_s is the specific volume collected per unit area and per NDP in time t. Disadvantages: It cannot be used as a fouling potential indicator of feedwater.
k_{fp} (L. Song et al. 2004 [59])	- Membrane: UF and RO - Foulant: colloids - Operating mode: constant pressure - Fouling mechanisms: cake filtration - Test: J vs. t	$k_{fp} = \frac{R_t^2 - R_0^t}{v_t}$	Characteristics: This normalization method has the objective of eliminating the effects of different operating parameters in the determination of the fouling rate. In this way, the fouling potential of feed water can be compared on a fair basis. Disadvantages: One of its results indicates that the fouling potential of large colloidal particles increases as the operating pressure increases. This is mainly due to the compressibility effect of the cake, which is strongly related to the nature of the colloid.

Table 1. Cont.

Methods, Indices and Parameters	Characteristics	Equation	Comments
Membrane Fouling Simulator (MFS) (J.S. Vrowenvelder et al. 2006 [60])		—	MFS uses the same membrane materials as spiral-wound RO/NF membrane, with the same dimensions and hydrodynamic behavior, and is equipped with a visor. Suitable for in situ observations in real time, non-destructive observations and parameters such as pressure drop can be monitored. It is mainly used as a biofouling monitor [61]. Disadvantages: There is no instant response of the fouling potential.
Feed Fouling Monitor (FFM) (A.H. Taheri et al., 2013 [17])		—	This technique uses a UF membrane to predict the increase of transmembrane pressure at constant fluxes in the presence of colloidal fouling. This prediction includes the developing hydraulic resistance and the CEOP components. Disadvantages: Lack of extension of this monitoring and modeling approach to real-world foulants and a full-scale RO desalination plant.
Feed Fouling Monitor-Salt Tracer Response (FFM-STRT) (A.H. Taheri et al., 2015 [16])		—	This method uses the FFM including an STRT to measure the development of concentration polarization in estimating (CEOP) the contribution. Foulants studied were humic acid and colloidal silica Disadvantages: There is no instant response of the fouling potential.

2.5. Fouling Potential Parameter (k_{fp})

L. Song et al. [59] defined a new standardization method for the determination of fouling potential in membrane processes. Initially, it was developed to evaluate the potential of colloidal fouling in UF membranes, but later was also applied in the characterization of fouling in large-scale RO processes [62–64].

Index k_{fp} (Pa s/m²) (called the fouling potential) is defined by Equation (9):

$$R_t = R_0 + k_{fp} \cdot \int_0^t J \, dt \tag{9}$$

In Equation (9), J (m/s) is the specific permeate flux, and R and R_t (Pa s/m) are the initial and final resistance of the membrane R_0. In this resistance, the resistive effect of the viscosity is included and is equivalent to multiplication of the resistance as is usually considered, R (m^{-1}), and the dynamic viscosity of the fluid μm (Pa s) (Equation (10)):

$$R_t = \mu \cdot R \tag{10}$$

If the parameter k_{fp} is assumed constant over time, it can be calculated using Equation (11):

$$k_{fp} = \frac{R_t - R_0}{v_t} \tag{11}$$

$$v_t = \int_0^t J \, dt \tag{12}$$

where v_t is the total specific volume of permeate over time t.

3. Predictive Models

These models are an alternative to fouling indices in the prediction of the fouling influence on RO systems. Some authors [26–29] have proposed equations to estimate the decline of the permeate flux (J_w) over time due to long-term variation of the water permeability coefficient (A). Generally, these correlations are applicable for the respective membrane type and for specific operating conditions.

One of the main drawbacks in the development of this type of model is the availability of long-term operating data for a wide range of operating conditions and different types of full-scale membranes. All models aim to describe the permeate flow decline over time or the variation of the normalized water permeability coefficient A_n due to compaction, fouling, etc.

A proposed model to predict the decline of J_w due to membrane compaction was used by M. Wilf et al. [26] to estimate the J_w decline in the long term (Equation (13)). Three years of experimental data from different Sea Water Reverse Osmosis (SWRO) desalination plants were used to identify the parameter of the model. They calculated the parameter for permeate flow decrements of 25% and 20%.

$$A_n = t^m \tag{13}$$

where m is a parameter with values between −0.035 and −0.041 [26] related to permeate flow decline of 20% and 25%, respectively, and t is the operating time in days.

Zhu et al. [27] also proposed a model (Equation (14)) to predict the coefficient A. This involves an exponential equation, but in this case, a hollow fiber membrane was utilized (Dupont^{TM} B-10, Wilmington, DE, USA) during one year of operating time. This correlation is not based on experiments, but on model-based simulation: variable feed pressure (6.28–7.09 MPa), constant feedwater concentration and temperature (35,000 mg/L and 27 °C, respectively). Belkacem et al. [65] used the Zhu model in terms of membrane resistance increase. The membrane used was the BW30LE-440 Filmtec^{TM} (Midland, MI, USA) in a two-stage desalination plant with re-circulation during one year of operation.

$$A_n = A_0 \cdot e^{\left(\frac{-t}{\tau}\right)} \tag{14}$$

where τ is a correlative parameter, and the value was 328 under the aforementioned operating conditions.

Abbas et al. [28] (Equation (15)) proposed a model to determine the variation of the normalized average water permeability coefficient $A_n = A/A_0$, where A_0 is the initial average water permeability coefficient. It was an exponential equation depending on three parameters and time, and the utilized membrane was the BW30-400 Filmtec^{TM}. Five years of operating data were used for the parameter identification. The feedwater temperature was between 28 and 30 °C, the concentration being in a range of 2540–2870 mg/L, and the feed pressure was around 1200 kPa.

$$A_n = \alpha \cdot e^{\left(\frac{\beta}{t+\gamma}\right)} \tag{15}$$

where $\alpha = 0.68$, $\beta = 79$ and $\gamma = 201.1$ for the aforementioned membrane and operating conditions.

A forth model was proposed by Ruiz-García et al. [29] (Equation (16)). They include the parameter k_{fp} in the model and gave specific information about the behavior of the performance decline in the long term. They proposed a two-stage pattern in the decline of A in RO systems: an initial Stage I, where a more pronounced decline than Stage II was shown. This is mainly due to membrane compaction, irreversible fouling (strongly adherent films) and k_{fp}. Stage II is related to a gradual decrease mostly due to irreversible fouling and the frequency and efficiency of the Chemical Cleaning (CC). The model described the mentioned stages by the superposition of two exponential functions. The used about 3300 operating days of a full-scale brackish water reverse osmosis (BWRO) desalination plant to fit the parameters of the model. They got three equations, one related to maximum values of the normalized water permeability coefficient (A_n) (Post-Chemical Cleaning (Post-CC)), average and minimum values (Pre-Chemical Cleaning (Pre-CC)). This allowed obtaining equations to estimate a range of values for the coefficient A_n in time.

$$A_n = \delta_1 \cdot e^{-\frac{t}{\tau_1} \cdot k_{fp}} + \delta_2 \cdot e^{-\frac{t}{\tau_2} \cdot k_{fp}} \tag{16}$$

The first exponential function is dependent on three parameters (δ_1, τ_1 and k_{fp}) and is related to the behavior in Stage I (Figure 3), while the second is dependent on two parameters (δ_2, τ_2 and k_{fp}) and is more related to Stage II (Figure 3). The first function gets closer to zero as Stage I ends. The δ are related to the weight of each exponential: the lower δ_1 is and the higher δ_2 is, the higher A_n is when the desalination plant is stabilized. τ concerns the decline in each stage (i.e., how fast is the irreversible effects (mainly fouling) affecting performance): the larger the value, the more constant is the function. Generally, the higher k_{fp} results in a faster decline of A_n in Stages I and II. They also carried out a comparison between the different models by using their experimental data.

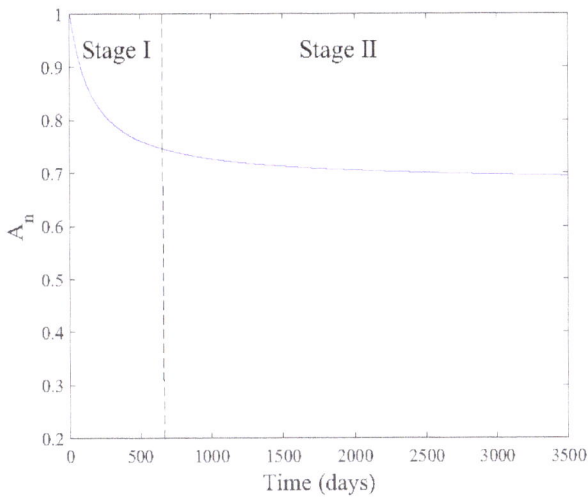

Figure 3. Schematic presentation of the two stages in A_n decline. (I) initial more pronounced drop due to compaction and irreversible fouling; (II) gradual decline mainly caused by irreversible fouling [29].

4. Conclusions and Perspective of Future

The analysis of the different techniques, parameters, indexes and models that have been developed to date in the characterization and evaluation of RO membrane fouling potential reveals the existence of gaps in effective methods for the characterization and evaluation of fouling. It seems that the efforts made to advance our knowledge have turned out to be ineffective in terms of the mitigation and control of membrane fouling due to gaps in effective methods for the characterization and evaluation of fouling. The task of developing reliable fouling prediction tools is extremely important for the desalination industry, since fouling is one of the main causes of performance decrease in full-scale RO desalination plants. There are different fouling rates that have been developed and used in this field, but there remains much work to be done to improve these methods, indices and evaluation parameters. Among the weaknesses or deficiencies observed in the current methods of fouling assessment are the following:

(a) Most conventional indexes, SDI and MFI are not appropriate.
(b) There are very few studies about indices or parameters applied directly to spiral wound membranes and feedwater with high salinity. Most of the studies are applied at the laboratory scale with well-controlled operating conditions, flat membrane systems and at low salinity. However, it is preferable for fouling potential to be determined with RO membranes and under operating conditions similar to those of full-scale desalination plants.
(c) Currently, the effect of Cake-Enhanced Osmotic Pressure (CEOP) has not been taken extensively into account in measuring fouling potential. However, CEOP can contribute to a significant loss of performance, even more than the hydraulic resistance brought about by cake formation.

The aforementioned prediction models are based on long-term data of full-scale RO desalination plants under full-scale operating conditions. Unfortunately, these models do not take into consideration important features of membranes such as the spacer geometry. The efficiency and frequency of chemical cleanings, which play an important role in the performance of this process, should also be considered in these models.

Conflicts of Interest: The authors declare no conflict of interest.

References

1. Misdan, N.; Ismail, A.; Hilal, N. Recent advances in the development of (bio)fouling resistant thin film composite membranes for desalination. *Desalination* **2016**, *380*, 105–111.
2. Zhang, R.; Liu, Y.; He, M.; Su, Y.; Zhao, X.; Elimelech, M.; Jiang, Z. Antifouling membranes for sustainable water purification: strategies and mechanisms. *Chem. Soc. Rev.* **2016**, *45*, 5888–5924.
3. Jiang, S.; Li, Y.; Ladewig, B.P. A review of reverse osmosis membrane fouling and control strategies. *Sci. Total Environ.* **2017**, *595*, 567–583.
4. She, Q.; Wang, R.; Fane, A.G.; Tang, C.Y. Membrane fouling in osmotically driven membrane processes: A review. *J. Membr. Sci.* **2016**, *499*, 201–233.
5. Tang, C.Y.; Chong, T.; Fane, A.G. Colloidal interactions and fouling of NF and RO membranes: A review. *Adv. Colloid Interface Sci.* **2011**, *164*, 126–143.
6. Karabelas, A.; Sioutopoulos, D. New insights into organic gel fouling of reverse osmosis desalination membranes. *Desalination* **2015**, *368*, 114–126.
7. Shirazi, S.; Lin, C.J.; Chen, D. Inorganic fouling of pressure-driven membrane processes—A critical review. *Desalination* **2010**, *250*, 236–248.
8. Matin, A.; Khan, Z.; Zaidi, S.; Boyce, M. Biofouling in reverse osmosis membranes for seawater desalination: Phenomena and prevention. *Desalination* **2011**, *281*, 1–16.
9. Hoek, E.M.; Allred, J.; Knoell, T.; Jeong, B.H. Modeling the effects of fouling on full-scale reverse osmosis processes. *J. Membr. Sci.* **2008**, *314*, 33–49.
10. Ruiz-García, A.; Ruiz-Saavedra, E. 80,000 h operational experience and performance analysis of a brackish water reverse osmosis desalination plant. Assessment of membrane replacement cost. *Desalination* **2015**, *375*, 81–88.
11. Kang, G.D.; Cao, Y.M. Development of antifouling reverse osmosis membranes for water treatment: A review. *Water Res.* **2012**, *46*, 584–600.
12. Shahkaramipour, N.; Tran, T.N.; Ramanan, S.; Lin, H. Membranes with Surface-Enhanced Antifouling Properties for Water Purification. *Membranes* **2017**, *7*, 13.
13. Lee, K.P.; Arnot, T.C.; Mattia, D. A review of reverse osmosis membrane materials for desalination—Development to date and future potential. *J. Membr. Sci.* **2011**, *370*, 1–22.
14. Lawler, J. Incorporation of Graphene-Related Carbon Nanosheets in Membrane Fabrication for Water Treatment: A Review. *Membranes* **2016**, *6*, 57.
15. Nady, N. PES surface modification using green chemistry: New generation of antifouling membranes. *Membranes* **2016**, *6*, 23.
16. Taheri, A.; Sim, L.; Chong, T.; Krantz, W.; Fane, A. Prediction of reverse osmosis fouling using the feed fouling monitor and salt tracer response technique. *J. Membr. Sci.* **2015**, *475*, 433–444.
17. Taheri, A.; Sim, S.; Sim, L.; Chong, T.; Krantz, W.; Fane, A. Development of a new technique to predict reverse osmosis fouling. *J. Membr. Sci.* **2013**, *448*, 12–22.
18. Ho, J.S.; Sim, L.N.; Webster, R.D.; Viswanath, B.; Coster, H.G.; Fane, A.G. Monitoring fouling behavior of reverse osmosis membranes using electrical impedance spectroscopy: A field trial study. *Desalination* **2017**, *407*, 75–84.
19. Esfahani, I.J.; Kim, M.; Yun, C.; Yoo, C. Proposed new fouling monitoring indices for seawater reverse osmosis to determine the membrane cleaning interval. *J. Membr. Sci.* **2013**, *442*, 83–96.
20. Koo, C.; Mohammad, A. Experimental Investigation on Performance of Fouling Prediction Devices for NF/RO System. *Int. J. Chem. Eng. Appl.* **2015**, *6*, 179.
21. Rabie, H.R.; Côté, P.; Adams, N. A method for assessing membrane fouling in pilot- and full-scale systems. *Desalination* **2001**, *141*, 237–243.
22. Jin, Y.; Lee, H.; Jin, Y.O.; Hong, S. Application of multiple modified fouling index (MFI) measurements at full-scale SWRO plant. *Desalination* **2017**, *407*, 24–32.
23. Wei, C.H.; Laborie, S.; Aim, R.B.; Amy, G. Full utilization of silt density index (SDI) measurements for seawater pre-treatment. *J. Membr. Sci.* **2012**, *405–406*, 212–218.
24. Jin, Y.; Ju, Y.; Lee, H.; Hong, S. Fouling potential evaluation by cake fouling index: Theoretical development, measurements, and its implications for fouling mechanisms. *J. Membr. Sci.* **2015**, *490*, 57–64.

25. Ju, Y.; Hong, S. Nano-colloidal fouling mechanisms in seawater reverse osmosis process evaluated by cake resistance simulator-modified fouling index nanofiltration. *Desalination* **2014**, *343*, 88–96.

26. Wilf, M.; Klinko, K. Performance of commercial seawater membranes. *Desalination* **1994**, *96*, 465–478.

27. Zhu, M.; El-Halwagi, M.M.; Al-Ahmad, M. Optimal design and scheduling of flexible reverse osmosis networks. *J. Membr. Sci.* **1997**, *129*, 161–174.

28. Abbas, A.; Al-Bastaki, N. Performance decline in brackish water Film Tec spiral wound RO membranes. *Desalination* **2001**, *136*, 281–286.

29. Ruiz-García, A.; Nuez, I. Long-term performance decline in a brackish water reverse osmosis desalination plant. Predictive model for the water permeability coefficient. *Desalination* **2016**, *397*, 101–107.

30. ASTM. *The Annual Book of ASTM Standard, Designation: D 4189-95. Standard Test Method for Silt Density Index (SDI) of Water*; American Society for Testing and Materials: West Conshohocken, PA, USA, 2010.

31. Schippers, J.; Verdouw, J. The modified fouling index, a method of determining the fouling characteristics of water. *Desalination* **1980**, *32*, 137–148.

32. Boerlage, S.F.; Kennedy, M.; Aniye, M.P.; Schippers, J.C. Applications of the MFI-UF to measure and predict particulate fouling in RO systems. *J. Membr. Sci.* **2003**, *220*, 97–116.

33. Yiantsios, S.G.; Karabelas, A.J. An assessment of the Silt Density Index based on RO membrane colloidal fouling experiments with iron oxide particles. *Desalination* **2003**, *151*, 229–238.

34. Boerlage, S. Understanding the SDI and Modified Fouling Indices (MFI0. 45 and MFI-UF). In Proceedings of the IDA World Congress On Desalination and Water Reuse 2007-Desalination: Quenching a Thirst, Maspalomas, Gran Canaria, Canary Islands, Spain, 21–26 October 2007.

35. Boerlage, S.F.; Kennedy, M.D.; Dickson, M.R.; El-Hodali, D.E.; Schippers, J.C. The modified fouling index using ultrafiltration membranes (MFI-UF): Characterisation, filtration mechanisms and proposed reference membrane. *J. Membr. Sci.* **2002**, *197*, 1–21.

36. ASTM. D 8002-15, Standard Test Method for Modified Fouling Index (MFI-0.45) of Water. In *Annual Book of ASTM Standards*; American Society for Testing and Materials: West Conshohocken, PA, USA, 2015.

37. Sanz Ataz, J.; Guerrero Gallego, L.; Taberna Camprubí, E.; Peña García, N.; Carulla Contreras, C.; Blavia Bergós, J. Prevención del ensuciamiento coloidal en sistemas de ósmosis inversa y nanofiltración. Aplicación del ánalisis de superficies con haces de electrones. *Tecnol. Agua* **2003**, *239*, 58–63.

38. Koo, C.H.; Mohammad, A.W.; Suja, F.; Talib, M.Z.M. Review of the effect of selected physicochemical factors on membrane fouling propensity based on fouling indices. *Desalination* **2012**, *287*, 167–177.

39. Sioutopoulos, D.C.; Karabelas, A.J. Correlation of organic fouling resistances in RO and UF membrane filtration under constant flux and constant pressure. *J. Membr. Sci.* **2012**, *407*, 34–46.

40. Sioutopoulos, D.C.; Karabelas, A.J. Evolution of organic gel fouling resistance in constant pressure and constant flux dead-end ultrafiltration: Differences and similarities. *J. Membr. Sci.* **2016**, *511*, 265–277.

41. Sioutopoulos, D.; Karabelas, A.; Yiantsios, S. Organic fouling of RO membranes: Investigating the correlation of RO and UF fouling resistances for predictive purposes. *Desalination* **2010**, *261*, 272–283.

42. Khirani, S.; Aim, R.B.; Manero, M.H. Improving the measurement of the Modified Fouling Index using nanofiltration membranes (NF-MFI). *Desalination* **2006**, *191*, 1–7.

43. Sioutopoulos, D.; Yiantsios, S.; Karabelas, A. Relation between fouling characteristics of RO and UF membranes in experiments with colloidal organic and inorganic species. *J. Membr. Sci.* **2010**, *350*, 62–82.

44. Alhadidi, A.; Kemperman, A.; Schippers, J.; Blankert, B.; Wessling, M.; van der Meer, W. SDI normalization and alternatives. *Desalination* **2011**, *279*, 390–403.

45. Koo, C.H.; Mohammad, A.W.; Suja, F.; Talib, M.Z.M. Use and Development of Fouling Index in Predicting Membrane Fouling. *Sep. Purif. Rev.* **2013**, *42*, 296–339.

46. Boerlage, S.F.; Kennedy, M.; Tarawneh, Z.; Faber, R.D.; Schippers, J.C. Development of the MFI-UF in constant flux filtration. *Desalination* **2004**, *161*, 103–113.

47. Chellam, S.; Wiesner, M.R. Evaluation of crossflow filtration models based on shear-induced diffusion and particle adhesion: Complications induced by feed suspension polydispersivity. *J. Membr. Sci.* **1998**, *138*, 83–97.

48. Adham, S.; Fane, A. *Crossflow Sampler Fouling Index*; National Water Research Institute: Fountain Valley, CA, USA, 2008.

49. Javeed, M.; Chinu, K.; Shon, H.; Vigneswaran, S. Effect of pre-treatment on fouling propensity of feed as depicted by the modified fouling index (MFI) and cross-flow sampler-modified fouling index (CFS-MFI). *Desalination* **2009**, *238*, 98–108.

50. Sim, L.N.; Ye, Y.; Chen, V.; Fane, A.G. Crossflow Sampler Modified Fouling Index Ultrafiltration (CFS-MFIUF)-An alternative Fouling Index. *J. Membr. Sci.* **2010**, *360*, 174–184.

51. Sim, L.N.; Ye, Y.; Chen, V.; Fane, A.G. Investigations of the coupled effect of cake-enhanced osmotic pressure and colloidal fouling in RO using crossflow sampler-modified fouling index ultrafiltration. *Desalination* **2011**, *273*, 184–196.

52. Hoek, E.M.V.; Elimelech, M. Cake-Enhanced Concentration Polarization: A New Fouling Mechanism for Salt-Rejecting Membranes. *Environ. Sci. Technol.* **2003**, *37*, 5581–5588, PMID:14717167.

53. Choi, J.S.; Hwang, T.M.; Lee, S.; Hong, S. A systematic approach to determine the fouling index for a RO/NF membrane process. *Desalination* **2009**, *238*, 117–127.

54. Yu, Y.; Lee, S.; Hong, K.; Hong, S. Evaluation of membrane fouling potential by multiple membrane array system (MMAS): Measurements and applications. *J. Membr. Sci.* **2010**, *362*, 279–288.

55. Naceur, M. The Modified Fouling Index Revisited: Proposal of A Novel Dimensionless Fouling Index for Membranes. *Int. J. Eng.* **2014**, *3*.

56. Hong, K.; Lee, S.; Choi, S.; Yu, Y.; Hong, S.; Moon, J.; Sohn, J.; Yang, J. Assessment of various membrane fouling indexes under seawater conditions. *Desalination* **2009**, *247*, 247–259.

57. Boerlag, S.F.; Kennedy, M.D.; Bonne, P.A.; Galjaard, G.; Schippers, J.C. Prediction of flux decline in membrane systems due to particulate fouling. *Desalination* **1997**, *113*, 231–233.

58. Sim, L.N.; Ye, Y.; Chen, V.; Fane, A.G. Comparison of MFI-UF constant pressure, MFI-UF constant flux and Crossflow Sampler-Modified Fouling Index Ultrafiltration (CFS-MFIUF). *Water Res.* **2011**, *45*, 1639–1650.

59. Song, L.; Chen, K.L.; Ong, S.L.; Ng, W.J. A new normalization method for determination of colloidal fouling potential in membrane processes. *J. Colloid Interface Sci.* **2004**, *271*, 426–433.

60. Vrouwenvelder, J.; van Paassen, J.; Wessels, L.; van Dam, A.; Bakker, S. The Membrane Fouling Simulator: A practical tool for fouling prediction and control. *J. Membr. Sci.* **2006**, *281*, 316–324.

61. Vrouwenvelder, J.; Bakker, S.; Wessels, L.; van Paassen, J. The Membrane Fouling Simulator as a new tool for biofouling control of spiral-wound membranes. *Desalination* **2007**, *204*, 170–174.

62. Tay, K.G.; Song, L. A more effective method for fouling characterization in a full-scale reverse osmosis process. *Desalination* **2005**, *177*, 95–107.

63. Zhao, Y.; Song, L.; Ong, S.L. Fouling behavior and foulant characteristics of reverse osmosis membranes for treated secondary effluent reclamation. *J. Membr. Sci.* **2010**, *349*, 65–74.

64. Chen, K.L.; Song, L.; Ong, S.L.; Ng, W.J. The development of membrane fouling in full-scale RO processes. *J. Membr. Sci.* **2004**, *232*, 63–72.

65. Belkacem, M.; Bekhti, S.; Bensadok, K. Groundwater treatment by reverse osmosis. *Desalination* **2007**, *206*, 100–106.

membranes

MDPI

Review

The Role of Ion Exchange Membranes in Membrane Capacitive Deionisation

Armineh Hassanvand [1], Kajia Wei [2], Sahar Talebi [1,3], George Q. Chen [1,3] and Sandra E. Kentish [1,3,*]

[1] Department of Chemical Engineering, University of Melbourne, Parkville, VIC 3010, Australia;
     ahassanvand@student.unimelb.edu.au (A.H.); stalebi@student.unimelb.edu.au (S.T.);
     gechen@unimelb.edu.au (G.Q.C.)
[2] Key Laboratory of Jiangsu Province for Chemical Pollution Control and Resources Reuse,
     School of Environment and Biological Engineering, Nanjing University of Science and Technology,
     Nanjing 210094, China; wkjzerone@hotmail.com
[3] The ARC Dairy Innovation Hub, Department of Chemical Engineering, University of Melbourne,
     Parkville, VIC 3010, Australia
* Correspondence: sandraek@unimelb.edu.au; Tel.: +61-8344-6682

Received: 7 July 2017; Accepted: 5 September 2017; Published: 14 September 2017

Abstract: Ion-exchange membranes (IEMs) are unique in combining the electrochemical properties of ion exchange resins and the permeability of a membrane. They are being used widely to treat industrial effluents, and in seawater and brackish water desalination. Membrane Capacitive Deionisation (MCDI) is an emerging, energy efficient technology for brackish water desalination in which these ion-exchange membranes act as selective gates allowing the transport of counter-ions toward carbon electrodes. This article provides a summary of recent developments in the preparation, characterization, and performance of ion exchange membranes in the MCDI field. In some parts of this review, the most relevant literature in the area of electrodialysis (ED) is also discussed to better elucidate the role of the ion exchange membranes. We conclude that more work is required to better define the desalination performance of the proposed novel materials and cell designs for MCDI in treating a wide range of feed waters. The extent of fouling, the development of cleaning strategies, and further techno-economic studies, will add value to this emerging technique.

Keywords: cation exchange membrane; anion exchange membrane; desalination; carbon

1. Introduction

An ion-exchange membrane consists of a polymer matrix in which ionic groups are fixed to the polymeric backbone. Depending upon the charge of the ionic group, IEMs are categorized as either anion-exchange membranes (AEMs) or cation-exchange membranes (CEMs). The most common positively charged groups fixed in the former are $-NH_3^+$, $-NRH_2^+$, $-NR_2H^+$, and NR_3^+, which allow the transport of anions and reject cations. On the other hand, CEMs contain $-SO_3^-$, $-COO^-$, PO_3^{2-}, and PO_3H^-, which allow the transport of cations and reject anions [1,2]. The combination of a cation and anion exchange membrane produces a third type of membrane known as a bipolar ion exchange membrane [3].

IEMs are used for mass separation, chemical synthesis, energy conversion, and storage processes [4]. The most commonly known applications are chloroalkali electrolysis and fuel cells [3,5,6], but these applications are not the focus of this paper. Rather, this paper focuses on separation processes where cations and/or anions are selectively transferred through the membrane upon applying an external electrical current [4]. This is the basis for electrodialysis (ED), Donnan dialysis, electrodialysis with bipolar membranes (EDBM), reverse electrodialysis (RED), and Membrane Capacitive Deionisation (MCDI).

As shown in Figure 1a, in electrodialysis (ED), the application of an electric field causes the cations and anions present in the feed stream to migrate toward the cathode and anode, respectively [7]. The cations pass through the CEM, but are then retained in the permeate channel due to the presence of an AEM. Similarly, anions are captured between an AEM and CEM. Hence, by placing the IEMs alternatively, the feed compartments (diluate side) become depleted of ions and the permeate compartments (concentrate side) become more concentrated [8].

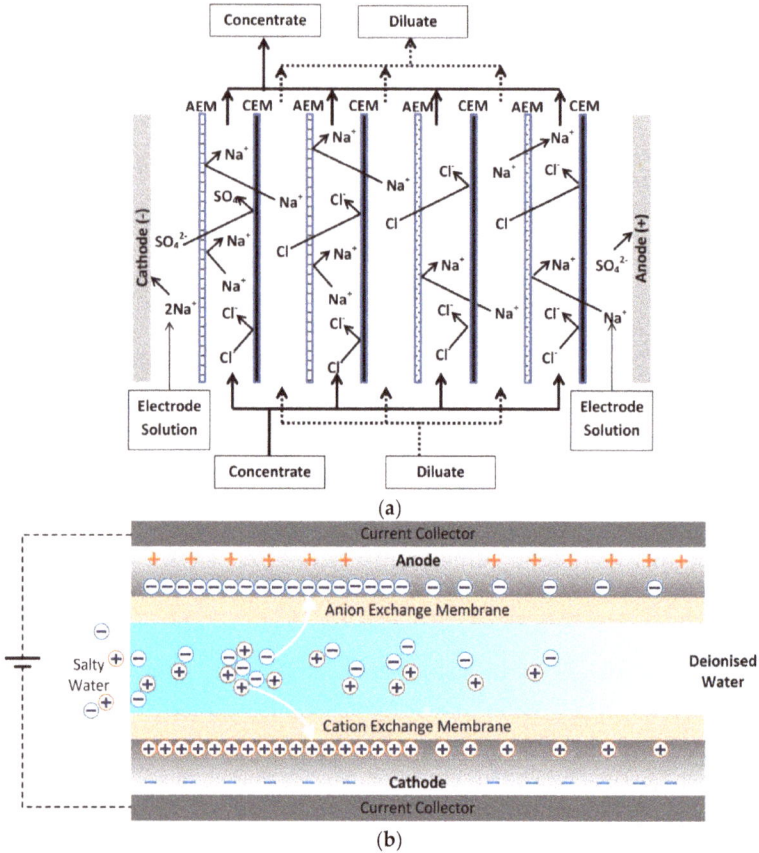

(a)

(b)

Figure 1. (**a**) Electrodialysis (adapted from [9]); and (**b**) Membrane Capacitive Deionisation (MCDI) during adsorption.

While ED is considered a mature technology in water desalination, Capacitive Deionisation (CDI) is an emerging technology. This approach has very low energy consumption at low salinity, is easy to operate and is low maintenance [10,11]. In CDI, an electric field is applied to carbon electrodes, causing charged species to be adsorbed within the carbon micropores [12]. During regeneration, the electric field is removed and the adsorbed ions are released back into a brine stream. One of the drawbacks limiting the charge efficiency in CDI is co-ion adsorption, i.e., the adsorption of ions to an electrode carrying the same surface charge [13]. In 2006, Lee et al. [14] suggested introducing ion-exchange membranes in front of the carbon electrodes to reduce this effect. As depicted in Figure 1b, in a MCDI cell, an anion-exchange membrane is placed in front of the anode to block the passage of cations and a cation-exchange membrane is placed in front of the cathode to reject the anions [15]. This approach significantly limits the co-ion adsorption. It also allows for the electrical charge to be reversed during the regeneration cycle, rather than being simply turned off. Zhao et al. [16] found that MCDI can be more energy efficient than Reverse Osmosis (RO) technology, when the feed water salinity is <2 g L^{-1} Total Dissolved Solids (TDS) and effluent water TDS >0.5 g L^{-1} (see Figure 2).

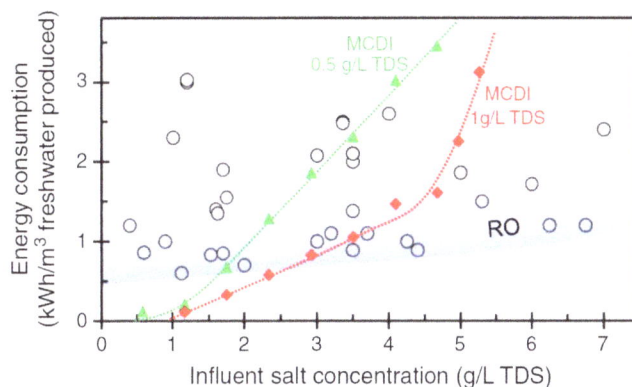

Figure 2. Comparison of energy consumption between MCDI and reverse osmosis. Triangles: energy consumption of MCDI to bring salt concentration to the level of 0.5 g L^{-1} Total Dissolved Solids (TDS). Diamonds: energy consumption of MCDI to reduce salt concentration to 1 g L^{-1} TDS. Circles: energy consumption of Reverse Osmosis (RO) collected from literature studies. Reproduced with permission from [16], copyright Elsvier, 2013.

This review provides a summary of recent accomplishments in the fast evolving area of MCDI and the novel IEMs being prepared for this purpose. Additionally, the effect of fouling and scaling in these systems is discussed. Recent progress in the area of ion transport through IEMs is also discussed briefly. In the final section, we describe the most recent MCDI stack designs where novel features are introduced to the conventional MCDI cell.

2. Ion Exchange Membranes for Membrane Capacitive Deionisation (MCDI) Applications

2.1. Homogeneous Ion Exchange Membranes

Within a homogeneous ion exchange membrane, the ionic groups are chemically bound to a polymeric backbone making a coherent ion-exchanger gel [1,2]. These functionalised polymers would swell greatly in water if not crosslinked; thus, crosslinking agents, most commonly divinyl benzene monomers, are also incorporated [17]. In addition to less swelling, crosslinking also results in an improved structure of the IEM that can facilitate the introduction of a greater density of fixed charge groups [18]. In a traditional production process, the divinyl monomers are mixed with a linear polymer, polymerisation initiators, and plasticisers to form a paste. This paste is then coated onto a backing fabric or net and the composite structure is heated to copolymerise the divinyl monomers. The ion exchange functionality is added after membrane formation by impregnation of chemicals such as chlorosulfonic acid, sulfuric acid, trimethylamine, or methyl iodide. However, the dangerous nature of these chemicals is resulting in an increasing tendency to instead use monomers that are already functionalised [2]. In this case, a water soluble crosslinking agent, typically formaldehyde, is used in place of divinyl benzene [19]. An alternate CEM can be prepared from a perfluorinated sulfonic acid polymer such as Nafion™ (DuPont, DE, USA).

IEMs that are commercially available for MCDI applications include Fumasep IEMs produced by FuMA-Tech GmbH (Bietigheim-Bissingen, Baden-Württemberg, Germany), Selemion IEMs produced by Asahi Glass Co., Ltd. (Tokyo, Japan), and Neosepta IEMs produced by ASTOM Co. (Tokyo, Japan) (Tables 1 and 2). These membranes benefit from high permselectivity, low electrical resistance, and high chemical and mechanical stability [3].

Several workers have developed IEMs specifically for MCDI. In this application, the membranes can be thinner than in ED applications, as they do not need to be self-supporting. This results in lower electrical resistance. Kwak et al. [20] developed a cation exchange membrane through heat crosslinking and esterification of three monomers, 4-styrenesulfonic acid sodium salt hydrate (NaSS), methacrylic acid (MAA), and methyl methacrylate (MMA). The synthesis of a poly(vinylidene fluoride)-sodium 4-vinylbenzene sulfonate copolymer (PVDF-g-PSVBS) CEM for deionisation applications was also reported by Kang et al. [21]. Jeong et al. [22] synthesized an aminated poly(vinylidene fluoride-g-4-vinylbenzyl chloride) (PVDF-g-VBC) anion exchange membrane, while Qiu et al. [23] used γ-irradiation to crosslink a thin sulfonated polystyrene film as a CEM.

2.2. Composite Electrodes for MCDI

The carbon electrodes used in CDI applications are generally prepared from activated carbon, or other forms of carbon such as graphene or carbon nanotubes. These carbon particles are bound together through the addition of a weakly hydrophilic polymer binder such as polyvinylidene fluoride (PVDF). Lee et al. [24] were the first to combine ion-exchange resin directly into a CDI electrode to improve the hydrophilicity of these structures. They achieved salt removal efficiencies 35% higher than a conductive carbon black/carbon composite electrode when Purolite anion exchange resin was added to both electrodes. Hou et al. [25] similarly used polyvinyl alcohol as a more hydrophilic binder.

However, later work has focused on incorporating ion exchange resin into the carbon electrode or coating it with an ion selective coating. This provides a composite structure within which the ion selectivity is comparable to separate IEM and electrode layers. Application of such electrodes can potentially result in a system that is lower in capital cost and electrical resistance than a conventional MCDI unit, but with the same current efficiency. This approach is referred to as modified MCDI (m-MCDI) [26].

Table 1. Fabrication and characterization properties of different cation-exchange membranes (CEMs) and composite electrodes for use in membrane capacitive deionisation (MCDI) processes.

Materials	Polymer or Polymer Coating	Water Uptake (wt %)	IEC (meq g^{-1})	Electrical Resistance (Ω cm^2)	Thickness of Polymer or Coating (μm)	Specific Capacitance (F g^{-1})	Ref.
Commercial Membranes	Fumasep FKS	12–15	0.8–1.2	2.0–4.5	120	-	Supplier
	Neosepta CMX	25–30	1.5–1.8	3.0	170	-	[9]
	Selemion CMV	25	2.4	-	120	-	[9]
	Dupont Nafion	16	0.9	1.5	117	-	[9]
IEMs for MCDI	NaSS-MAA-MMA	121	0.99	0.7	90–140	-	[20]
	PVDF-g-PSVBS	61	1.14	2	-	-	[21]
	Crosslinked Sulfonated polystyrene	30	0.9	0.37	25	-	[23]
Composite Electrodes	Polyethyleneimine (PEI)	-	-	-	-	52	[26]
	PVA/SSA-coated	-	-	0.67–1.17	-	97–116	[27]
	SG-CNFs [a]	-	-	-	<1	117	[28]
	PVA/SSA-coated	-	-	0.64	10	0.74 F cm^{-2}	[29]
	Sulfonated PPO [b]	-	0.93	-	4.9	-	[30]
	Sulfonated BPPO [c] coated	-	-	-	-	-	[31]
	Sulfonated graphene	-	-	-	-	108	[32]
	PVA/SSA/SSA-MA	44	2.8	-	6.4	-	[33]

[a] Sulfonated graphene-carbon nanofibres; [b] poly(phenylene oxide); [c] Bromomethylated poly (2,6-dimethyl-1,4-phenylene oxide).

Table 2. Fabrication and characterization properties of different anion-exchange membranes (AEMs) and composite electrodes for MCDI process.

Materials	Polymer or Polymer Coating	Water Uptake (wt %)	IEC (meq g^{-1})	Electrical Resistance (Ω cm^2)	Thickness of Polymer or Coating (μm)	Specific Capacitance (F g^{-1})	Ref.
Commercial Membranes	Fumasep FAS	15–30 13–23	1.6–2.0 1.0–1.3	0.3–0.6 1.7–3.0	30 135	-	Supplier
	Neosepta AMX	25–30	1.4–1.7	2.4	140	-	[9]
	Selemion AMV	19	1.9	2.8	120	-	[9]
IEMs for MCDI	Aminated PVDF-g-VBC	25	1.0	4.8	-	-	[34]
Composite Electrodes	Purolite	-	-	-	-	19	[24]
	dimethyldiallyl ammonium chloride	-	-	-	-	53	[26]
	Anion exchange resin	-	1.20	-	-	-	[35]
		-	2.9	-	~40	-	[36]
	Aminated PVA	-	1.76	-	20–40	180	[37]
	Aminated PVA	-	-	-	-	184	[38]
	Aminated PSF [a]	-	1.0	5.2	-	-	[30]
	Aminated BPPO	-	-	-	-	-	[31]
	Aminated Graphene	-	-	-	-	91	[32]
	Aminated PSF	37	2.1	-	10.6	-	[33]

[a] Polysulfone.

Kim and Choi [27,29] were the first to try this approach and coated a mixture of poly(vinyl alcohol) (PVA) and sulfosuccinic acid (SSA) directly onto a carbon electrode to form a cation selective electrode (Figure 3a). The other electrode was left uncoated. This approach gave a 15 to 30% increase in specific capacitance for the electrode, while giving less electrical resistance than commercial IEMs. The same group [35,36] later fabricated nitrate-selective composite electrodes (Figure 3b) by coating anion exchange resin powder onto a carbon electrode. In MCDI experiments, they coupled this novel electrode with a standard Neosepta CMX membrane attached to the other electrode. Tian et al. [37] synthesised an anion exchange membrane electrode that combined a cross-linked PVA layer functionalised with quaternary amine groups above an activated carbon layer, while Gu et al. [38] used a similar coating on a graphene sponge electrode (Figure 3c). Qian et al. [28] used sulfonated graphene to prepare an ultrathin cation-selective coating achieving enhanced electrochemical specific capacitance and charge efficiency.

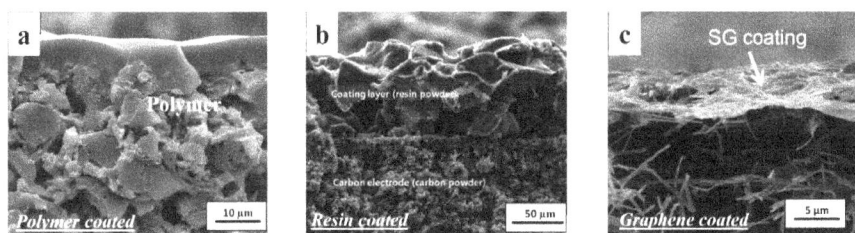

Figure 3. SEM images of three typical composite electrodes. (**a**) Carbon electrode coated with PVA/SSA polymer solution, copyright Elsvier, 2010; (**b**) composite carbon electrode coated with resin powder, copyright Elsvier, 2012; (c) Sulfonated graphite (SG) carbon nanofibre composite. Reproduced with permission from [27,28,35], copyright John Wiley and Sons, 2015.

While these researchers developed only one ion selective electrode in isolation, other researchers have developed electrode pairs, generally by sulfonation and amination of a base polymer, which is then sprayed or coated onto the carbon electrode [30,31]. Liu et al. [32] similarly directly sulfonated and aminated 3D graphene. Kim et al. [33] blended and crosslinked PVA with sulfur succinic acid (SSA) and poly(styrene sulfonic acid-co-maleic acid) (PSSA-MA) to provide a CEM and used animated polysulfone for the AEM. Liu et al. [26] introduced the anion exchange polymer dimethyldiallyl ammonium chloride and cation exchange polymer polyethyleneimine (PEI) as a binder in carbon nanotube electrodes to form such a pair.

These composite electrodes appear a promising approach to reducing electrical resistance. However, more work is required to determine if they are indeed lower in cost than the traditional arrangement and to evaluate their long term performance in desalination of a wide range of charged species.

3. MCDI Performance Parameters

There are several key parameters that indicate MCDI or CDI performance. The salt removal efficiency is the percentage of salt removed from the feed stream in each adsorption cycle and reflects the cell geometry, feed flowrates, charge applied, and carbon properties. In general, MCDI provides better salt removal efficiency compared with CDI due to the almost complete elimination of co-ion adsorption (Figure 4). Such efficiencies are mainly above 70% for MCDI and m-MCDI, while for CDI cells, these are mostly below 70%.

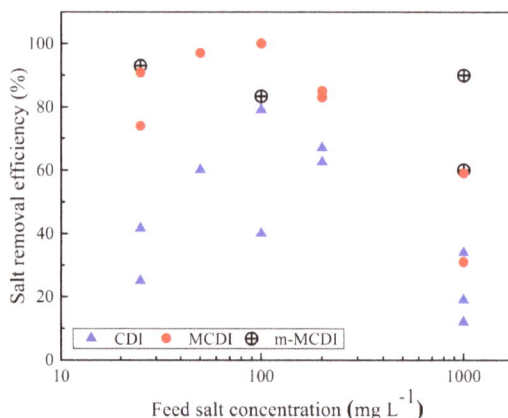

Figure 4. Salt removal efficiencies as a function of feed salt concentration from recent research reports on Capacitive Deionisation (CDI), MCDI, and modified MCDI (m-MCDI) [14,24,26,29–31,39–42].

The adsorption capacity is the amount of salt adsorbed per gram of capacitive materials within the adsorption cycle. This capacity depends most strongly upon the type of electrode used, but is generally higher in direct experimental comparisons for m-MCDI or MCDI versus CDI (Table 3).

Table 3. A comparison of salt adsorption capacities from parallel studies of m-MCDI or MCDI versus Capacitive Deionisation (CDI).

Feed Salt Concentration	Adsorption Capacity (mg g^{-1})			
(mg L^{-1})	m-MCDI	MCDI	CDI	Reference
50	-	2.04	1.05	[41]
100	2.09	-	2.0	[31]
200	-	5.3	3.7	[40]
200	-	10.2	8.1	[29]
300	-	3.5	1	[43]
400	-	4.3	3	[44]
400	9.5	-	5.0	[28]
500	9.3	6.5	-	[26]
750	-	45.6	30.3	[45]
1000	9	6	1.9	[46]

CDI and MCDI cells are operated either at constant current (CC) where the voltage varies during the cycle, or constant voltage (CV) where the current is varied. The charge efficiency (Λ) is used to characterise the CV mode and is defined as the ratio of adsorbed salt (Γ, mol g^{-1}) per cycle to the charge (Σ, C g^{-1}) transferred in this cycle [47]:

$$\Lambda = \frac{\Gamma \times F}{\Sigma} \qquad (1)$$

where F represents Faraday constant (96,485 C mol^{-1}), Γ is the adsorption capacity, and Σ is calculated by integrating the corresponding current [48].

The current efficiency, λ, describes the CC mode and is similarly defined as the amount of ions adsorbed over current applied.

$$\lambda = \frac{(C_{in} - C_{out})VF}{It_{ads}} \qquad (2)$$

where C_{in} and C_{out} are the salt concentrations (mol L^{-1}) of influent and effluent, I is the applied electrical current, t_{ads} is the adsorption duration, and V is the solution volume (L) [49].

The charge efficiencies and current efficiencies for MCDI are mostly above 50%, while those for CDI are lower (see Figure 5). The use of composite electrodes (m-MCDI) shows even higher charge efficiencies. Choi [50] found that the charge efficiency in CV mode was greater than the comparable current efficiency in CC mode, although this trend is less evident when a range of data is considered such as in Figure 5 [51].

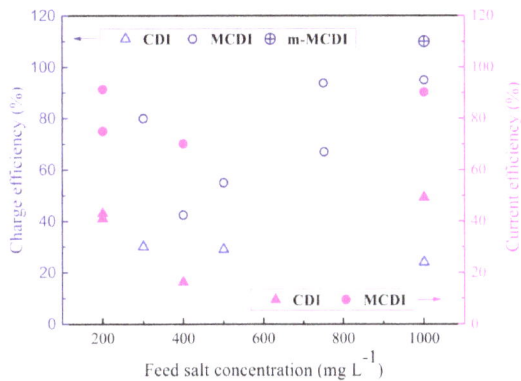

Figure 5. Charge and current efficiencies as a function of feed salt concentration for a range of recent studies [25,28,29,40,42–45,52–54].

The energy consumption in both CDI and MCDI is a direct function of the feed salt concentration. In a pilot scale CDI study, Mossad and Zou [55] proposed a minimum energy consumption of 1.85 kWh m^{-3} (10 kJ g^{-1}) for CDI at the highest flow rate. In lab-scale tests, MCDI has been shown to provide lower energy consumption, below 1 kWh m^{-3} or 6 kJ g^{-1} [44,49]. Dlugolecki et al. [56] were able to achieve a value of 0.26 kWh m^{-3} by recovering the energy during the discharge part of the MCDI cycle. Choi [50] found that the MCDI energy consumption was much lower while operated under CC mode against CV mode.

It should be noted that most work to date has considered only the separation of NaCl using MCDI. Choi et al. [57] considered the selective removal application of nitrate, while Ryu et al. [58] suggested a novel recovery system for lithium with a modified MCDI cell. Yoon et al. [59] proposed the use of calcium alginate as a cation exchange coating material on a negative electrode and used this approach effectively for calcium removal. In this case, the salt sorption capacity and charge efficiency were 15.6 mg g^{-1} and 95%, respectively; much higher than that for CDI (9.8 mg g^{-1} and 55%).

In another study, Choi et al. [60] used a commercial monovalent cation permselective membrane to produce calcium-rich solutions from an MCDI process, by selectively removing sodium. They mixed NaCl and $CaCl_2$ at similar mass ratios and achieved a selectivity of 1.8 (removal of Na^+ to that of Ca^{2+}). This selectivity fell at higher feed TDS concentrations, lower pH values, lower applied voltage, and longer operation time. The authors compared this approach to the use of nanofiltration (NF) and argued that the NF approach was unable to produce a divalent-rich solution. However, this result is not consistent with other literature sources that do indicate NF can provide divalent selectivity [61–64]. The authors also compared the electrical energy consumption in MCDI to that of NF. With specific energy consumption of around 0.2 KWh m^{-3}, MCDI was comparable with NF only at low salinity and water recoveries. However, it is worth mentioning that Choi et al. [60] considered the electrical energy consumption during both during adsorption and desorption. The use of energy recovery systems during MCDI regeneration can reduce this energy requirement [56]. Overall, there is limited data on MCDI with such alternative salts, making this a fruitful area for further research.

The research outcomes discussed in this section are also mostly collected from lab scale MCDI units. More work is required to test these membranes at a larger scale. van Limpt et al. [65] are one of the few groups to report results from commercial MCDI systems. They monitored the operation of two MCDI modules in series (each providing 6 m^2 cell area) on cooling water recycle streams. The modules were operated under constant current conditions for more than six months at two different site locations. MCDI operation resulted in 70% conductivity removal, 83% water recovery, and savings of 85% in chemical usage. High water recovery was achieved by lowering the flow rate during regeneration. An energy consumption of 0.1 to 0.2 kWh m^{-3} was reported, which is significantly lower than that of RO in brackish water deioniziation (0.85–1.55 kWh m^{-3}).

4. Fouling

Fouling is an unfavorable phenomenon caused by the attachment of a substance or a living organism to the surface of the membrane. Mikhaylin et al. [66] classified IEM fouling into three categories—colloidal fouling, organic fouling, and biofouling. Non-dissolved suspended solids such as aluminium silicate clays contribute to colloidal fouling, while organic fouling is caused by organic materials such as oils, proteins, and humic acid. Biofouling is initiated by deposition of bacteria or algae onto the membrane, where these microorganisms proliferate and exude extracellular polymeric substances [67,68]. Alternatively, scaling occurs when a dissolved species precipitates onto the membrane surface.

4.1. Fouling in Capacitive Deionisation

In CDI mode, fouling directly affects the carbon electrodes. Mossad et al. [69] used the sodium salt of humic acid as a model foulant of CDI electrodes and reported a fall in salt removal efficiency and feed flow rate over 30 h of operation. They attributed this deterioration to organic fouling blocking the carbon pores. Energy consumption increased by 39% with the introduction of only 10 mg L^{-1} of organic matter to the feed composition. For a feed containing NaCl and humic acid, neither hydraulic or acid cleaning was effective in recovering the flow rate; an alkaline solution, however, could recover 86% of the initial flow rate [69]. For a feed containing both humic acid and a range of divalent salts, hydraulic cleaning was less effective, with only 62.5% of the flow rate recovered with alkaline cleaning. These researchers also observed that the amount of iron in the effluent during cleaning was significantly higher than the iron concentration in the feed, suggesting that Fe ions accumulated more readily than Ca or Mg ions on the carbon electrodes.

Zhang et al. [70] similarly reported an increase in energy consumption during the long-term operation of two inland brackish water desalination CDI units in Australia. It was suggested that dissolved organic matter should be removed prior to the CDI unit to maintain the unit sustainability and efficiency [69]. These researchers [70] cleaned their CDI unit with 0.01 M citric acid for calcium and magnesium scaling, followed by 0.01 M NaOH solution to remove organic fouling to recover the cell performance.

Wang et al. [71] observed a deterioration in CDI desalination and regeneration performance in the presence of humic acid in domestic wastewater biotreated feed. Using impedance spectra, they calculated the resistance of fouled and virgin electrodes as 1.7 and 1.2 Ω cm^{-2}, respectively. They reported that cleaning with water was capable of removing protein-like substances, while 0.01 M NaOH was most effective in the removal of humic-like substances from the carbon surface.

4.2. Fouling of Ion Exchange Membranes

To the best of our knowledge, Kim et al. [52] is the only research group who have studied the use of MCDI technology in the presence of foulants. They observed a decrease in salt removal efficiency in the first 6 cycles of operation when using a solution of 17 mM NaCl containing 5 and 10 mg L^{-1} of octane as the model foulant. However, this efficiency reached a stable value in the four subsequent cycles. They attributed this trend to octane reaching adsorption equilibrium on the IEM surface. They assumed octane fouling interrupted the access of ions to the carbon electrodes, lessening the salt removal efficiency.

It is anticipated that fouling of IEMs in the MCDI arrangement should generally be less damaging to process operations than fouling of the electrodes in CDI arrangements. Further, we anticipate that fouling in these systems should replicate trends observed in electrodialysis studies, so the following section summarizes literature in this field.

In most electrodialysis studies, peptide and protein fouling is observed on the AEM [1,72,73]. This reflects the fact that these species are generally negatively charged and hence accumulate in the boundary layer near the surface of this membrane. Conversely, Langevin et al. [74] reported that CEM membranes are doubly more prone to peptide fouling in comparison to AEMs. This contrasting result probably arises from the fact that these authors completed their investigations in the absence of electrical current. Therefore, the interaction between the charged amino acids and the membrane's fixed charged groups became the only governing factor. With systems involving proteins, the process should also be maintained at temperatures lower than 40 °C to avoid protein denaturation and agglomeration [75].

Lee et al. [76] used zeta potential to obtain the surface charges of three organic foulants, humate, bovine serum albumin (BSA), and sodium dodecylbenzenesulfonate (SDBS). They then used batch equilibrium experiments to determine the adsorption capacity of these negatively charged organic foulants on a Neosepta AMX membrane and fitted the results to Langmuir isotherms. Among the three model foulants, SDBS showed the highest adsorption capacity, even though the humate was more negatively charged. In subsequent ED trials, the SDBS (at 52 mg L^{-1}) also caused a greater increase in membrane resistance and decline in current efficiency, relative to 1.0 g L^{-1} of humate and BSA. Lee et al. [76] then examined the reversibility of the fouling step by running three experiments in a row, comparing the ED performance before and after fouling without the application of any cleaning agent. They observed that the organic fouling on AMX was not reversible; it was adsorbed chemically to the membrane.

Lindstrand et al. [77] investigated the fouling of a surface active fatty acid (octanoic acid), a surfactant (SDBS) and an alkaline bleach plant filtrate on Selemion AMV and CMV membranes. While the CEM resistance increased marginally due to scaling with the bleach plant filtrate, that of the AEM increased significantly in the presence of octanoic acid, due to electrostatic attractive forces. The authors argued that this fouling was more significant at surfactant concentrations approaching the critical micelle concentration. The AEM fouling was also more severe for octanoic acid at pH 3–3.5 than at pH 9. Since octanoic acid is not dissociated at the low pH values, they were expecting to observe similar fouling on both the AEM and CEM, but the CEM was not affected. They argued that the repulsion of the electrical dipole in the acid molecules by the CEM fixed charges outweighed the hydrophobic forces. However, it is also likely that the electric field caused the dissociation of the octanoic acid even at this lower pH, as has been observed by our own group and others for similar weak acids [78].

The pH of the feed solution promotes different types of fouling to take place. Diblíková et al. [79] reported protein precipitation on the AEM when processing cheese whey at a diluate pH of 4–5. It is known that at such pH, β-lactoglobulin has low solubility [80]. In addition, the negatively charged whey proteins cannot pass through the AEM due to their large size and thus precipitate on the surface of the membrane [81,82]. Under acidic conditions, the presence of Ca^{2+} and CO_3^{2-} in the feed solution also promotes the precipitation of proteins on the CEM [82,83]. Several researchers reported that the gel-like protein layer noticed on AEM during whey demineralization is reversible, as it detaches easily during equipment disassembly [81,84]. However, Bleha et al. [85] argued that protein fouling becomes irreversible once it reacts with functional groups of the IEMs.

Mineral fouling (scaling) occurs most readily on the CEM in an acidic environment and upon the AEM in a neutral or basic environment where the solubility of hydroxides and carbonates reduces [8,66,75,82,86,87]. Increases in temperature also exacerbate the precipitation of calcium salts due to their well-known reverse solubility [88]. The presence of certain ions in the feed solution can further promote the precipitation of other components. Bazinet and Araya-Farias [86] noted that when a solution containing $CaCl_2$ and Na_2CO_3 were used, calcium hydroxide was the main deposit, while no calcium carbonate was detected. This was explained by the absence of magnesium in the solution, as magnesium is required to induce calcium carbonate nucleation.

Fouling in Electrodialysis Reversal (EDR) systems may provide a better analogue to MCDI, as in both operations the DC current is periodically reversed [89]. Vermass et al. [90] demonstrated the effectiveness of polarity reversal in EDR to remove multivalent ions and organic fouling. However, they reported that not all fouling is reversible and a combination of antifouling strategies must be employed for long-term charge efficiency stability [90].

4.3. Alterations to Membrane Chemistry

Salts in the feed solution can alter the counter-ion concentration within the membrane itself and thus affect performance. For example, Ayala-Bribiesca et al. [84] showed that calcium present in the feed solution could replace the counter-ion of the membrane and hence change its electrical conductivity. Shee et al. [91] similarly observed changes in the membrane composition in response to changes in the feed. Langevin et al. [74] observed that a CEM pretreated with acid (1 M HCl) was more prone to fouling in a soy protein hydrolysate solution than membrane pretreated in an alkali or neutral solution. This was attributed to the reaction of the H^+ counter-ion within the membrane with this negatively charged species.

A number of researchers have shown that strongly alkaline solutions will damage ion exchange membranes [92,93]. Sata et al. [92] investigated the effect of NaOH concentrations ranging from 3.0 to 9.0 N on common AEMs. They showed that while all membrane showed a loss of anion exchange functionality, the N-methyl-pyridinium groups were more rapidly degraded, compared with benzyl trimethylammonium groups. The degradation reactions were slower at low concentrations of NaOH and lower temperatures. Membranes based on a polysulfone backbone weakened mechanically upon alkali exposure, while those made with a polyethylene fabric retained their mechanical strength. Similarly, Vega et al. [94] reported that the extent of damage to a Neosepta AMX membrane was less under low KOH concentrations (pH 10) than at higher concentrations (pH 14). This membrane is reinforced by ploy(vinyl chloride) fabric and alkali exposure that caused both the color to darken and mechanical strength to weaken due to dehydrochlorination of this fabric. Garcia-Vasuez at al. [95] showed that while AEMs exhibit both loss of functional groups and polymer backbone integrity, the sulfonate groups on the CEM are not directly affected, with only backbone chain scission occurring.

Ghalloussi et al. [96] analysed the structure and physiochemical properties of four different ion exchange membranes after 2 years of ED operation with a food industry solution containing organic acids. They found that the AEMs were most severely damaged, with evidence of organic colloidal particles being sorbed into the membrane structure itself. This led to increased water content and thickness, greater surface hydrophilicity, as well as structural damage (see Figure 6). Conversely, there was evidence of a loss of sulfonic acid functional sites from the CEM, leading to a decrease in water content, greater hydrophobicity, and a more dense structure. Both membranes suffered a loss in specific conductivity and ion exchange capacity, with the CEM also experiencing a loss in permselectivity.

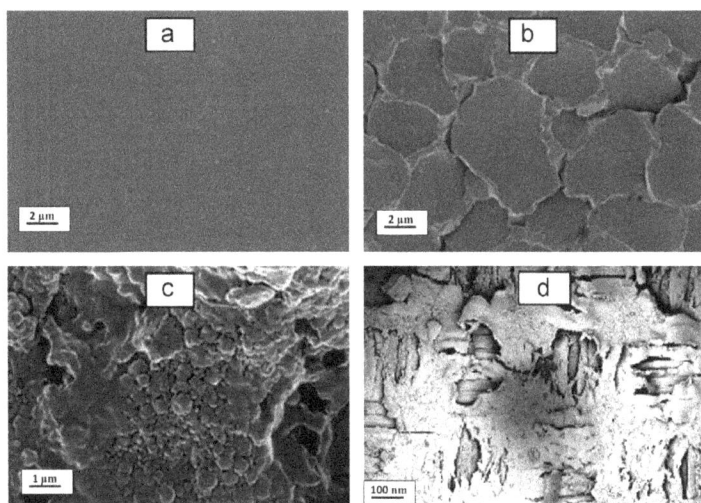

Figure 6. SEM micrographs of Neosepta® membranes. New CEM (**a**); used CEM (**b**) and used AEM (**c**,**d**). Reproduced with permission from [96], copyright Elsvier, 2013.

4.4. Cleaning Solutions

Physical cleaning of membranes involves forward or backwashing, air sparging, or vibration. Those processes have proven to be effective for pressure-driven membrane processes, but not for dense non-porous ion-exchange membranes such as used in MCDI processes [97]. For these operations, chemical membrane cleaning is preferred as it utilizes chemical reactions to weaken the bonds between the foulants themselves and the foulant-membrane surface [88]. There are five groups of chemicals that can be used as cleaning agents for membrane processes, namely: alkalis, acids, metal chelating agents, surfactants, and enzymes—each targeting different types of deposits. The type of cleaning chemical to be utilized is usually selected based upon the nature of the foulant, the compatibility of the chemical with the processing equipment, chemical availability, cost, and safety [88,98]. However, chemical cleaning has many disadvantages, namely: membrane damage, usage of large quantities of chemicals, and waste generation.

Generally, alkaline cleaning agents are known for their ability to remove organic foulants, such as peptides and proteins, as the functional groups of those foulants deprotonate at pH 11 [74], while an acid clean can remove mineral deposits. Similarly, chlorinating agents are often used to remove biological fouling from membranes. However, Garcia-Vasuez at al. [95] has shown that both AEM and CEM membranes degrade upon exposure to sodium hydochlorite, losing electrical conductivity, ion exchange capacity, and mechanical strength. Furthermore, although the use of sequestering agents, such as EDTA, are found useful for mineral deposit removal from pressure driven membranes [88], they should be used with care for ion-exchange membranes, as they may tend to remove the charged ions from the membrane structure.

Cleaning thus needs to be carried out with some care for ion exchange membranes, as they are readily degraded by extremes of pH, particularly alkalinity, as discussed in Section 4.3. The American Water Works Association claims that 2–5% HCl solution, 3–5% NaCl solution adjusted to a pH of 8–10 using NaOH, and 10–50 mg L^{-1} chlorine solution are the only chemicals that should be utilized for ion-exchange membranes in ED stacks [99]. Langevin and Bazinet [74] investigated the effect of 2% and 5% NaCl solutions as cleaning agents. They found that such solutions were able to remove larger foulant particles (>900 Da) from a Neosepta CMX membrane, but was less effective on smaller particles.

The interaction between membrane charge and foulant should be taken into account [100], as alkali and acid cleaning agents have the tendency to replace the equilibrated counter-ions in the membrane with H^+ and OH^-, respectively. This can cause problems in later operation as these ions are replaced with other salts. Research has shown that AEMs are more sensitive to cleaning chemicals compared to CEM. Garcia-Vasquez et al. [101] repeated an ex-situ clean-in-place (CIP) cycle for 400 h on an AEM and a CEM both having the same materials as the support and binder, but different ion-exchange groups. The cycle consisted of soaking for 30 min in 0.1 M HCl; rinsing with deionized water; soaking for 30 min in 0.1 M NaOH and rinsing again with deionized water. The cycle resulted in an increase in the hydrophilicity of the AEM causing higher water uptake, greater membrane electrical conductivity, formation of pores and cavities in the membrane, loss of toughness and flexibility, and modification of the binder material. Less damage was noted for the CEM, which was attributed to the minimal change in water uptake as the charged groups changed from Na^+ to H^+.

5. Modelling MCDI Behaviour

An MCDI stack can be modelled by considering the resistance to flow of ions from the flow channel, through the IEMs, through the macropores in the carbon, and into the carbon micropores (Figure 7). The modified Donnan theory proposed by Biesheuvel et al. [102] can be used to describe the ion storage within these pores. This theory specifies the relationship between the concentration in the micropores with the Donnan potential drop of these micropores ($\Delta\varphi_{electrode,donnan}$) and the Stern layer potential drop ($\Delta\varphi_{St}$) within the electrical double layer (see Figure 7). However, in this Section, we focus on the role of IEMs, as these are the defining difference between CDI and MCDI.

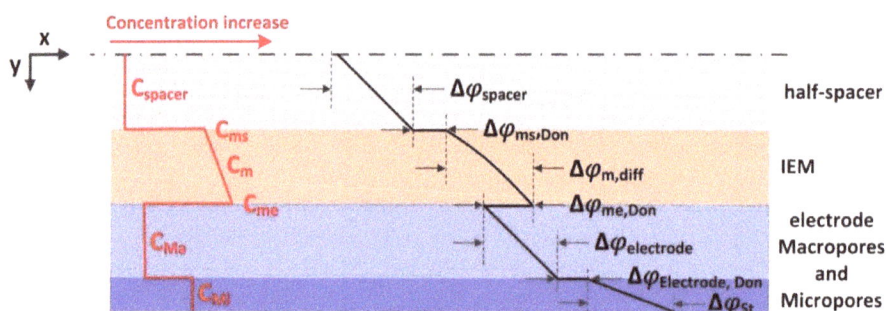

Figure 7. Schematic illustration of counter-ion concentration (C) and dimensionless voltage (φ) distribution over a half-cell within an MCDI stack. Subscripts ms, m, me, Ma, and Mi stand for membrane/spacer interface, membrane phase, membrane/electrode interface, carbon macropores, and micropores, respectively. Reproduced with permission from [103], copyright Elsvier, 2017.

The flux of counter-ions $(J_{+,m})$ through the IEM can be described by the Nernst-Planck equation. In the absence of convective flow, this equation is composed of a diffusion term driven by a chemical potential gradient, and an electromigration term driven by an electrical potential gradient:

$$J_{+,m} = -\mathcal{D}_{+,m}\left(\frac{dC_{+,m}}{dy} + C_{+,m}\frac{dln\gamma_{+,m}}{dy}\right) - \mathcal{D}_{i,m}\frac{z_iC_{+,m}F}{RT}\frac{d\psi_{m,diff}}{dy} \tag{3}$$

where $C_{+,m}$ is the concentration, $\mathcal{D}_{+,m}$ is the diffusion coefficient and $\gamma_{+,m}$ is the activity coefficient of the counter-ion, $C_{fix,m}$ is the concentrations of fixed charge groups in the membrane, $\psi_{m,diff}$ is the diffusion electrical potential, z_i is the charge of the ion, R is the universal gas constant (8.314 J mol^{-1} K^{-1}), T is the temperature (K), and y is the distance perpendicular to the membrane surface.

In general, the fixed charge group concentration [104–106], affinity of the competing ions [107–109], and water uptake of the membrane [106,110] are known to affect the ultimate ion concentration within the membrane $(C_{i,m})$. The ion activity coefficients $(\gamma_{+,m})$ in the membrane cannot be measured directly and in most studies, these are either assumed to be unity, or equal to that in the solution (i.e., $\gamma_+/\gamma_{+,m} = 1$) [111–113]. However, the activity coefficient of the counter-ions is indeed expected to be significantly lower than that of the free aqueous solution [114], due to the high concentration of the counter-ions $(C_{+,m})$ within this phase [115]. In contrast, the concentration of co-ions in the membrane $(C_{-,m})$ is much lower hence this activity coefficient is expected to be closer to unity. Recent advancements have been made by Kamcev et al. [116] to relate the ion activity coefficients to the concentration of fixed charges in the membrane $(C_{fix,m})$ and a dimensionless linear charge density (ζ) of the polymer chains, based on Manning's counter-ion condensation theory for polyelectrolyte solutions [15]. For a 1:1 electrolyte $\gamma_{+,m}\gamma_{-m}$ is:

$$\gamma_{+,m}\gamma_{-m} = [\frac{X/\zeta+1}{X+1}]\cdot\exp[-X/(X+2\zeta)] \tag{4}$$

where X is the ratio of fixed charge group to mobile ion concentration (i.e., $X = \frac{C_{fix,m}}{C_{-,m}}$).

The Meares model is widely used to describe the co-ion diffusion coefficient through the swollen polymer network $(D_{-,m})$ [114]:

$$\frac{D_{-,m}}{D_{-,s}} = \left(\frac{1-\Phi_p}{1+\Phi_p}\right)^2 \tag{5}$$

where $D_{-,s}$ is the diffusion coefficient of the co-ion in the solution and Φ_p is the polymer volume fraction in the membrane [117–119]. For counter-ions, the electrostatic effect from the fixed charge groups cannot be neglected. To account for this effect, Manning's model [120] can again be applied, by assuming that ion diffusion is affected by the local electric field and that condensed counter-ions have no mobility. This results in the extended Meares/Manning Model [121] for the counter-ion diffusion coefficient of 1:1 electrolytes $D_{+,m}$:

$$\frac{D_{+,m}}{D_{+,s}} = \left(\frac{\frac{X}{\zeta}+1}{X+1}\right)\left(1-\frac{1}{3}A(1;\frac{X}{\zeta})\right)(\frac{\Phi_w}{2-\Phi_w})^2 \tag{6}$$

where Φ_w is the water volume fraction in the membrane, the parameter A is dependent upon measurable parameters X and ζ.

While a range of research groups have established techniques to evaluate the concentration, activity, and diffusivity coefficients in IEMs, this knowledge has not been widely employed in MCDI models. The application of more accurate membrane related parameters will further improve the ion transport models evolving for MCDI.

6. MCDI Novel Stack Developments and Outlook

The search for novel MCDI configurations has led to the introduction of various technologies where researchers aimed to overcome the restrictions and drawbacks associated with conventional MCDI systems. In this section, recent developments in this area are described to frame future directions in this field.

6.1. Radial Deionisation (RDI)

Atlantis Technologies are currently commercialising an MCDI cell which has been modified to provide up to 100 electrode pairs in a cylindrical arrangement [122]. This approach has been tested on phosphate mining wastewater, shale gas produced water and landfill leachate and been shown to effectively remove trace metals such as selenium, mercury, arsenic, and uranium, in addition to more common salts [123].

6.2. Flow Chamber Modification

In an approach similar to electrodeionisation, Liang et al. [42], filled the flow channel in an MCDI cell with ion exchange resin to reduce the electrical resistance at low salinities. This approach gave a salt removal efficiency of 90%, higher than that of either MCDI (60%) or CDI (19%) at comparable conditions. Similarly, Bian et al. [124] filled the flow chamber with granular activated carbon (GAC) (as shown in Figure 8). They observed an enhanced desalination rate compared with a conventional MCDI cell, which was attributed to a lower electrical resistance within the flow channel, as well as the introduction of additional storage sites within the GAC.

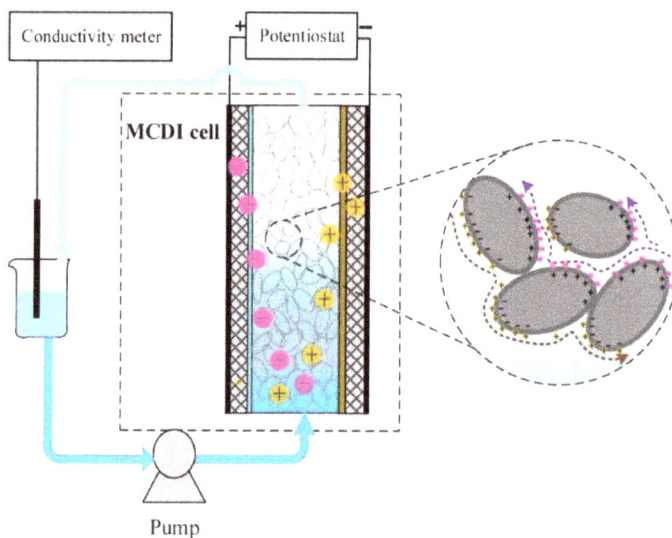

Figure 8. Schematic diagram of GAC (granular activated carbon)-MCDI as depicted by Bian et al. [124]. Reproduced with permission, copyright Elsvier, 2015.

The approach was most effective for NaCl solutions of <200 mg L^{-1}. A smaller GAC particle size (0.4–0.8 mm compared with 2–5 mm) showed a better desalination rate, which was attributed to a lower energy barrier for ion transport from the bulk to the GAC pores [124]. Using non-conductive glass beads was less effective, indicating that the GAC was not just acting to increase turbulence. Conversely, while porous graphite granules showed similar ohmic resistance to the activated carbon, these could not match the desalination rate of regular MCDI due to the higher ionic resistance.

6.3. Flow-Electrodes Capacitive Deionisation

In 2013, Jeon et al. [125] introduced a novel configuration called flow-electrode capacitive deionisation (FCDI) in which a flowing carbon suspension replaced the solid carbon electrodes. As depicted in Figure 9, in FCDI, the ions are adsorbed into the suspended carbon materials flowing between the current collectors and ion-exchange membranes.

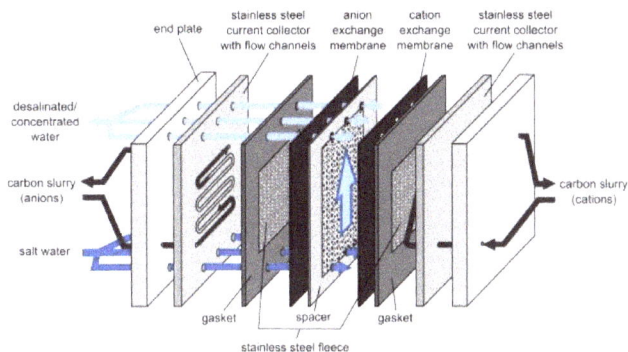

Figure 9. Schematic diagram of Flow-electrode capacitive deionisation (FCDI). Reproduced with permission from Gendel et al. [126], copyright Elsvier, 2014.

This approach allows capacitive deionisation to operate in a continuous mode, rather than in adsorption and desorption cycles [125,126]. Further, whereas in a conventional MCDI process, the ion sorption capacity is restricted to the size of the fixed carbon electrodes, the ion exchange capacity of an FCDI system can be enhanced by altering the flow path size and flow rate of carbon slurry [127,128].

To regenerate the carbon suspensions, some research groups simply mix the suspension collected at the anode with that of the cathode and then let the carbon materials settle [125,127]. Alternatively, Gendel et al. [126] coupled two FCDI cells with opposing electrical potentials in which desalination and regeneration occurred separately. Later, Rommerskirchen et al. [129] developed a smart design by combing the desalination and regeneration cells into a single electrosorption module. In this case, one type of ions transfers through the ion-exchange membrane placed in the middle of the cell from a diluate compartment to a concentrate compartment. The opposite ion is adsorbed into the carbon slurry flowing on the diluate side (see Figure 10). Then, by circulating the carbon suspension into the concentrate side, the adsorbed ion is desorbed back in to the brine stream [129].

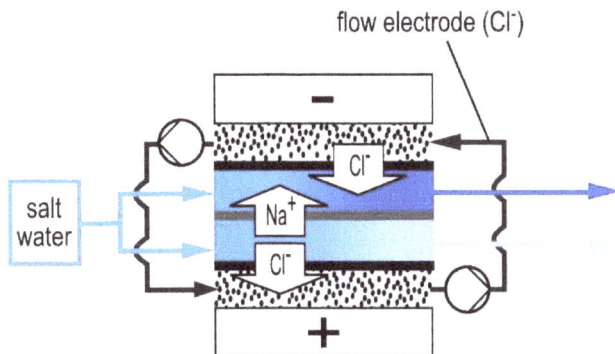

Figure 10. Single module flow-electrode capacitive deionisation concept as depicted by Rommerskirchen et al. [129]. Reproduced with permission, copyright Elsvier, 2015.

Gendel el al. [126] examined the effect of flow rate and water splitting ratio between the desalination and regeneration cells in FCDI, while Porada et al. [127] studied the effect of carbon content and water residence time. The latter reported that while salt removal rate increased with carbon content, beyond 20 wt % the flowing slurry was prone to clogging. Yang et al. [130] demonstrated that dispersing carbon particles in 2 wt % NaCl solution resulted in significantly improved desalination efficiency compared to the use of deionized water. Ma et al. [131] developed redox-active flow electrodes with the addition of aqueous hydroquinone and benzoquinone to increase the charge transfer efficiency. Further research and long-term performance data will help to determine the optimum composition of these suspensions, with high conductivity and low viscosity to achieve high charge transfer and avoid clogging.

The authors believe that the introduction of flow-electrodes is a major step forward towards the commercialization of MCDI since it enables continuous operation and provides higher water recovery. However, Hoyt et al. [132] are among the few who have modelled the flowing carbon slurry. We believe this area requires more attention to justify the significantly higher salt adsorption and to predict the performance of FCDI under different operational conditions.

6.4. Hybrid Capacitive Deionisation (CDI)

Lee et al. [133] replaced the cathode with a sodium manganese oxide electrode that benefited from higher specific capacity (Figure 11). In this approach, sodium ions are adsorbed into the sodium manganese electrode, while chloride ions adsorb in a standard format MCDI anode.

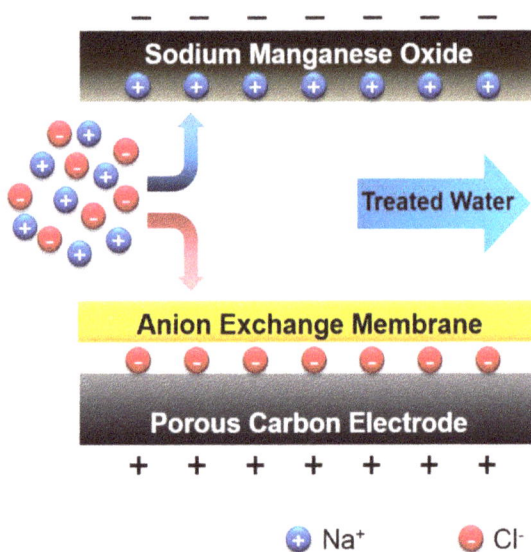

Figure 11. Schematic diagram of hybrid capacitive deionisation (HCDI). Reproduced with permission from Lee et al. [133], copyright Royal Society of Chemistry, 2014.

Using this approach, Lee et al. [133] could improve the ion removal capacity from 13.5 mg g^{-1} in CDI and 22.4 mg g^{-1} of MCDI to 31.2 mg g^{-1} (per total mass of sodium manganese oxide and carbon). While the adsorption rate was higher than that of CDI, the performance was slower compared with regular MCDI. Further, the cost of these electrodes is likely to be well above those of simple activated carbon. As noted by Suss et al. [134], the ultimate performance-normalized costs associated with such novel CDI configurations requires assessment.

7. Conclusions and Future Research Directions

This article has provided an overview of the current status of MCDI, with a focus on the role of the ion exchange membrane in this process. Recent developments in the area of composite electrodes, novel IEMs, performance metrics, fouling and cleaning, and innovative configurations of MCDI were discussed in details. MCDI is a promising technique for desalination at low feedwater salinities, where the energy consumption appears lower than in comparable membrane processes, at least at the lab scale. However, further techno-economic studies comparing this approach to other desalination technologies including electrodialysis, nanofiltration, and reverse osmosis are of great importance to encourage more pilot-plant development and larger scale investments in the area. Additionally, while many research groups are enhancing the performance of MCDI by the fabrication of novel IEMs or composite electrode-IEM electrodes, or by introducing novel MCDI configurations, more attention must be paid to operational issues. Specifically, the performance of these materials must be evaluated with a much wider range of monovalent and multivalent salts, not just NaCl. Their stability must also be assessed in fouling systems where regular cleaning will be necessary. Such fouling is an important issue that can jeopardize the efficiency of MCDI at a commercial scale and is thus worthy of further investigation. Finally, while ion transport through IEMs was only briefly described in this paper, this research field requires in-depth analysis and mathematical modelling to better distinguish the role of the membrane in MCDI applications.

Acknowledgments: Funding from the Dairy Innovation Research Hub, an Industrial Transformation Research Hub (IH120100005) of the Australian Research Council is acknowledged.

Conflicts of Interest: The authors declare no conflict of interest.

References

1. Xu, T. Ion exchange membranes: State of their development and perspective. *J. Membr. Sci.* **2005**, *263*, 1–29. [CrossRef]
2. Sata, T. *Ion Exchange Membranes: Preparation, Characterization, Modification and Application*; Royal Society of Chemistry: London, UK, 2004.
3. Ran, J.; Wu, L.; He, Y.; Yang, Z.; Wang, Y.; Jiang, C.; Ge, L.; Bakangura, E.; Xu, T. Ion exchange membranes: New developments and applications. *J. Membr. Sci.* **2017**, *522*, 267–291. [CrossRef]
4. Strathmann, H. *Ion-Exchange Membrane Separation Processes*; Elsevier: Amsterdam, The Netherlands, 2004; Volume 9.
5. Nagarale, R.K.; Gohil, G.S.; Shahi, V.K. Recent developments on ion-exchange membranes and electro-membrane processes. *Adv. Colloid Interface Sci.* **2006**, *119*, 97–130. [CrossRef] [PubMed]
6. Merle, G.; Wessling, M.; Nijmeijer, K. Anion exchange membranes for alkaline fuel cells: A review. *J. Membr. Sci.* **2011**, *377*, 1–35. [CrossRef]
7. Strathmann, H. Electrodialysis and its application in the chemical process industry. *Sep. Purif. Methods* **1985**, *14*, 41–66. [CrossRef]
8. Bazinet, L. Electrodialytic phenomena and their applications in the dairy industry: A review. *Crit. Rev. Food Sci. Nutr.* **2005**, *45*, 307–326. [CrossRef] [PubMed]
9. Kentish, S.E.; Kloester, E.; Stevens, G.W.; Scholes, C.A.; Dumée, L.F. Electrodialysis in aqueous-organic mixtures. *Sep. Purif. Rev.* **2015**, *44*, 269–282. [CrossRef]
10. Demirer, O.N.; Naylor, R.M.; Rios Perez, C.A.; Wilkes, E.; Hidrovo, C. Energetic performance optimization of a capacitive deionization system operating with transient cycles and brackish water. *Desalination* **2013**, *314*, 130–138. [CrossRef]
11. Li, Y.; Zhang, C.; Jiang, Y.; Wang, T.-J.; Wang, H. Effects of the hydration ratio on the electrosorption selectivity of ions during capacitive deionization. *Desalination* **2016**, *399*, 171–177. [CrossRef]
12. Porada, S.; Zhao, R.; van der Wal, A.; Presser, V.; Biesheuvel, P.M. Review on the science and technology of water desalination by capacitive deionisation. *Prog. Mater. Sci.* **2013**, *58*, 1388–1442. [CrossRef]

13. Han, L.; Karthikeyan, K.G.; Anderson, M.A.; Wouters, J.J.; Gregory, K.B. Mechanistic insights into the use of oxide nanoparticles coated asymmetric electrodes for capacitive deionisation. *Electrochim. Acta* **2013**, *90*, 573–581. [CrossRef]

14. Lee, J.-B.; Park, K.-K.; Eum, H.-M.; Lee, C.-W. Desalination of a thermal power plant wastewater by membrane capacitive deionization. *Desalination* **2006**, *196*, 125–134. [CrossRef]

15. Biesheuvel, P.M.; Zhao, R.; Porada, S.; van der Wal, A. Theory of membrane capacitive deionisation including the effect of the electrode pore space. *J. Colloid Interface Sci.* **2011**, *360*, 239–248. [CrossRef] [PubMed]

16. Zhao, R.; Porada, S.; Biesheuvel, P.M.; van der Wal, A. Energy consumption in membrane capacitive deionisation for different water recoveries and flow rates, and comparison with reverse osmosis. *Desalination* **2013**, *330*, 35–41. [CrossRef]

17. McRae, W.A. Electroseparations, Electrodialysis. In *Kirk-Othmer Encyclopedia of Chemical Technology*; John Wiley & Sons, Inc.: Hoboken, NJ, USA, 2000.

18. He, S.; Liu, L.; Wang, X.; Zhang, S.; Guiver, M.D.; Li, N. Azide-assisted self-crosslinking of highly ion conductive anion exchange membranes. *J. Membr. Sci.* **2016**, *509*, 48–56. [CrossRef]

19. Kariduraganavar, M.Y.; Nagarale, R.K.; Kittur, A.A.; Kulkarni, S.S. Ion-exchange membranes: Preparative methods for electrodialysis and fuel cell applications. *Desalination* **2006**, *197*, 225–246. [CrossRef]

20. Kwak, N.-S.; Koo, J.S.; Hwang, T.S.; Choi, E.M. Synthesis and electrical properties of NaSS–MAA–MMA cation exchange membranes for membrane capacitive deionisation (MCDI). *Desalination* **2012**, *285*, 138–146. [CrossRef]

21. Kang, K.W.; Hwang, C.W.; Hwang, T.S. Synthesis and properties of sodium vinylbenzene sulfonate-grafted poly(vinylidene fluoride) cation exchange membranes for membrane capacitive deionisation process. *Macromol. Res.* **2015**, *23*, 1126–1133. [CrossRef]

22. Jeong, K.S.; Hwang, W.C.; Hwang, T.S. Synthesis of an aminated poly(vinylidene fluride-g-4-vinyl benzyl chloride) anion exchange membrane for membrane capacitive deionisation(MCDI). *J. Membr. Sci.* **2015**, *495*, 316–321. [CrossRef]

23. Qiu, Q.; Cha, J.-H.; Choi, Y.-W.; Choi, J.-H.; Shin, J.; Lee, Y.-S. Preparation of stable polyethylene membranes filled with crosslinked sulfonated polystyrene for membrane capacitive deionisation by γ-irradiation. *Macromol. Res.* **2017**, *25*, 92–95. [CrossRef]

24. Lee, J.-B.; Park, K.-K.; Yoon, S.-W.; Park, P.-Y.; Park, K.-I.; Lee, C.-W. Desalination performance of a carbon-based composite electrode. *Desalination* **2009**, *237*, 155–161. [CrossRef]

25. Hou, C.-H.; Liu, N.-L.; Hsu, H.-L.; Den, W. Development of multi-walled carbon nanotube/poly(vinyl alcohol) composite as electrode for capacitive deionisation. *Sep. Purif. Technol.* **2014**, *130*, 7–14. [CrossRef]

26. Liu, Y.; Pan, L.; Xu, X.; Lu, T.; Sun, Z.; Chua, D.H. Enhanced desalination efficiency in modified membrane capacitive deionisation by introducing ion-exchange polymers in carbon nanotubes electrodes. *Electrochim. Acta* **2014**, *130*, 619–624. [CrossRef]

27. Kim, J.-S.; Choi, J.-H. Fabrication and characterization of a carbon electrode coated with cation-exchange polymer for the membrane capacitive deionisation applications. *J. Membr. Sci.* **2010**, *355*, 85–90. [CrossRef]

28. Qian, B.; Wang, G.; Ling, Z.; Dong, Q.; Wu, T.; Zhang, X.; Qiu, J. Sulfonated Graphene as Cation-Selective Coating: A New Strategy for High-Performance Membrane Capacitive Deionization. *Adv. Mater. Interfaces* **2015**, *2*. [CrossRef]

29. Kim, Y.-J.; Choi, J.-H. Improvement of desalination efficiency in capacitive deionization using a carbon electrode coated with an ion-exchange polymer. *Water Res.* **2010**, *44*, 990–996. [CrossRef] [PubMed]

30. Kim, J.; Kim, C.; Shin, H.; Rhim, J. Application of synthesized anion and cation exchange polymers to membrane capacitive deionization (MCDI). *Macromol. Res.* **2015**, *23*, 360–366. [CrossRef]

31. Lee, J.-Y.; Seo, S.-J.; Yun, S.-H.; Moon, S.-H. Preparation of ion exchanger layered electrodes for advanced membrane capacitive deionization (MCDI). *Water Res.* **2011**, *45*, 5375–5380. [CrossRef] [PubMed]

32. Liu, P.; Wang, H.; Yan, T.; Zhang, J.; Shi, L.; Zhang, D. Grafting sulfonic and amine functional groups on 3D graphene for improved capacitive deionization. *J. Mater. Chem. A* **2016**, *4*, 5303–5313. [CrossRef]

33. Kim, J.S.; Jeon, Y.S.; Rhim, J.W. Application of poly(vinyl alcohol) and polysulfone based ionic exchange polymers to membrane capacitive deionization for the removal of mono- and divalent salts. *Sep. Purif. Technol.* **2016**, *157*, 45–52. [CrossRef]

34. Jeong, J.S.; Kim, H.S.; Cho, M.D.; Kang, H.R. Regeneration Methods of Capacitive Deionization Electrodes in Water Purification. U.S. Patent 20,160,289,097 A1, 6 October 2016.

35. Kim, Y.-J.; Choi, J.-H. Selective removal of nitrate ion using a novel composite carbon electrode in capacitive deionization. *Water Res.* **2012**, *46*, 6033–6039. [CrossRef] [PubMed]

36. Yeo, J.-H.; Choi, J.-H. Enhancement of nitrate removal from a solution of mixed nitrate, chloride and sulfate ions using a nitrate-selective carbon electrode. *Desalination* **2013**, *320*, 10–16. [CrossRef]

37. Tian, G.; Liu, L.; Meng, Q.; Cao, B. Preparation and characterization of cross-linked quaternised polyvinyl alcohol membrane/activated carbon composite electrode for membrane capacitive deionisation. *Desalination* **2014**, *354*, 107–115. [CrossRef]

38. Gu, X.; Deng, Y.; Wang, C. Fabrication of anion-exchange polymer layered graphene-melamine electrodes for membrane capacitive deionisation. *ACS Sustain. Chem. Eng.* **2017**, *5*, 325–333. [CrossRef]

39. Li, H.; Gao, Y.; Pan, L.; Zhang, Y.; Chen, Y.; Sun, Z. Electrosorptive desalination by carbon nanotubes and nanofibres electrodes and ion-exchange membranes. *Water Res.* **2008**, *42*, 4923–4928. [CrossRef] [PubMed]

40. Kim, Y.-J.; Choi, J.-H. Enhanced desalination efficiency in capacitive deionisation with an ion-selective membrane. *Sep. Purif. Technol.* **2010**, *71*, 70–75. [CrossRef]

41. Li, H.; Zou, L. Ion-exchange membrane capacitive deionisation: A new strategy for brackish water desalination. *Desalination* **2011**, *275*, 62–66. [CrossRef]

42. Liang, P.; Yuan, L.; Yang, X.; Zhou, S.; Huang, X. Coupling ion-exchangers with inexpensive activated carbon fiber electrodes to enhance the performance of capacitive deionisation cells for domestic wastewater desalination. *Water Res.* **2013**, *47*, 2523–2530. [CrossRef] [PubMed]

43. Omosebi, A.; Gao, X.; Landon, J.; Liu, K. Asymmetric electrode configuration for enhanced membrane capacitive deionisation. *ACS Appl. Mater. Interfaces* **2014**, *6*, 12640–12649. [CrossRef] [PubMed]

44. Zhao, Y.; Wang, Y.; Wang, R.; Wu, Y.; Xu, S.; Wang, J. Performance comparison and energy consumption analysis of capacitive deionization and membrane capacitive deionization processes. *Desalination* **2013**, *324*, 127–133. [CrossRef]

45. Ding, M.; Shi, W.; Guo, L.; Leong, Z.Y.; Baji, A.; Yang, H.Y. Bimetallic metal–organic framework derived porous carbon nanostructures for high performance membrane capacitive desalination. *J. Mat. Chem. A* **2017**, *5*, 6113–6121. [CrossRef]

46. Duan, F.; Li, Y.; Cao, H.; Wang, Y.; Zhang, Y.; Crittenden, J.C. Activated carbon electrodes: Electrochemical oxidation coupled with desalination for wastewater treatment. *Chemosphere* **2015**, *125*, 205–211. [CrossRef] [PubMed]

47. Zhao, R.; Biesheuvel, P.M.; Miedema, H.; Bruning, H.; van der Wal, A. Charge efficiency: A functional tool to probe the double-layer structure inside of porous electrodes and application in the modeling of capacitive deionisation. *J. Phys. Chem. Lett.* **2010**, *1*, 205–210. [CrossRef]

48. Porada, S.; Bryjak, M.; van der Wal, A.; Biesheuvel, P.M. Effect of electrode thickness variation on operation of capacitive deionisation. *Electrochim. Acta* **2012**, *75*, 148–156. [CrossRef]

49. Huyskens, C.; Helsen, J.; de Haan, A.B. Capacitive deionisation for water treatment: Screening of key performance parameters and comparison of performance for different ions. *Desalination* **2013**, *328*, 8–16. [CrossRef]

50. Choi, J.-H. Comparison of constant voltage (CV) and constant current (CC) operation in the membrane capacitive deionisation process. *Desalination Water Treat.* **2014**, *56*, 921–928. [CrossRef]

51. Qu, Y.; Campbell, P.G.; Gu, L.; Knipe, J.M.; Dzenitis, E.; Santiago, J.G.; Stadermann, M. Energy consumption analysis of constant voltage and constant current operations in capacitive deionisation. *Desalination* **2016**, *400*, 18–24. [CrossRef]

52. Kim, Y.-J.; Hur, J.; Bae, W.; Choi, J.-H. Desalination of brackish water containing oil compound by capacitive deionisation process. *Desalination* **2010**, *253*, 119–123. [CrossRef]

53. Li, H.; Nie, C.; Pan, L.; Sun, Z. The study of membrane capacitive deionisation from charge efficiency. *Desalin. Water Treat.* **2012**, *42*, 210–215. [CrossRef]

54. Wimalasiri, Y.; Mossad, M.; Zou, L. Thermodynamics and kinetics of adsorption of ammonium ions by graphene laminate electrodes in capacitive deionisation. *Desalination* **2015**, *357*, 178–188. [CrossRef]

55. Mossad, M.; Zou, L. A study of the capacitive deionisation performance under various operational conditions. *J. Hazard. Mater.* **2012**, *213–214*, 491–497. [CrossRef] [PubMed]

56. Dlugolecki, P.; van der Wal, A. Energy recovery in membrane capacitive deionisation. *Environ. Sci. Technol.* **2013**, *47*, 4904–4910. [CrossRef] [PubMed]

57. Kim, Y.-J.; Kim, J.-H.; Choi, J.-H. Selective removal of nitrate ions by controlling the applied current in membrane capacitive deionisation (MCDI). *J. Membr. Sci.* **2013**, *429*, 52–57. [CrossRef]

58. Ryu, T.; Lee, D.-H.; Ryu, J.C.; Shin, J.; Chung, K.-S.; Kim, Y.H. Lithium recovery system using electrostatic field assistance. *Hydrometallurgy* **2015**, *151*, 78–83. [CrossRef]

59. Yoon, H.; Lee, J.; Kim, S.-R.; Kang, J.; Kim, S.; Kim, C.; Yoon, J. Capacitive deionisation with Ca-alginate coated-carbon electrode for hardness control. *Desalination* **2016**, *392*, 46–53. [CrossRef]

60. Choi, J.; Lee, H.; Hong, S. Capacitive deionisation (CDI) integrated with monovalent cation selective membrane for producing divalent cation-rich solution. *Desalination* **2016**, *400*, 38–46. [CrossRef]

61. Rice, G.; Barber, A.R.; O'Connor, A.J.; Stevens, G.W.; Kentish, S.E. Rejection of dairy salts by a nanofiltration membrane. *Sep. Purif. Technol.* **2011**, *79*, 92–102. [CrossRef]

62. Garcia-Aleman, J.; Dickson, J.M. Permeation of mixed-salt solutions with commercial and pore-filled nanofiltration membranes: Membrane charge inversion phenomena. *J. Membr. Sci.* **2004**, *239*, 163–172. [CrossRef]

63. Labbez, C.; Fievet, P.; Szymczyk, A.; Vidonne, A.; Foissy, A.; Pagetti, J. Retention of mineral salts by a polyamide nanofiltration membrane. *Sep. Purif. Technol.* **2003**, *30*, 47–55. [CrossRef]

64. Schaep, J.; Vandecasteele, C.; Mohammad, A.W.; Bowen, W.R. Analysis of the salt retention of nanofiltration membranes using the Donnan-steric partitioning pore model. *Sep. Sci. Technol.* **1999**, *34*, 3009–3030. [CrossRef]

65. Van Limpt, B.; van der Wal, A. Water and chemical savings in cooling towers by using membrane capacitive deionisation. *Desalination* **2014**, *342*, 148–155. [CrossRef]

66. Mikhaylin, S.; Bazinet, L. Fouling on ion-exchange membranes: Classification, characterization and strategies of prevention and control. *Adv. Colloid Interface Sci.* **2016**, *229*, 34–56. [CrossRef] [PubMed]

67. Ong, C.S.; Goh, P.S.; Lau, W.J.; Misdan, N.; Ismail, A.F. Nanomaterials for biofouling and scaling mitigation of thin film composite membrane: A review. *Desalination* **2016**, *393*, 2–15. [CrossRef]

68. Piyadasa, C.; Ridgway, H.F.; Yeager, T.R.; Stewart, M.B.; Pelekani, C.; Gray, S.R.; Orbell, J.D. The application of electromagnetic fields to the control of the scaling and biofouling of reverse osmosis membranes—A review. *Desalination* **2017**, *418*, 19–34. [CrossRef]

69. Mossad, M.; Zou, L. Study of fouling and scaling in capacitive deionisation by using dissolved organic and inorganic salts. *J. Hazard. Mater.* **2013**, *244–245*, 387–393. [CrossRef] [PubMed]

70. Zhang, W.; Mossad, M.; Zou, L. A study of the long-term operation of capacitive deionisation in inland brackish water desalination. *Desalination* **2013**, *320*, 80–85. [CrossRef]

71. Wang, C.; Song, H.; Zhang, Q.; Wang, B.; Li, A. Parameter optimization based on capacitive deionisation for highly efficient desalination of domestic wastewater biotreated effluent and the fouled electrode regeneration. *Desalination* **2015**, *365*, 407–415. [CrossRef]

72. Korngold, E.; De Korosy, F.; Rahav, R.; Taboch, M.F. Fouling of anion-selective membranes in electrodialysis. *Desalination* **1970**, *8*, 195–220. [CrossRef]

73. Lee, H.-J.; Moon, S.-H.; Tsai, S.-P. Effects of pulsed electric fields on membrane fouling in electrodialysis of NaCl solution containing humate. *Sep. Purif. Technol.* **2002**, *27*, 89–95. [CrossRef]

74. Langevin, M.-E.; Bazinet, L. Ion-exchange membrane fouling by peptides: A phenomenon governed by electrostatic interactions. *J. Membr. Sci.* **2011**, *369*, 359–366. [CrossRef]

75. Fidaleo, M.; Moresi, M. Electrodialysis applications in the food industry. *Adv. Food Nutr. Res.* **2006**, *51*, 265–360. [PubMed]

76. Lee, H.-J.; Hong, M.-K.; Han, S.-D.; Cho, S.-H.; Moon, S.-H. Fouling of an anion exchange membrane in the electrodialysis desalination process in the presence of organic foulants. *Desalination* **2009**, *238*, 60–69. [CrossRef]

77. Lindstrand, V.; Sundström, G.; Jönsson, A.-S. Fouling of electrodialysis membranes by organic substances. *Desalination* **2000**, *128*, 91–102. [CrossRef]

78. Chen, G.Q.; Eschbach, F.I.I.; Weeks, M.; Gras, S.L.; Kentish, S.E. Removal of lactic acid from acid whey using electrodialysis. *Sep. Purif. Technol.* **2016**, *158*, 230–237. [CrossRef]

79. Diblíková, L.; Čurda, L.; Kinčl, J. The effect of dry matter and salt addition on cheese whey demineralisation. *Int. Dairy J.* **2013**, *31*, 29–33. [CrossRef]

80. Sienkiewicz, T.; Riedel, C.-L. *Whey and Whey Utilization*; VEB Fachbuchverlag: Leipzig, Germany, 1986.

81. Ayala-Bribiesca, E.; Pourcelly, G.; Bazinet, L. Nature identification and morphology characterization of cation-exchange membrane fouling during conventional electrodialysis. *J. Colloid Interface Sci.* **2006**, *300*, 663–672. [CrossRef] [PubMed]

82. Ayala-Bribiesca, E.; Pourcelly, G.; Bazinet, L. Nature identification and morphology characterization of anion-exchange membrane fouling during conventional electrodialysis. *J. Colloid Interface Sci.* **2007**, *308*, 182–190. [CrossRef] [PubMed]

83. Bazinet, L.; Montpetit, D.; Ippersiel, D.; Mahdavi, B.; Amiot, J.; Lamarche, F. Neutralization of hydroxide generated during skim milk electroacidification and its effect on bipolar and cationic membrane integrity. *J. Membr. Sci.* **2003**, *216*, 229–239. [CrossRef]

84. Ayala-Bribiesca, E.; Araya-Farias, M.; Pourcelly, G.; Bazinet, L. Effect of concentrate solution pH and mineral composition of a whey protein diluate solution on membrane fouling formation during conventional electrodialysis. *J. Membr. Sci.* **2006**, *280*, 790–801. [CrossRef]

85. Bleha, M.; Tishchenko, G.; Šumberová, V.; Kůdela, V. Characteristic of the critical state of membranes in ED-desalination of milk whey. *Desalination* **1992**, *86*, 173–186. [CrossRef]

86. Bazinet, L.; Araya-Farias, M. Effect of calcium and carbonate concentrations on cationic membrane fouling during electrodialysis. *J. Colloid Interface Sci.* **2005**, *281*, 188–196. [CrossRef] [PubMed]

87. Casademont, C.; Farias, M.A.; Pourcelly, G.; Bazinet, L. Impact of electrodialytic parameters on cation migration kinetics and fouling nature of ion-exchange membranes during treatment of solutions with different magnesium/calcium ratios. *J. Membr. Sci.* **2008**, *325*, 570–579. [CrossRef]

88. Trägårdh, G. Membrane cleaning. *Desalination* **1989**, *71*, 325–335. [CrossRef]

89. Katz, W.E. The electrodialysis reversal (EDR) process. *Desalination* **1979**, *28*, 31–40. [CrossRef]

90. Vermaas, D.A.; Kunteng, D.; Veerman, J.; Saakes, M.; Nijmeijer, K. Periodic feedwater reversal and air sparging as antifouling strategies in reverse electrodialysis. *Environ. Sci. Technol.* **2014**, *48*, 3065–3073. [CrossRef] [PubMed]

91. Shee, F.L.T.; Angers, P.; Bazinet, L. Microscopic approach for the identification of cationic membrane fouling during cheddar cheese whey electroacidification. *J. Colloid Interface Sci.* **2008**, *322*, 551–557. [CrossRef] [PubMed]

92. Sata, T.; Tsujimoto, M.; Yamaguchi, T.; Matsusaki, K. Change of anion exchange membranes in an aqueous sodium hydroxide solution at high temperature. *J. Membr. Sci.* **1996**, *112*, 161–170. [CrossRef]

93. Komkova, E.N.; Stamatialis, D.F.; Strathmann, H.; Wessling, M. Anion-exchange membranes containing diamines: Preparation and stability in alkaline solution. *J. Membr. Sci.* **2004**, *244*, 25–34. [CrossRef]

94. Vega, J.A.; Chartier, C.; Mustain, W.E. Effect of hydroxide and carbonate alkaline media on anion exchange membranes. *J. Power Sources* **2010**, *195*, 7176–7180. [CrossRef]

95. Garcia-Vasquez, W.; Ghalloussi, R.; Dammak, L.; Larchet, C.; Nikonenko, V.; Grande, D. Structure and properties of heterogeneous and homogeneous ion-exchange membranes subjected to ageing in sodium hypochlorite. *J. Membr. Sci.* **2014**, *452*, 104–116. [CrossRef]

96. Ghalloussi, R.; Garcia-Vasquez, W.; Chaabane, L.; Dammak, L.; Larchet, C.; Deabate, S.V.; Nevakshenova, E.; Nikonenko, V.; Grande, D. Ageing of ion-exchange membranes in electrodialysis: A structural and physicochemical investigation. *J. Membr. Sci.* **2013**, *436*, 68–78. [CrossRef]

97. Wang, Q.; Yang, P.; Cong, W. Cation-exchange membrane fouling and cleaning in bipolar membrane electrodialysis of industrial glutamate production wastewater. *Sep. Purif. Technol.* **2011**, *79*, 103–113. [CrossRef]

98. Haddad, M.; Mikhaylin, S.; Bazinet, L.; Savadogo, O.; Paris, J. Electrochemical acidification of kraft black liquor: Effect of fouling and chemical cleaning on ion exchange membrane integrity. *ACS Sustain. Chem. Eng.* **2016**, *5*, 168–178. [CrossRef]

99. Association, A.W.W. *Electrodialysis and Electrodialysis Reversal: M38*; American Water Works Association: Washington, DC, USA, 1995; Volume 38.

100. Guo, H.; You, F.; Yu, S.; Li, L.; Zhao, D. Mechanisms of chemical cleaning of ion exchange membranes: A case study of plant-scale electrodialysis for oily wastewater treatment. *J. Membr. Sci.* **2015**, *496*, 310–317. [CrossRef]

101. Garcia-Vasquez, W.; Dammak, L.; Larchet, C.; Nikonenko, V.; Grande, D. Effects of acid-base cleaning procedure on structure and properties of anion-exchange membranes used in electrodialysis. *J. Membr. Sci.* **2016**, *507*, 12–23. [CrossRef]

102. Biesheuvel, P.M.; Porada, S.; Levi, M.; Bazant, M.Z. Attractive forces in microporous carbon electrodes for capacitive deionisation. *J. Solid State Electrochem.* **2014**, *18*, 1365–1376. [CrossRef]
103. Hassanvand, A.; Chen, G.Q.; Webley, P.A.; Kentish, S.E. Improvement of MCDI operation and design through experiment and modelling: Regeneration with brine and optimum residence time. *Desalination* **2017**, *417*, 36–51. [CrossRef]
104. Xie, W.; Cook, J.; Park, H.B.; Freeman, B.D.; Lee, C.H.; McGrath, J.E. Fundamental salt and water transport properties in directly copolymerized disulfonated poly(arylene ether sulfone) random copolymers. *Polymer* **2011**, *52*, 2032–2043. [CrossRef]
105. Geise, G.M.; Freeman, B.D.; Paul, D.R. Characterization of a sulfonated pentablock copolymer for desalination applications. *Polymer* **2010**, *51*, 5815–5822. [CrossRef]
106. Geise, G.M.; Paul, D.R.; Freeman, B.D. Fundamental water and salt transport properties of polymeric materials. *Prog. Polym. Sci.* **2014**, *39*, 1–42. [CrossRef]
107. Sata, T.; Sata, T.; Yang, W. Studies on cation-exchange membranes having permselectivity between cations in electrodialysis. *J. Membr. Sci.* **2002**, *206*, 31–60. [CrossRef]
108. Van der Bruggen, B.; Koninckx, A.; Vandecasteele, C. Separation of monovalent and divalent ions from aqueous solution by electrodialysis and nanofiltration. *Water Res.* **2004**, *38*, 1347–1353. [CrossRef] [PubMed]
109. Galama, A.H.; Daubaras, G.; Burheim, O.S.; Rijnaarts, H.H.M.; Post, J.W. Seawater electrodialysis with preferential removal of divalent ions. *J. Membr. Sci.* **2014**, *452*, 219–228. [CrossRef]
110. Ju, H.; Sagle, A.C.; Freeman, B.D.; Mardel, J.I.; Hill, A.J. Characterization of sodium chloride and water transport in crosslinked poly(ethylene oxide) hydrogels. *J. Membr. Sci.* **2010**, *358*, 131–141. [CrossRef]
111. Crank, J.; Park, G.S. *Diffusion in Polymers*; Academic Press: Cambridge, MA, USA, 1968.
112. Pintauro, P.N.; Bennion, D.N. Mass transport of electrolytes in membranes. 2. Determination of sodium chloride equilibrium and transport parameters for Nafion. *Ind. Eng. Chem. Fundam.* **1984**, *23*, 234–243. [CrossRef]
113. Helfferich, F.G. *Ion Exchange*; Dover Publications: New York, NY, USA, 1995.
114. Mackie, J.; Meares, P. The diffusion of electrolytes in a cation-exchange resin membrane I. Theoretical. In *Proceedings of the Royal Society of London A: Mathematical, Physical and Engineering Sciences*; The Royal Society: London, UK, 1955.
115. Kamcev, J.; Paul, D.R.; Freeman, B.D. Ion activity coefficients in ion exchange polymers: Applicability of Manning's counterion condensation theory. *Macromolecules* **2015**, *48*, 8011–8024. [CrossRef]
116. Kamcev, J.; Galizia, M.; Benedetti, F.M.; Jang, E.-S.; Paul, D.R.; Freeman, B.; Manning, G.S. Partitioning of Mobile Ions Between Ion Exchange Polymers and Aqueous Salt Solutions: Importance of Counter-ion Condensation. *Phys. Chem. Chem. Phys.* **2016**, *18*, 6021–6031. [CrossRef] [PubMed]
117. Mackay, D.; Meares, P. The electrical conductivity and electro-osmotic permeability of a cation-exchange resin. *Trans. Faraday Soc.* **1959**, *55*, 1221–1238. [CrossRef]
118. Kamo, N.; Toyoshima, Y.; Kobatake, Y. Fixed charge density effective to membrane phenomena. *Colloid Polym. Sci.* **1971**, *249*, 1069–1071.
119. Ueda, T.; Kamo, N.; Ishida, N.; Kobatake, Y. Effective fixed charge density governing membrane phenomena. IV. Further study of activity coefficients and mobilities of small ions in charged membranes. *J. Phys. Chem.* **1972**, *76*, 2447–2452. [CrossRef]
120. Manning, G.S. Limiting laws and counterion condensation in polyelectrolyte solutions II. Self-Diffusion of the small ions. *J. Phys. Chem.* **1969**, *51*, 934–938. [CrossRef]
121. Kamcev, J.; Paul, D.R.; Manning, G.S.; Freeman, B.D. Predicting salt permeability coefficients in highly swollen, highly charged ion exchange membranes. *ACS Appl. Mater. Interfaces* **2017**, *9*, 4044–4056. [CrossRef] [PubMed]
122. Curran, P.M. Concentric Layer Electric Double Layer Capacitor Cylinder, System, and Method of Use. U.S. Patent 9,193,612, 24 November 2015.
123. RDI™ Desalination System. 2017. Available online: http://www.atlantis-water.com/rdi-desalination-system-2/ (accessed on 7 July 2017).
124. Bian, Y.; Yang, X.; Liang, P.; Jiang, Y.; Zhang, C.; Huang, X. Enhanced desalination performance of membrane capacitive deionisation cells by packing the flow chamber with granular activated carbon. *Water Res.* **2015**, *85*, 371–376. [CrossRef] [PubMed]

125. Jeon, S.-I.; Park, H.-R.; Yeo, J.-G.; Yang, S.C.; Cho, C.H.; Han, M.H.; Kim, D.K. Desalination via a new membrane capacitive deionisation process utilizing flow-electrodes. *Energy Environ. Sci.* **2013**, *6*, 1471–1475. [CrossRef]

126. Gendel, Y.; Rommerskirchen, A.K.E.; David, O.; Wessling, M. Batch mode and continuous desalination of water using flowing carbon deionisation (FCDI) technology. *Electrochem. Commun.* **2014**, *46*, 152–156. [CrossRef]

127. Porada, S.; Weingarth, D.; Hamelers, H.V.; Bryjak, M.; Presser, V.; Biesheuvel, P. Carbon flow electrodes for continuous operation of capacitive deionisation and capacitive mixing energy generation. *J. Mater. Chem. A* **2014**, *2*, 9313–9321. [CrossRef]

128. Yang, S.; Jeon, S.-I.; Kim, H.; Choi, J.; Yeo, J.-G.; Park, H.-R.; Kim, D.K. Stack design and operation for scaling up the capacity of flow-electrode capacitive deionisation technology. *ACS Sustain. Chem. Eng.* **2016**, *4*, 4174–4180. [CrossRef]

129. Rommerskirchen, A.; Gendel, Y.; Wessling, M. Single module flow-electrode capacitive deionisation for continuous water desalination. *Electrochem. Commun.* **2015**, *60*, 34–37. [CrossRef]

130. Yang, S.-C.; Choi, J.; Yeo, J.-G.; Jeon, S.-I.; Park, H.-R.; Kim, D.K. Flow-electrode capacitive deionisation using an aqueous electrolyte with a high salt concentration. *Environ. Sci. Technol.* **2016**, *50*, 5892–5899. [CrossRef] [PubMed]

131. Ma, J.; He, D.; Tang, W.; Kovalsky, P.; He, C.; Zhang, C.; Waite, T.D. Development of redox-active flow electrodes for high-performance capacitive deionisation. *Environ. Sci. Technol.* **2016**, *50*, 13495–13501. [CrossRef] [PubMed]

132. Hoyt, N.C.; Savinell, R.F.; Wainright, J.S. Modeling of flowable slurry electrodes with combined Faradaic and nonfaradaic currents. *Chem. Eng. Sci.* **2016**, *144*, 288–297. [CrossRef]

133. Lee, J.; Kim, S.; Kim, C.; Yoon, J. Hybrid capacitive deionisation to enhance the desalination performance of capacitive techniques. *Energy Environ. Sci.* **2014**, *7*, 3683–3689. [CrossRef]

134. Suss, M.; Porada, S.; Sun, X.; Biesheuvel, P.; Yoon, J.; Presser, V. Water desalination via capacitive deionisation: What is it and what can we expect from it? *Energy Environ. Sci.* **2015**, *8*, 2296–2319. [CrossRef]

![membranes logo] *membranes*

MDPI

Review

Electro-Conductive Membranes for Permeation Enhancement and Fouling Mitigation: A Short Review

Patrizia Formoso [1], Elvira Pantuso [1], Giovanni De Filpo [2,*] and Fiore Pasquale Nicoletta [1]

[1] Department of Pharmacy, Health and Nutritional Sciences, University of Calabria, I-87036 Rende (CS), Italy; patrizia.formoso@unical.it (P.F.); elvirapnt.ep@gmail.com (E.P.); fiore.nicoletta@unical.it (F.P.N.)
[2] Department of Chemistry and Chemical Technologies, University of Calabria, I-87036 Rende (CS), Italy
* Correspondence: giovanni.defilpo@unical.it; Tel.: +39-0984-492-095

Received: 22 June 2017; Accepted: 20 July 2017; Published: 28 July 2017

Abstract: The research on electro-conductive membranes has expanded in recent years. These membranes have strong prospective as key components in next generation water treatment plants because they are engineered in order to enhance their performance in terms of separation, flux, fouling potential, and permselectivity. The present review summarizes recent developments in the preparation of electro-conductive membranes and the mechanisms of their response to external electric voltages in order to obtain an improvement in permeation and mitigation in the fouling growth. In particular, this paper deals with the properties of electro-conductive polymers and the preparation of electro-conductive polymer membranes with a focus on responsive membranes based on polyaniline, polypyrrole and carbon nanotubes. Then, some examples of electro-conductive membranes for permeation enhancement and fouling mitigation by electrostatic repulsion, hydrogen peroxide generation and electrochemical oxidation will be presented.

Keywords: membrane fouling; membrane cleaning; stimuli responsive polymer membranes; electro-responsive membranes; fouling mitigation; permeation enhancement

---

1. Introduction

The most important property for assessing the quality of a separation process through a membrane is its selectivity for a compound on another compound, also known as permselectivity.

Higher permeability values require lower membrane areas to separate a given compound, and a very good selectivity leads to higher purity products and, accordingly, to optimized values of rejection [1,2]. Nevertheless, the onset of fouling, due to the deposition/adsorption of particulate and soluble materials on membrane surfaces with time, causes a decrease in the permeability and selectivity with detrimental effects on membrane processes.

The present review summarizes recent developments in the preparation of electro-conductive membranes and the mechanisms of their response to external electric voltages in order to obtain an improvement in permeation and mitigation in the fouling growth. In particular, this paper deals with the properties of electro-conductive polymers and the preparation of electro-conductive polymer membranes with a focus to responsive membranes based on polyaniline, polypyrrole and carbon nanotubes, which represent the most used electro-conductive polymers. Then, some examples of electro-conductive membranes for permeation enhancement and mitigation of membrane fouling by electrostatic repulsion, hydrogen peroxide generation and electrochemical oxidation will be presented.

2. Electro-Conductive Membranes

Membranes are selective barriers able to separate components with different sizes or physical/chemical properties. The efficiency of a membrane separation process depends on the selectivity and permeability of used membranes. Selectivity, i.e., the ability to separate solutes, contaminants and particles with different sizes or physical/chemical properties is determined by the rejection of the unwanted compound and the permeation of desired compound. Generally, the membrane selectivity depends on the affinity between the substances and membrane porous surface, effective pore size and distribution. The permeability of a membrane is typically quantified by the trans-membrane flux and influenced by the pore size and surface properties of membranes [3]. Therefore, the performance of porous membranes can be weakened upon adsorption and deposition of foulants, present in the feed mixtures, on the porous surface.

Thus, for various size-based membrane separation applications, tunable/switchable pore sizes are required to achieve adjustable selectivity and permeability in response to external stimuli (single or multiple) or to environmental changes in feed conditions.

All responsive membranes have channels able to self-regulate their permselectivity in response to environmental stimuli, such as temperature, pH, specific molecules/ions, light, electric/magnetic fields, ionic strength, and redox reactions [4–18]. A common method to add responsiveness to a membrane is the use of stimuli-responsive polymers, copolymers and mixtures of polymers and additives during or after the membrane formation [19].

Such responsive membranes can act as smart valves, allowing an on demand flux control by dynamic modification of their structure and transport properties (e.g., permselectivity and hydrophilicity). In such a way, it is possible to enable a fouling mitigation and tunable self-cleaning membrane surfaces without the use of physical/chemical cleaning methods required for membranes under normal operating conditions.

Electro-Responsive Polymer Membranes (ERPMs), i.e., polymer membranes able to respond to an electric potential, can be obtained by membrane functionalization with custom-designed electrically conductive polymers.

According such a rationale several biosensors, electronic devices, and biomimetic devices have been prepared from electrically conductive polymer membranes [20–23]. The specific reactivity/polarity/conformation of used conducting polymers, virgin or properly functionalized for a better integration in the pore structure, enables an electro-responsiveness in filtration membranes usually adopted for water treatment [24,25].

Porous membranes used for microfiltration (MF) and ultrafiltration (UF) processes are characterized by a pore size ranging from 0.1 μm to 10 μm and 2 nm to 100 nm, respectively. The coating with a thin and selective polymer layer enables their use in nanofiltration (NF), reverse osmosis (RO), desalinization [26], and crystallization [27] processes. Some of such membranes can gain a responsive behavior by grafting electro-responsive polymers onto their surface or inside pore walls [3,28].

Unfortunately, only a limited number of electrically conductive polymers are suitable to be easily integrated into the most common production process of composite membranes, such as Non-Solvent Induced Phase Separation for MF and UF membranes, and interfacial polymerization or coating for NF and RO membranes. Recently, a different method based on using responsive polymer self-assembly has been proposed for functionalizing commercial membranes (either by a post-processing procedure or in a single step process) and improving permselectivity and fouling potential [29,30].

2.1. Electro-Conductive Polymers

In water treatment applications, it is important that membranes show opportune surface structure (e.g., pore size and distribution), purposeful hydrophilicity, adequate chemical-physical properties, high mechanical stability, and long term durability. ERPMs can contain specific organic and/or inorganic solid nanofillers in their porous structure (nanocomposite or hybrid membranes) to give or enhance the membrane properties that would otherwise not be met by the conducting polymer alone [31].

Electro-conductive polymers can be classified according to the movement of electric charges [32] in:

1. Intrinsic electro-conductive polymers, characterized by conjugated π-π or p-π systems;
2. Redox polymers that possess redox potentials within their structure groups (reduction/oxidation capacity).

The electronic transport in intrinsic electro-conductive polymers is due to the electron transfer from π type bonds to nearby simple σ bonds, due to the repulsion effect of same type charges. In presence of heteroatoms (N, S or O type) within the macromolecular polymer chains, the electron transfer is from π type bonds to non-participating p electrons of the heteroatoms that, moving to σ single bonds, further induces the movement of π electrons from the nearby double bonds by electrostatic repulsion effect. The conductivity of intrinsic conductive polymers significantly increases by oxidative and reductive doping (p-doping and n-doping, respectively).

The electronic movement in the case of redox polymers is gained through "donor-acceptor" reversible chemical reactions, according to Equation (1):

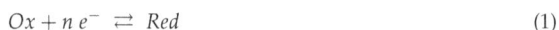

$$Ox + n\,e^- \rightleftarrows Red \tag{1}$$

if the chemical groups, with their redox potential distributed within the macromolecular structure, enable electronic jumps between groups [32].

Table 1 shows the most representative electro-conductive polymers, but not their numerous derivatives. Nevertheless, there are few stable conducting polymers in harsh aqueous environment and most of recent studies on ERPM are limited to the use of commercially available conducting polymers.

Table 1. Some electro-conductive polymers and their abbreviations.

Chemical Name	Abbreviation
Polyacetylene	PAc
Polyaniline	PANI
Polyazulene	PAZ
Polybutadiene	PBD
Polyisopren	PIP
Poly(isothianaphtene)	PITN
Polyfuran	PFu
Poly(α-naphthylamine)	PNA
Poly(p-phenylene)	PPP
Polythiophene	PTh

2.2. Preparation of Electro-Responsive Polymer Membranes

Surface material research has recently led to the manufacturing of many smart membranes by either chemical bonds or physical incorporation of electro-responsive materials on porous membrane substrates. It is well known that in typical processes for preparation of conventional membranes, the enrichment of membranes with conductive polymers is limited by the doping amounts used to improve membrane electro-responsiveness without loss of the mechanical properties. Electro-responsive membranes can be obtained from casting of conductive polymer thin films or self-assembling of monolayers onto the membrane surface by different methods such as plasma deposition, chemical vapour deposition, spin coating, chemical and electrochemical reactions, and layer-by-layer assembly [3]. The physical coating with a conductive polymer generally leads to variations in membrane swelling degree and changes in membrane permselectivity [33,34]. Alternatively, functional polymers, as well known as polymer brushes, can be attached in a controlled manner on the membrane surfaces or within membrane pores by physical adsorption or covalent bonds [35–38]. Polymer brushes can be covalently attached to membrane surfaces and pores either by 'grafting-from' methods or by 'grafting-to' techniques. In the 'grafting-from' methods, functional monomers are polymerized onto active sites present on the membrane pores and surface. The 'grafting-from' method is considered very advantageous because the presence of linear polymers or crosslinked networks in the pores can reduce the steric hindrance of neighbouring bonded polymer chains [39,40]. On the contrary, in the 'grafting-to' methods responsive membranes are fabricated by chemical/physical incorporation of opportunely end-functionalized polymers that can react onto the desired surfaces. In both grafting methods, the presence of an electro-responsiveness in the grafted polymer brushes can be used to alter the chain conformation and lead to responsive surfaces with an electro-tuneable permselectivity. Grafting can be usually induced by plasma, photo-irradiation, redox reactions, temperature and controlled radical reactions such as reversible addition fragmentation chain transfer polymerization and atom transfer radical polymerization. Polymer self-assembly methods to prepare ERPM membrane in a single step are limited by the difficulties generally found in the synthesis of conductive copolymers. Recently, polymer self-assembly methods have used amphiphilic copolymers in order to prevent membrane fouling and retain permeability [41–45]. Barghi et al. [46] synthesized a flexible, biocompatible, semi-hydrophilic, and electro-conductive membrane by crosslinking copolymerization of a highly electro-conductive monomer (hydroxymethyl-3,4-ethylenedioxy thiophene, HMEDOT) with a highly mechanical resistant polyamide (polytetramethylene-N-hydroxyethyl adipamine, PTMHEA) opportunely hydrophilized with acetaldehyde and in situ polymerized by an oxidative plasma treatment. The PHMEDOT homopolymer grafted onto the PHMEDOT-*co*-PTMHEA surface reduced considerably the copolymer electrical resistance both in dry and wet conditions (105 kΩ cm^{-2} and 2 kΩ cm^{-2}, respectively). Pore size and distribution, roughness, and water flux were finely controlled by changing the thickness of PHMEDOT homopolymer.

2.2.1. Responsive Membranes Based on Polyaniline

Polyaniline (PANI) is one of the most investigated conductive polymers because of its long term environmental stability, high conductivity, and relative low cost. The high chemical selectivity of PANI and its composites makes them particularly attractive as sensors for the detection of a number of gases and vapours, including methanol, ammonia, HCl, CHCl$_3$, NO$_2$, and CO [47,48]. Distinctive drawbacks, such as low solubility in the majority of solvents commonly used for membranes preparation, low mechanical flexibility, and thermal instability at temperatures above 160 °C, do not allow obtaining pure PANI membranes. Therefore, PANI-based responsive membranes are blends of PANI with other polymers suitable for membrane preparation. Polysulphone (PSF), polystyrene (PS), polypropylene (PP), cellulose (CEL) and its derivatives are some chemically inert polymers used for PANI-based membrane preparation. PANI-based membranes are mainly used in selective separation processes of gases and some chemical species from complex liquid solutions, in antistatic textile materials, biosensors, anticorrosive films, and electric and electronic devices (e.g., light emitting diodes and photovoltaic cells).

PSF/PANI-based membranes are designed for advanced separation of polar compounds from various mixtures and obtained from simultaneous formation of a PSF-based membrane and aniline polymerization within membranes in oxidative conditions. PANI results generally well distributed in the whole microporous structure and not only on membrane surface [49]. In addition, PS/PANI-based membranes can be obtained via phase inversion processes, whereas phase changes take place through precipitation in vapour phase [50]. Obviously, conductive properties of PS/PANI-based membranes depend on PS/PANI weight ratio within the composites. PP/PANI-based membranes maintain the microporous structure of the supporting polypropylene. Different pore diameters can be obtained as long as PANI is formed within the PP pores through soaking of supporting polymer films in aniline, followed by aniline oxidative polymerization using ammonium peroxydisulphate and HCl [51].

PP/PANI-based composites can be used for selective separation of chemical species from various liquid media through reverse osmosis, microfiltration, ultrafiltration, and nanofiltration processes. CEL/PANI nanocomposites can be prepared by in situ chemical oxidative polymerization routes of aniline within the fibre microstructure of CEL [52]. An enhancement in the PANI content inside membrane nanocomposites and a consequent increase in their electric conductivity is observed by increasing the reaction time. Longer reaction times give rise to PANI aggregation and formation of discontinuities within the nanocomposite structure with a consequent decrease of electric conductivity. Other CEL/PANI-based membranes can be prepared by deposition of a thin layer of PANI onto membrane interface (cellulose or its esters) by in situ oxidative polymerization of aniline [53]. Therefore, the electric conductivities of cellulose acetate/PANI membranes increase from 10^{-3} to 11 S m^{-1} and 98 S m^{-1} using liquid [54] and vapour [55] phase polymerization, respectively.

2.2.2. Responsive Membranes Based on Polypyrrole

Polypyrrole (PPy) is characterized by very low conductivity and a low processability due to its poor mechanical strength. An appropriated doping with anions such as dodecylsulphate, chloride, sulphate and perchlorate can easily increase conductivity. Doped PPy shows a good chemical and thermal stability, a higher conductivity compared with other conductive polymers, and improved plasticity and elasticity by inclusion within polymer structure. PPy-based composite membranes are frequently used in concentration- (gas separation from complex mixtures and pervaporation) and electric potential-gradient processes (electro-dialysis). PPy is polymerized by electrochemical and chemical oxidative polymerization [56–61]. The vapour-phase polymerization of pyrrole is an additional method to form conducting PPy films on membranes [62–65]. Addition of surfactants such as sodium dodecylbenzensulphonate, sodium alkylnaphtalenesulphonate and sodium alkylsulphonate within the chemical oxidation reaction media gives higher PPy electro-polymerization efficiencies, larger electric conductivities, better fluxes and selectivity control in the composite membranes [60]. The resulting nanoporous membranes are able to tune their pore sizes by application of an electrical potential, whose strength is less than 1.1 V.

Tsai et al. [60] prepared a PPy-based nanoporous membrane with tuneable wettability from a polypyrrole film doped with dodecylbenzenesulfonate anions (DBS) and electropolymerized on a coated Si wafer. Due to the reorientation of DBS dopant molecules, the membrane surface wettability was tuned from a more hydrophobic behaviour (with a contact angle of 134°) to a less hydrophobic behaviour (with a contact angle of 107°) by application of low electrical potentials (from 0.7 to −1.0 V).

2.2.3. Responsive Membranes Based on Carbon Nanotubes

The use of metal nanoparticles and carbon nanotubes (CNTs) as conducting elements is a valuable approach for the preparation of effective electro-sensitive materials to be used in several fields including drug delivery [66,67], liquid crystal displays [68], solar energy cells [69], conductive devices [70–72]. CNTs are commonly employed in hybrid polymer membranes to improve their performance in terms of fouling potential, permselectivity, and flux. The specific features of CNTs, such as well-defined structure, chemical bonding properties and high aspect ratio, concur to their interesting electro-mechanical properties that can improve the morphological, rheological, thermal, mechanical, and electrical properties of the host polymers. Commercially available and laboratory-scale produced single-walled (SWCNT), double-walled (DWCNT) and multi-walled (MWCNT) CNTs can be incorporated into final and/or intermediate polymer materials. The most important challenges in the preparation and effective utilization of CNTs in polymer membranes are an adequate interfacial adhesion between polymer matrix and CNTs and a homogeneous distribution of CNTs throughout the composite matrix in order to prevent their agglomeration [73]. Moreover, CNTs concentration limits and inhomogeneous orientation in membranes represent additional issues to be overcome. Some approaches to face these challenges include the use of surfactant molecules, polymer wrapping, long sonication times and chemical sidewall-functionalization in order to favour debundling and enhance hydrophilicity [74].

2.3. Electro-Conductive Membranes for Permeation Enhancement

In recent years, the use of stimuli responsive membranes has become a promising method for reducing fouling potential. Treatments with stimuli responsive molecules in the form of thin films and nano-brushes give surface functionality to conventional membranes and reduce their fouling potential [75–83] (Figure 1). For example, it is possible to increase the membrane permeability and solve the problem of foulant deposition within pores by opening gates and enlarging pore size [84–86].

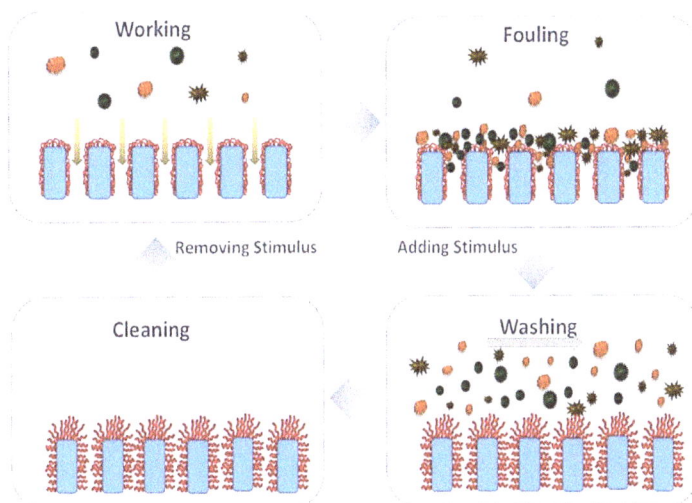

Figure 1. Examples of Stimuli Responsive Membranes. Reprinted from [83], with permission from Royal Society of Chemistry.

Lalia et al. [87] proposed self-cleaning PVDF membranes by using highly tangled carbon nanostructures (CNS) with an average diameter of 7–8 nm. Membranes were characterized by improved processability, high electrical conductivity and large surface area [88] (Figure 2).

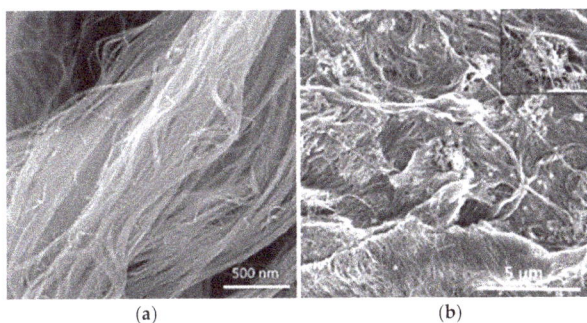

Figure 2. SEM images of carbon nanotube structures: (**a**) pure and (**b**) cast on PVDF membrane. Reprinted from [87], with permission from Elsevier.

These membranes were prepared via vacuum filtration, followed by a heat treatment at 160 °C to melt PVDF and provide binding sites inside the entangled CNS structure with the aim to improve the membrane mechanical strength. Then, membrane performance was tested for in situ surface cleaning in a cross-flow filtration using a yeast suspension as feed. In the electrolytic cleaning the CNS/PVDF surface acted as the cathode, a platinum foil was used as the anode and Ag/AgCl was employed as the reference electrode in 0.5 M H_2SO_4 solution. Electrolysis led to the generation of hydrogen micro-bubbles on the membrane surface, which removed foulants and recovered flux in successive cycles. Permeation fluxes exponentially decreased with time in absence of periodic electrolysis, while they increased of about 40% respect to their original values after 4.6 h of filtration in the presence of periodic electrolysis (Figure 3).

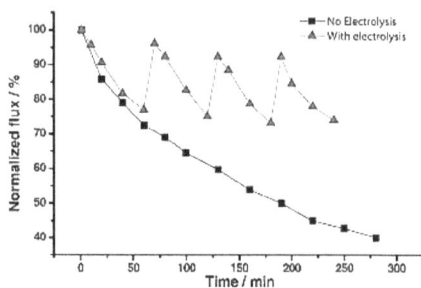

Figure 3. Time behaviour of normalized flux for CNS/PVDF membrane with and without electrolysis. Reprinted from [87], with permission from Elsevier.

Recently, Duan et al. [89] used a polyvinyl alcohol and carboxylated MWCNTs (PVA/MWCNT-COOH) membrane to remove Cr(VI) from water through a combined process of electrostatic repulsion, electrochemical reduction, and precipitation. The overall removal efficiency exceeded 95%, a very high value if compared with the maximum rejection of 20% by commercial UF polysulfone membranes with a cut-off of 10 kDa. An electrochemical treatment of Cr(VI) is usually conducted in a mass-transport limited batch process that needs long contact times, making the process hard to scale up. These mass transfer restrictions can be overcome by electrochemical filtration, where the contaminated water is forced through a porous electrode, as a PVA/MWCNT-COOH membrane, capable of supporting electrochemical reactions, such as oxygen reduction, chlorine oxidation, and water splitting [90,91]. The removal mechanism resulted to be highly dependent on solution conductivity: higher solution conductivities involved electrochemical reduction and precipitation of Cr(III) on the membrane surface, while very low conductivities led to electrostatic repulsions accounting for Cr(VI) rejection from the permeate. The increase of membrane surface charge density by application of an external potential (3, 5 and 7 V, with membrane as cathode), increased the Cr(VI) removal from 45.0% (for non-polarized PVA/MWCNT-COOH membrane) to 86.5% (at the highest cell potential). The membrane contact time and background ionic strength of the feed solution influenced significantly the Cr(VI) removal. Electrostatic repulsive forces between the negatively charged PVA/MWCNT-COOH membrane and CrO_4^{2-} could prevent chromium ions permeation under low salinity conditions without applying external potentials. At high electrolyte concentrations, soluble Cr(VI) is reduced to insoluble Cr(III) and precipitates on the membrane surface primarily as $Cr(OH)_3$ by reaction with hydroxide ions generated by the water splitting on the MWCNT network, and can be removed from the treated water stream. Moreover, thicker membranes (6 μm-tick) showed superior performance with better rejection/removal and higher current densities, also when PVA/MWCNT-COOH membranes were used to treat tap water spiked with 1 ppm Cr (VI) by application of 7 V to the membrane counter electrode (Figure 4).

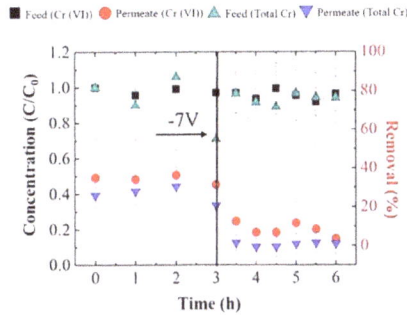

Figure 4. Removal of chromium from tap water spiked with 1 ppm Cr(VI) using a 6 μm-tick membrane. Reprinted from [89], with permission from Elsevier.

3. Membrane Fouling

Fouling can be considered the "Achilles heel" of membrane processes. It is essentially due to the deposition/adsorption of particulate and soluble materials on membrane surface and, in case of porous membranes, to the entrapment of foulant molecules inside membrane pores [92]. Several factors can influence fouling: the feed conditions (e.g., ionic strength, pH and presence of cations), membrane surface morphology and properties such as roughness, charge and hydrophilicity [93].

Membrane fouling can be essentially classified into three main categories: reversible, irreversible and irrecoverable, depending on the nature of foulant attachment onto membrane surface. Reversible fouling is caused by external deposition of material and can be removed by simple cleaning methods, while irreversible fouling refers to foulants, which can only be removed by harsh chemical and/or thermal treatments. The term irrecoverable fouling refers to fouling that cannot be removed by any cleaning method, but only by a long operational period [92] (Figure 5).

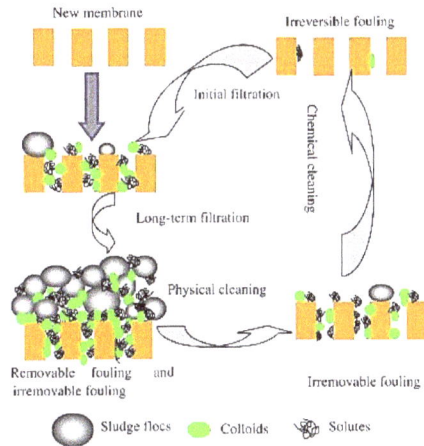

Figure 5. Reversible, irreversible and irrecoverable fouling in membrane processes. Reprinted from [92], with permission from Elsevier.

Another important classification divides fouling in abiotic fouling and biofouling. Abiotic fouling is responsible of the formation of a "cake layer" consisting of rejected material, while biofouling is the accumulation of microorganisms onto the surfaces and within the pores of membranes [94].

Fouling can significantly reduce membrane performance by:

1. a lowering in productivity because of longer filtration times,
2. an alteration of membrane selectivity as a consequence of a change in pore size,
3. a shortening of membrane life because of the severe chemical cleaning [95,96],
4. an increase of operational costs [97].

The material accumulated onto surface or within pores may reduce the membrane permeability and results in a general reduction of the permeate flux over time [94]. For constant pressure operations, where the transmembrane pressure is maintained at a constant value during filtration, fouling causes an increase in filtration resistance, that leads to a flux decline, *FD*, over time defined as [98]:

$$FD = \frac{F_i - F_f}{F_i} \times 100 \qquad (2)$$

where F_i and F_f are the initial and final fluxes, respectively.

The characteristics and the position of deposited materials determinate the extent and reversibility of permeate flux decline. A partial restore of permeate flux can be obtained by membrane cleaning (either by hydraulic or chemical methods) in order to remove some/all the accumulated material [94] (Figure 5).

Conventional cleaning methods include back-flushing, feed pulsing, permeate back-pulsing and air sparging [99]. However, these methods have some limits because they can provide only a temporary relief to flux losses (Figure 6) and damage membranes [100] causing significant changes in their properties (e.g., surface charge, hydrophobicity and permeability) [101]. Further drawbacks are the increased operational costs, reduction of membrane lifetime, and need to interrupt the production to activate such cleaning procedures [99].

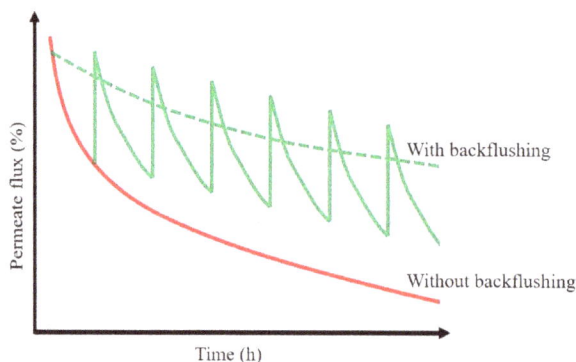

Figure 6. Effects of periodic back-flushings on permeate flux over time.

Obviously, a highly fouling-resistant membrane requires infrequent cleanings, reduces operating and disposal costs, increases the operational life and provides consistent permeate quality over time [94].

3.1. Novel Approaches to Mitigate Fouling

Extensive work has been done on developing methods to mitigate the negative effects of fouling on membrane performance including optimisation of the membrane composition to minimise attractive interactions between foulants and surface [102], pre-treatment to remove the most reactive foulants [103] and enhanced module design and operation that reduce fouling through a more effective hydrodynamics [104]. A novel approach suggested to prevent fouling is the formulation of membranes characterised by an active layer with a low surface energy so that attached foulants can be readily washed away with reduced changes in water fluxes and permeations [105].

Surface modification of commercial membranes by post-treatment is one of the most frequently method to decrease membrane fouling potential [106,107]. Some post-synthesis modifications include [94]:

1. a decrease in the membrane hydrophobicity,
2. a reduction of the surface roughness,
3. an increase in the membrane selectivity,
4. a modification of the surface charge.

Since main polymer membranes are hydrophobic, a frequent problem in membrane processes is the hydrophobic interactions between membrane surface and hydrophobic solutes present in the feed solution. The use of a hydrophilic membrane could decrease the fouling potential [92]. Numerous attempts have been made to improve anti-fouling performance by increasing membrane surface hydrophilicity and smoothness [6,108]. Du et al. [109] proposed a new post treatment process to increase membrane surface hydrophilicity and smoothness by a surface microstructure reassembly.

Recently, the attention of researchers has shifted to Electrically Responsive Polymer Membranes (ERPMs) characterised by high electrical conductivities. These self-cleaning membranes can be used to mitigate the effects of fouling in several types of separation processes [87]. The cleaning mechanism in ERPMs can be based on electrostatic repulsion, electrochemical oxidation, hydrogen peroxide production, surface morphology changes, piezoelectric vibrations, electro-chemical bubble generation [32,110–113].

3.1.1. Electrostatic Repulsion and Hydrogen Peroxide Generation

A possible method to mitigate fouling consists in the generation of electrostatic repulsion between charged surfaces and foulants because most of the membrane foulants are negatively charged such as sludge flocs, soluble microbial products and polymer substances [114]. In addition, electrically charged membranes have been used as electro-catalytic platforms in order to transform various aqueous contaminants through electrochemical reactions [115].

Huang et al. [116] proposed a simple method to control fouling introducing a stainless steel mesh between the supporting layer and active layer of a MF polymer membrane without changing its surface physical-chemical properties. A homogeneous conducting polyvinylidene difluoride (PVDF) solution was cast on a stainless steel mesh (pore size 96 μm, thickness 43 μm) assembled on a polyester nonwoven fabric. The composite membrane was made by immersion precipitation in a non-solvent bath. Experiments were performed applying an electrical field of 2 V cm^{-1} with the membrane acting as cathode. A high water flux and low electrical resistance were found (66 L m^{-2} h^{-1} and around 200 Ω, respectively). The antifouling performance of these membranes was attributed to the combination of electrostatic repulsive forces between charged membranes and tested foulants, as well as to the organic oxidation by electrochemically generated hydrogen peroxide at the cathode (in situ membrane cleaning), leading to a decreased fouling potential (Figure 7). The electrical potential decreased the fouling rates for all tested model foulants (bovine serum albumin, sodium alginate, humic acid, and silicon dioxide).

CNTs are frequently used as additives in view of improving membrane performance [73,117]. In particular, electrically conductive membranes obtained by CNT entrapment in a crosslinked network have been demonstrated to mitigate several forms of fouling through the application of electrical potentials [118–123].

Dudchenko et al. [124] used a sequential pressure/deposition process to set up robust and electrically conductive thin films made of glutaraldehyde-based cross-linked PVA and MWCNTs-COOH on a polysulfone UF support. This membrane exhibited high electrical conductivity

(2500 S m^{-1}), excellent robustness and permeability. PVA/MWCNT-COOH were used in cross-flow devices (Figure 8) for electro-filtration process and showed separation properties similar to the commercially available PSF-35 UF membranes.

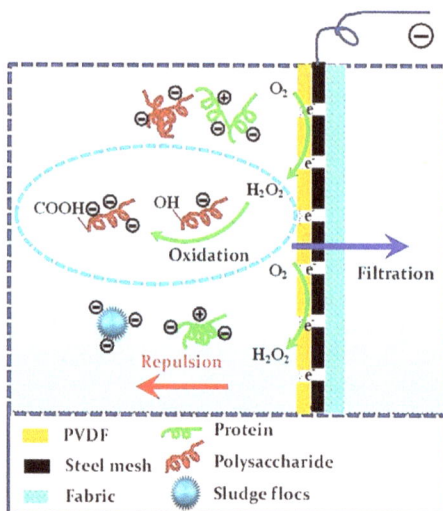

Figure 7. Schematic illustration of anti-fouling mechanism. Reprinted from [116], with permission from Nature Publishing Group.

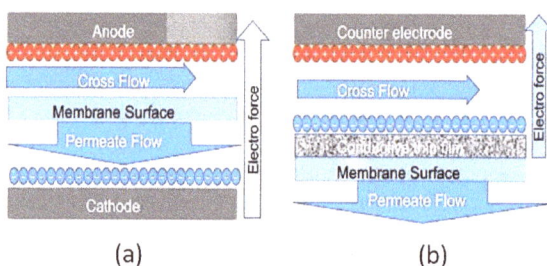

Figure 8. Electro-filtration set-up with: (**a**) conventional and (**b**) conductive membranes. Reprinted from [124], with permission from Elsevier.

When an electric potential was applied, PVA/MWCNT-COOH membranes were able to inhibit fouling at very high concentrations (3.0–5.0 g/L) of alginic acid, which was used as a negative charged model foulant. After 100 min of operation with the PVA/MWCNT-COOH membrane working as a cathode element (−3.4 V vs. Ag/AgCl reference electrode), the change in operating pressure was reduced by 51% compared with the control membrane working without voltage. Fouling mitigation was explained using a modified Poisson-Boltzmann equation and a DLVO-type theory, indicating that electrostatic interactions gave significant repulsive forces between the membrane surface and charged organic foulant molecules.

ERPMs have been demonstrated to be efficient in solving fouling problems in anaerobic bioreactors, when vigorous air scouring cannot be used to clean membrane surfaces [125,126].

Duan et al. [127] prepared a UF conductive membrane by deposition of CNT–COOH on a PSF support followed by the crosslinking of a PVA layer. The PVA/CNT–COOH network deposited on PSF surface created a smooth (46 ± 2 nm) electrically conducting (1132 ± 32 S m^{-1}) layer. The application of an electric voltage (5 V) using membrane as cathode, led to a significant reduction of membrane fouling because the main degradation products of the anaerobic processes are negatively charged small molecules. When the system operated at a constant flux of 30 L m^{-2} h^{-1} with no applied potential, pressure increased from 1.5 to 3 psi over the time. On the contrary, when the membrane was used as a cathode, the pressure increased from 1.5 psi to 2.4 and 2.2 psi, when an electric potential of 3 V and 5 V was applied, respectively (Figure 9).

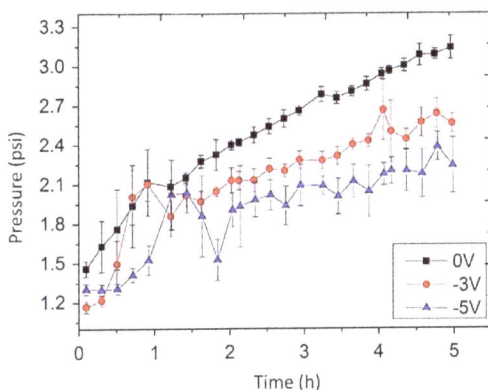

Figure 9. Impact of applied electrical potential on transmembrane pressure. Reprinted from [127], with permission from Elsevier.

Interestingly, during back-flushing, when the membrane was anodically switched (1.5 V), a rapid and irreversible fouling was recorded confirming that most of foulants were negatively charged.

Another effective cleaning method to mitigate fouling without membrane damage is the generation of microbubbles on the membrane surface through electro-reduction [128].

3.1.2. Electrochemical Oxidation

The electrically conducting form of PANI is emeraldine, which is obtained through the electrochemical polymerization of PANI under acidic condition (Figure 10) [32,129–132].

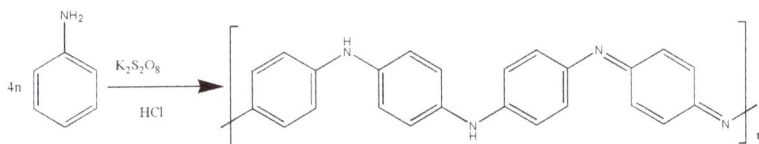

Figure 10. Emeraldine synthesis and structure.

PANI/CNT electrically conducting membranes were designed to evaluate their capacity for in situ electrochemical cleaning via electro-oxidation, without any external chemical addition [122].

Recently, Duan et al. [122] made an highly conductive and anodically stable polyaniline/ carboxylated carbon nanotubes (PANI/CNT–COOH) UF membrane by electro-polymerization of aniline on a PSF/CNT substrate under different acidic conditions (sulfuric, hydrochloric, and oxalic acid). Electrochemical polymerization under acidic conditions forms PANI in the emeraldine form. In addition, hydrophilic PANI-based membranes are usually more resistant to organic and biologic fouling as well as more conductive than PVA/CNT membranes. The PANI/CNT–COOH membranes obtained from sulfuric acid exhibited the best stability, conductivity and hydrophilicity with no impact on selectivity and permeability (Figure 11) and resulted ideal membranes for water treatment applications.

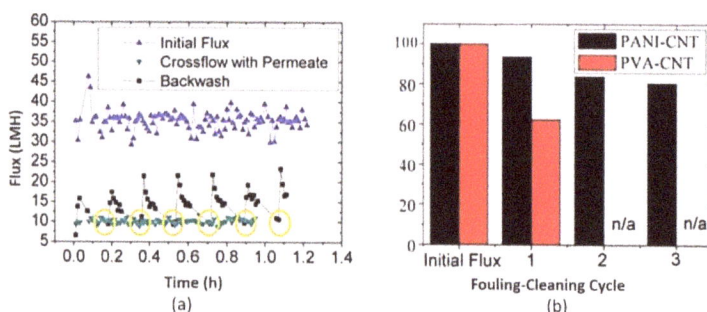

Figure 11. (**a**) Flux and (**b**) fouling behavior in PANI/CNT-COOH membranes. Reprinted from [122], with permission from American Chemical Society.

Moreover, PANI/CNT–COOH membranes showed enhanced resistance to anodic oxidation, with little degradation observed up to 3 V vs. Ag/AgCl (Figure 12) under neutral pH conditions.

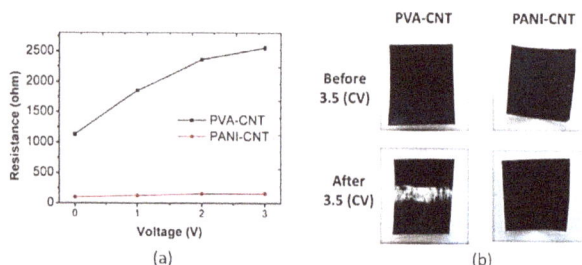

Figure 12. (**a**) Resistance and (**b**) stability in PANI/CNT–COOH and PVA/CNT-COOH membranes. Reprinted from [122], with permission from American Chemical Society.

Experiments conducted with bovine serum albumin showed an easy fouling cleaning of PANI/CNT–COOH membrane surfaces by in situ oxidation and fluxes restored to their initial values by application of a 3 V potential. Moreover, a methylene blue (MB) solution was easily electrochemically oxidized with 90% efficiency in a single pass through an anodically charged ERPM (3 V, 1 μm thick membrane, membrane residence time lower than 0.2 s), avoiding the need for additional and expensive chemical cleaning agents [133]. The electro-oxidation of 5 ppm of MB on PANI/CNT–COOH ERPM required only 2.5 kW m^{-3} with a contact time lower than 1 s. In contrast, typical photocatalytic processes for MB on titanium dioxide require up to 40 kW m^{-3} and contact times ranging from 30 to 60 min [134,135].

Recently, many researches underlined the huge potential of CNT/polymer composites in water treatment such as desalination. Shawky et al. [73] synthesized polyamide/MWCNT nanocomposite membranes (PA/ MWCNT) by a polymer grafting process and investigated the NaCl and humic acid rejection. The strong interactions between MWCNTs and PA matrix resulted in a remarkable structural compactness and significant improvement of mechanical properties of the obtained membranes. In addition, the salt rejection considerably increased, even if the permeate flux was slightly reduced.

De Lannoy et al. [136] evaluated the effects of MWCNTs-COOH on a hydrophobic PSF membrane, widely used in UF processes in spite of its relatively low tensile strength. Surface hydrophilicity, membrane permeability and tensile strength of PSF/MWCNTs-COOH composites increased as a function of CNT carboxylation. However, a decreased MWCNT retention within the membranes and an increased leaching during membrane cleaning were observed at higher carboxylation efficiencies.

A highly conducting and flexible composite membrane was realized with a thin layer of PVA, covalently cross-linked to MWCNTs–COOH and succinic acid, onto a cellulose nitrate membrane (Figure 13) [137].

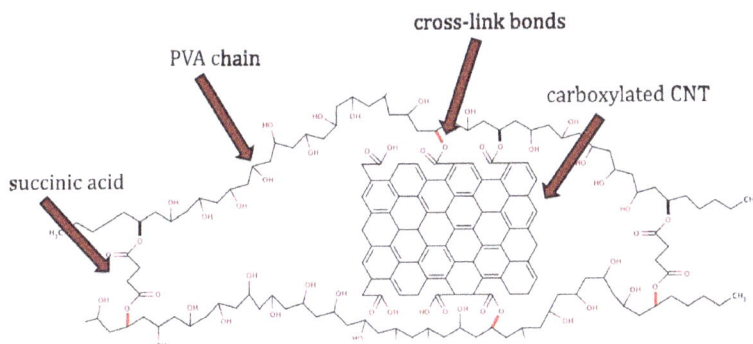

Figure 13. Thin layer of PVA covalently cross-linked to MWCNTs–COOH and succinic acid onto a cellulose nitrate membrane. The succinic acid molecules and MWCNTs–COOH cross-linked the PVA strands, immobilizing MWCNTs and altering the spacing between PVA strands. Reprinted from [137], with permission from Elsevier.

This PVA/MWCNT–COOH composite showed high electrical conductivity and permeate flux with low polymer crystallinity and surface tension. Membranes prepared with 20 wt % MWCNT–COOH and 20 min curing time exhibited conductivities as high as 3.6×10^3 S m^{-1}, pure water flux of 1440 L/m^2 h at pressures of 550 kPa, and triple-point initial contact angles as low as 40° with high hysteresis. Better separation characteristics were achieved in PVA/MWCNT–COOH membranes by incorporating smaller amounts of MWCNT–COOH (2 and 5 wt %), but at the expense of the membrane permeability (Figure 14). The authors suggest that the MWCNT-COOH loading could be easily employed to control the molecular weight cut-off.

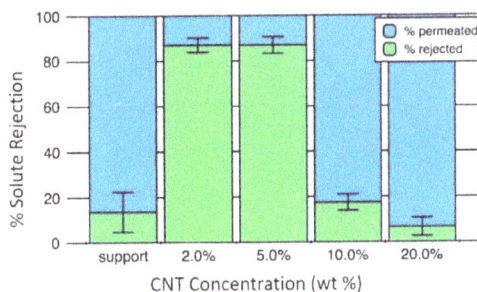

Figure 14. Solute rejection as a function of MWCNT–COOH wt % content with respect to PVA in PVA/MWCNT–COOH composites. Reprinted from [137], with permission from Elsevier.

CNT-based ERPMs show long-term stability (no notable change in their conductivity over time is observed when they are used as cathodes) and interesting electro-cleaning properties, as previously reported. Nevertheless, CNT-based ERPMs result to be unstable under elevated anodic potentials in aqueous environments due to CNT oxidation and breakdown when exposed to hydroxyl radicals produced on their surfaces [122]. Coating and anchoring of stable metal nanoparticles (e.g., bismuth-doped tin oxide and cobalt oxide) on CNT surfaces increase stability up to 2.2 V vs. Ag/AgCl reference electrode.

Graphene is a two-dimensional, one-atom-thick layer of graphite with tunable size and structure and can be engineered for different filtration processes, ranging from ultrafiltration to reverse osmosis. Graphene shows enhanced physical-chemical properties such as electrical and thermal conductivity, mechanical strength, optical transparency, solution processability, and specific surface area up to $2630\ \mathrm{m^2\ g^{-1}}$. Therefore, graphene has been widely used in flexible transparent electrodes, energy storage devices, solar cells, and electronics and optoelectronics applications. Usually, graphene can be obtained by exfoliation of highly pure graphite and, therefore, does not retain the CNTs metal impurities deriving from their metal-catalysis-driven growth process. Graphene nano-platelets (GNP) can lead to highly ordered membranes or films by means of different routes (filtration-assisted assembly, chemical vapor deposition, electrochemical deposition, and layer-by-layer methods). Moreover, the rich surface chemistry of bidimensional graphene favors the fine-tuning of the interfacing properties with numerous porous supporting materials, such as PSF, PES (polyethersulfone) and PTFE (polytetrafluoroethylene).

Liu et al. [138] developed a novel electrochemical filter for water purification by graphene nano-platelets enabled by carbon nanotubes (GNP:CNT) in a polytetrafluoroethylene membrane (PTFE/GNP:CNT). CNTs were the conductive binders for graphene nano-platelets (Figure 15).

Figure 15. Schematic representation of PTFE/GNP:CNT electrochemical filter. Reprinted from [138], with permission from Royal Society of Chemistry.

In particular, the researchers dispersed different weight ratios (from 50:50 to 100:0) of GNPs and CNTs in N-methyl-2-pyrrolidone, and vacuum filtered the stable suspension onto a PTFE membrane. Anodic oxidation of the PTFE/GNP:CNT electrode was tested using ferrocyanide ($\mathrm{Fe(CN)_6^{-4}}$) as a model electron donor. When the anodic filter was used in batch mode, electro-oxidation rates increased linearly with applied potential and plateaued because of mass transfer limitations. When the PTFE/GNP:CNT filter was evaluated as part of a flow-through system, no plateau was observed for high concentrations of ($\mathrm{Fe(CN)_6^{-4}}$) (10 mmol $\mathrm{L^{-1}}$) as a result of increased mass transfer rates. Overall, electro-oxidation rates increased up to 15-fold due to convection enhanced transfer of the target molecule to the electrode surface and reduction of mass transfer over potential.

Moreover, the efficiency of PTFE/GNP:CNT filters for anodic degradation was evaluated with three selected organic pollutants (tetracycline, phenol and oxalate). For all three organic compounds, electro-oxidation kinetics increased with increasing anode potential until a maximum removal rate (0.010, 0.064, and 0.050 mol $\mathrm{h^{-1}\ m^{-2}}$ for tetracycline, phenol, and oxalate, respectively) achieved at 0.8 V (Figure 16).

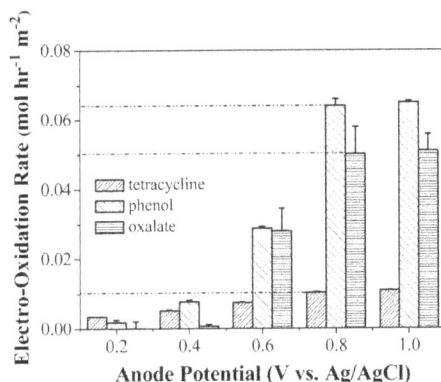

Figure 16. Electro-oxidative filtration of tetracycline, phenol and oxalate as a function of anode potential. Reprinted from [138], with permission from Royal Society of Chemistry.

3.2. Biofouling Mitigation with Electro-Responsive Membranes

Membrane processes are vulnerable to bacterial adhesion and biofilm growth on the membrane surface. Development of biofouling leads to a dramatic decrease of productivity, especially when in the feed solution are present organic matter and nutrients, as in the case of wastewater effluents. Most efforts to prevent biofilm development are based on limiting the initial bacterial attachment or increasing detachment. Methods to develop surfaces with anti-biofouling properties include linking and embedding of antimicrobial nanoparticles [139–142], grafting of polymer brushes, that form a hydrated gel layer that prevents bacteria from interacting with surfaces [143], and electrically charged surfaces [144,145]. These coating methods have been demonstrated to be effective, but the coating material loss causes the decline of their performance over time [146–148]. Ronen et al. [120] studied the bacterial deposition and detachment rates as a function of the electrical potential applied to the membrane surface. In the experiment, the authors used a conducting PVA/CNT composite UF membrane and ITO electrodes positioned on both sides of a modified cuvette containing *Escherichia Coli* suspension at a concentration of 10^8 cells mL^{-1}. The microbial attachment was investigated using a direct observation cross-flow membrane system mounted on a fluorescent microscope. Different electrical potentials (0, 0.5, 1, and 1.5 V) were applied to the electrodes and the impact of the electric voltage was investigated by measuring cell integrity and cell viability using propidium iodide and 5-cyano-2,3-ditolyltetrazolium chloride as fluorescent indicators, respectively.

SEM images of membrane surface after detachment phase showed that cells had a regular shape on membranes without applied potential, while the irregular structure of cells, that were remained attached to membrane subjected to a potential of ±1.5 V, was indicative of cell damage (Figure 17).

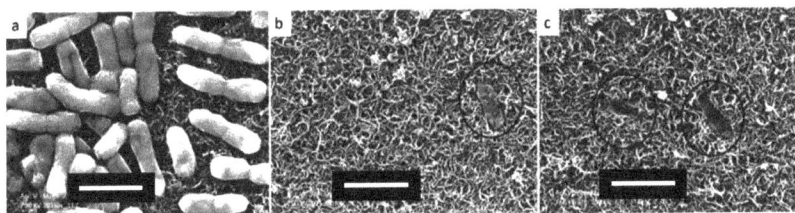

Figure 17. SEM images of membranes after detachment with: (**a**) no applied potential, (**b**) 1.5 V applied with membrane as anode, and (**c**) 1.5 V applied with membrane as cathode. Scale bars are 2 µm. Reprinted from [120], with permission from American Chemical Society.

The main mechanism proposed to explain the antifouling properties of these membranes was the generation of hydrogen peroxide due to electro-reduction of oxygen, when a low electrical potential was applied. The production of hydrogen peroxide on membrane surfaces caused the reduction of microbial cell viability, increased cellular permeability and prevented bacterial attachment.

Another interesting anti-biofouling method based on the electrostatic repulsion between membrane surface and attached bacteria was investigated by Baek et al. [149]. They produced an electro-conductive feed spacer (ECFS) in a lab-scale cross flow membrane system, in which low electric potentials were applied to minimize chlorine gas generation. A titanium mesh and a stainless steel mesh were used as model ECFS, on which an electrical polarization was induced. After 24 h from biofouling occurrence, the permeate flux was decreased to about 47%, while was recovered to 80%, 89%, and 91% when the ECFS was polarized for 30 min with +1.0 V, −1.0 V, and alternating electrical potentials (cycles of +1.0 V for 1 min and −1.0 V for 1 min), respectively.

The electrically conductive PA/MWCNT nanocomposites by de Lannoy et al. [74] showed higher biofilm-preventing capabilities, larger electrical conductivity ($\sim$400 S m^{-1}), better monovalent ion rejection (greater than 95%), and higher water permeability than commercially available membranes. Biofilm induced a non-reversible flux decline, while the flux decrease for membranes with an applied electric potential (1.5 V) was lower (due just to bacteria deposition) and fully recoverable with a short rinse with the feed solution without added cleaning agents. The prevention of microbial biofilms was probably due to local pH instabilities and unfavourable conditions for bacterial growth arising from the electrical potential.

4. Conclusions

This paper reviewed the recent progresses in electro-conductive membranes as smart devices able to respond to the application of an electric signal. Different classes of electro-conductive membranes were examined and the advantages in their using were discussed with particular emphasis to the beneficial effects on membrane transport properties and fouling mitigation.

Acknowledgments: MIUR, the Italian Ministry for University and Research, is acknowledged for financial support (EX-60/2016).

Author Contributions: Patrizia Formoso and Elvira Pantuso collected literature and wrote the manuscript. Giovanni De Filpo and Fiore Pasquale Nicoletta gave suggestions on the organization of the manuscript, supervised the work and revised the manuscript. All authors read and approved the final manuscript.

Conflicts of Interest: The authors declare no conflict of interest.

References

1. Giorno, L. Permselectivity. In *Encyclopedia of Membranes*; Drioli, E., Giorno, L., Eds.; Springer: Berlin/Heidelberg, Germany, 2016; ISBN 978-3-662-44323-1.
2. Mulder, M. *Basic Principles of Membrane Technology*, 2nd ed.; Kluwer Academic Publishers: Dordrecht, The Netherlands, 1996; ISBN 978-0-7923-4247-2.
3. Wandera, D.; Wickramasinghe, S.R.; Husson, S.M. Stimuli-responsive membranes. *J. Membr. Sci.* **2010**, *357*, 6–35. [CrossRef]
4. Yu, S.; Liu, X.; Liu, J.; Wu, D.; Liu, M.; Gao, C. Surface modification of thin-film composite polyamide reverse osmosis membranes with thermoresponsive polymer (TRP) for improved fouling resistance and cleaning efficiency. *Sep. Purif. Technol.* **2011**, *76*, 283–291. [CrossRef]
5. Bhattacharyya, D.; Schäfer, T.; Wickramasinghe, S.R.; Daunert, S. *Responsive Membranes and Materials*; Wiley: New York, NY, USA, 2012; ISBN 978-0-470-97430-8. [CrossRef]
6. Nicoletta, F.P.; Cupelli, D.; Formoso, P.; De Filpo, G.; Colella, V.; Gugliuzza, A. Light responsive polymer membranes: A review. *Membranes* **2012**, *2*, 134–197. [CrossRef] [PubMed]
7. Wang, R.; Xiang, T.; Yue, W.; Li, H.; Liang, S.; Sun, S.; Zhao, C. Preparation and characterization of pH-sensitive polyethersulfone hollow fiber membranes modified by poly(methyl methylacrylate-*co*-4-vinylpyridine) copolymer. *J. Membr. Sci.* **2012**, *423–424*, 275–283. [CrossRef]

8. Wang, Z.; Yao, X.; Wang, Y. Swelling-induced mesoporous block copolymer membranes with intrinsically active surfaces for size-selective separation. *J. Mater. Chem.* **2012**, *22*, 20542. [CrossRef]
9. Frost, S.; Ulbricht, M. Thermoresponsive ultrafiltration membranes for the switchable permeation and fractionation of nanoparticles. *J. Membr. Sci.* **2013**, *448*, 1–11. [CrossRef]
10. Gugliuzza, A. Intelligent membranes: Dream or reality? *Membranes* **2013**, *3*, 151–154. [CrossRef] [PubMed]
11. Katsuno, C.; Konda, A.; Urayama, K.; Takigawa, T.; Kidowaki, M.; Ito, K. Pressure-responsive polymer membranes of slide-ring gels with movable cross-links. *Adv. Mater.* **2013**, *25*, 4636–4640. [CrossRef] [PubMed]
12. Li, H.; Liao, J.; Xiang, T.; Wang, R.; Wang, D.; Sun, S.; Zhao, C. Preparation and characterization of pH- and thermo-sensitive polyethersulfone hollow fiber membranes modified with P(NIPAAm-MAA-MMA)terpolymer. *Desalination* **2013**, *309*, 1–10. [CrossRef]
13. Qiu, X.; Yu, H.; Karunakaran, M.; Pradeep, N.; Nunes, S.P.; Peinemann, K-V. Selective separation of similarly sized proteins with tunable nanoporous block copolymer membranes. *ACS Nano* **2013**, *7*, 768–776. [CrossRef] [PubMed]
14. Yang, Q.; Himstedt, H.M.; Ulbricht, M.; Qian, X.; Wickramasinghe, S.R. Designing magnetic field responsive nanofiltration membranes. *J. Membr. Sci.* **2013**, *430*, 70–78. [CrossRef]
15. Chen, P.R.; Chuang, Y.J. The development of conductive nanoporous chitosan polymer membrane for selective transport of charged molecules. *J. Nanomater.* **2013**, *2013*. [CrossRef]
16. He, Y.; Chen, X.; Bi, S.; Fu, W.; Shi, C.; Chen, L. Conferring pH-sensitivity on poly (vinylidene fluoride) membrane by poly (acrylic acid-co-butyl acrylate) microgels. *React. Funct. Polym.* **2014**, *74*, 58–66. [CrossRef]
17. Ma, S.; Meng, J.; Li, J.; Zhang, Y.; Ni, L. Synthesis of catalytic polypropylene membranes enabling visible-light-driven photocatalytic degradation of dyes in water. *J. Membr. Sci.* **2014**, *453*, 221–229. [CrossRef]
18. Xiao, L.; Isner, A.; Waldrop, K.; Saad, A.; Takigawa, D.; Bhattacharyya, D. Development of bench and full-scale temperature and pH responsive functionalized PVDF membranes with tunable properties. *J. Membr. Sci.* **2014**, *457*, 39–49. [CrossRef] [PubMed]
19. Kaner, P.; Bengani-Lutz, P.; Sadeghi, I.; Asatekin, A. Responsive filtration membranes by polymer self-assembly. *Technology* **2016**, *4*, 1–12. [CrossRef]
20. Meng, H.; Hu, J. A brief review of stimulus—Active polymers responsive to thermal, light, magnetic, electric, and water/solvent stimuli. *J. Intell. Mater. Syst. Struct.* **2010**, *21*, 859–885. [CrossRef]
21. Otero, T.F.; Martinez, J.G.; Arias-Pardilla, J. Biomimetic electrochemistry from conducting polymers. A review: Artificial muscles, smart membranes, smart drug delivery and computer/neuron interfaces. *Electrochim. Acta* **2012**, *84*, 112–128. [CrossRef]
22. Ates, M. A review study of (bio) sensor systems based on conducting polymers. *Mater. Sci. Eng. C* **2013**, *33*, 1853–1859. [CrossRef] [PubMed]
23. Guo, B.; Glavas, L.; Albertsson, A-C. Biodegradable and electrically conducting polymers for biomedical applications. *Prog. Polym. Sci.* **2013**, *38*, 1263–1286. [CrossRef]
24. Asatekin, A.; Mayes, A.M. Polymer filtration membranes. In *Encyclopedia of Polymer Science and Technology*; Mark, H.F., Bikales, N.M., Eds.; Wiley: New York, NY, USA, 2009; ISBN 9780471440260. [CrossRef]
25. Baker, R.W. *Membrane Technology and Applications*, 2nd ed.; Wiley: New York, NY, USA, 2004; ISBN 9780470743720. [CrossRef]
26. Majidi Salehi, S.; Di Profio, G.; Fontananova, E.; Nicoletta, F.P.; Curcio, E.; De Filpo, G. Membrane distillation by novel hydrogel composite membranes. *J. Membr. Sci.* **2016**, *504*, 220–229. [CrossRef]
27. Di Profio, G.; Polino, M.; Nicoletta, F.P.; Belviso, B.D.; Caliandro, R.; Fontananova, E.; De Filpo, G.; Curcio, E.; Drioli, E. Tailored hydrogel membranes for efficient protein crystallization. *Adv. Funct. Mater.* **2014**, *24*, 1582–1590. [CrossRef]
28. Hernandez, S.; Saad, A.; Ormsbee, L.; Bhattacharyya, D. Nanocomposite and responsive membranes for water treatment. In *Emerging Membrane Technology For Sustainable Water Treatment*; Hankins, N.P., Singh, R., Eds.; Elsevier: Amsterdam, The Netherlands, 2016; pp. 389–431. ISBN 978-0-444-63312-5. [CrossRef]
29. Asatekin, A.; Vannucci, C. Self-assembled polymer nanostructures for liquid filtration membranes: A review. *Nanosci. Nanotechnol. Lett.* **2015**, *7*, 21–32. [CrossRef]
30. Bengani, P.; Kou, Y.; Asatekin, A. Zwitterionic copolymer self-assembly for fouling resistant, high flux membranes with size-based small molecule selectivity. *J. Membr. Sci.* **2015**, *493*, 755–765. [CrossRef]
31. Yin, J.; Deng, B.L. Polymer-matrix nanocomposite membranes for water treatment. *J. Membr. Sci.* **2015**, *479*, 256–275. [CrossRef]

32. Batrinescu, G.; Constantin, L.A.; Cuciureanu, A.; Constantin, M.A. Conductive polymer-based membranes. In *Conducting Polymers*; Yilmaz, F., Ed.; Intech (Open Science): Rijeka, Croatia, 2016; ISBN 978-953-51-2691-1. [CrossRef]

33. Wesson, R.D.; Dow, E.S.; Williams, S.R. Responsive membranes/material-based separations: Research and development needs. In *Responsive Membranes and Materials*; Bhattacharyya, D., Schäfer, T., Wickramasinghe, S.R., Daunert, S., Eds.; Wiley: New York, NY, USA, 2013; pp. 385–393. ISBN 978-0-470-97430-8. [CrossRef]

34. Chen, H.; Hsieh, Y.L. Dual temperature- and pH-sensitive hydrogels from interpenetrating networks and copolymerization of N-isopropylacrylamide and sodium acrylate. *J. Polym. Sci. Part A: Polym. Chem.* **2004**, *42*, 3293–3301. [CrossRef]

35. Chun, Y.; Mulcahy, D.; Zou, L.; Kim, I.S. A short review of membrane fouling in forward osmosis processes. *Membranes* **2017**, *7*, 30. [CrossRef] [PubMed]

36. Azzaroni, O.; Brown, A.A.; Huck, W.T.S. Tunable wettability by clicking counterions into polyelectrolyte brushes. *Adv. Mater.* **2007**, *19*, 151–154. [CrossRef]

37. Husson, S.M. Synthesis aspects in the design of responsive membranes. In *Responsive Membranes and Materials*; Bhattacharyya, D., Schäfer, T., Wickramasinghe, S.R., Daunert, S., Eds.; John Wiley & Sons, Ltd.: Chichester, UK, 2013; pp. 73–96. ISBN 978-0-470-97430-8. [CrossRef]

38. Yang, Q.; Wickramasinghe, S.R. Responsive membranes for water treatment. In *Responsive Membranes and Materials*; Bhattacharyya, D., Schäfer, T., Wickramasinghe, S.R., Daunert, S., Eds.; John Wiley & Sons, Ltd.: Chichester, UK, 2013; ISBN 978-0-470-97430-8. [CrossRef]

39. Mendes, P.M. Stimuli-responsive surfaces for bio-applications. *Chem. Soc. Rev.* **2008**, *37*, 2512–2529. [CrossRef] [PubMed]

40. Kastantin, M.; Tirrell, M. Helix formation in the polymer brush. *Macromolecules* **2011**, *44*, 4977–4987. [CrossRef] [PubMed]

41. Hester, J.F.; Banerjee, P.; Mayes, A.M. Preparation of protein-resistant surfaces on poly(vinylidene fluoride) membranes via surface segregation. *Macromolecules* **1999**, *32*, 1643–1650. [CrossRef]

42. Hester, J.F.; Banerjee, P.; Won, Y.Y.; Akthakul, A.; Acar, M.H.; Mayes, A.M. ATRP of amphiphilic graft copolymers based on PVDF and their use as membrane additives. *Macromolecules* **2002**, *35*, 7652–7661. [CrossRef]

43. Jung, B. Preparation of hydrophilic polyacrylonitrile blend membranes for ultrafiltration. *J. Membr. Sci.* **2004**, *229*, 129–136. [CrossRef]

44. Asatekin, A.; Kang, S.; Elimelech, M.; Mayes, A.M. Anti-fouling ultrafiltration membranes containing polyacrylonitrile-graft-poly(ethylene oxide) as an additive. *J. Membr. Sci.* **2007**, *298*, 136–146. [CrossRef]

45. Wang, Y.Q.; Su, Y.L.; Sun, Q.; Ma, X.L.; Jiang, Z.Y. Generation of anti-biofouling ultrafiltration membrane surface by blending novel branched amphiphilic polymers with polyethersulfone. *J. Membr. Sci.* **2006**, *286*, 228–236. [CrossRef]

46. Barghi, H.; Taherzadeha, M.J. Synthesis of an electroconductive membrane using poly(hydroxymethyl-3,4-ethylenedioxythiophene-cotetramethylene-N-hydroxyethyl adipamide). *J. Mater. Chem. C* **2013**, *1*, 6347–6354. [CrossRef]

47. Bai, H.; Shi, G. Gas sensors on conducting polymers. *Sensors* **2007**, *7*, 267–307. [CrossRef]

48. Nicolas-Debarnot, D.; Poncin-Epaillard, F. Polyaniline as a new sensitive layer for gas sensors. *Anal. Chim. Acta* **2003**, *475*, 1–15. [CrossRef]

49. Cuciureanu, A.; Batrinescu, G.; Badea, N.N.; Radu, D.A. The influence of changing the polyaniline and polysulphone ratio on composite Psf–PANI membranes performances. *Mater. Plast.* **2010**, *47*, 416–420. Available online: http://www.revmaterialeplastice.ro/article_eng.asp?ID=2758 (accessed on 16 June 2017).

50. Guimaraes, I.S.; Hidalgo, A.A.; Cunha, H.N.; Santos, L.M.; Santos, J.A.V.; Santos, J.R., Jr. Thermal and morphological characterization of conducting polyaniline/polystyrene blends. *Synth. Met.* **2012**, *162*, 705–709. [CrossRef]

51. Yang, J.; Hou, J.; Zhu, W.; Xu, M.; Wan, M. Substituted polyaniline-polypropylene film composite: Preparation and properties. *Synth. Met.* **1996**, *80*, 283–289. [CrossRef]

52. Mattoso, L.H.; Medeiros, E.S.; Baker, D.A.; Avloni, J.; Wood, D.F.; Orts, W.J. Electrically conductive nanocomposite made from cellulose nanofibrils and polyaniline. *J. Nanosc. Nanotech.* **2009**, *9*, 2017–2022. [CrossRef]

53. Blinova, N.V.; Stejskal, J.; Trchova, M.; Cirici-Marjanovic, G.; Sapurina, I. Polymerization of aniline on aniline membranes. *J. Phys. Chem. B* **2007**, *111*, 2440–2448. [CrossRef] [PubMed]

54. Qaiser, A.A.; Hyland, M.M.; Patterson, D.A. Effects of various polymerization techniques on PANI deposition at the surface of cellulose ester microporous membrane: XPS and electrical studies. *Synth. Met.* **2012**, *162*, 958–967. [CrossRef]

55. Shehzad, M.A.; Qaiser, A.A.; Javaid, A.; Saeed, F. In situ solution-phase polymerization and chemical vapor deposition of polyaniline on microporous cellulose ester membranes: AFM and electrical conductivity studies. *Synth. Met.* **2015**, *200*, 164–171. [CrossRef]

56. Wang, L.X.; Li, X.G.; Yang, Y.L. Preparation, properties and applications of polypyrroles. *React. Funct. Polym.* **2001**, *47*, 125–139. [CrossRef]

57. Liu, J.; Liu, L.; Gao, B.; Yang, F. Integration of bio-electrochemical cell in membrane bioreactor for membrane cathode fouling reduction through electricity generation. *J. Membr. Sci.* **2013**, *430*, 196–202. [CrossRef]

58. Liu, L.; Liu, J.; Bo, G.; Yang, F.; Crittenden, C.Y. Conductive and hydrophilic polypyrrole modified membrane cathodes and fouling reduction in MBR. *J. Membr. Sci.* **2013**, *429*, 252–258. [CrossRef]

59. Liu, L.; Zhao, F.; Liu, J.; Yang, F. Preparation of highly conductive cathodic membrane with graphene (oxide)/PPy and the membrane antifouling property in filtrating. *J. Membr. Sci.* **2013**, *437*, 99–107. [CrossRef]

60. Tsai, Y.T.; Choi, C.H.; Gao, N.; Yang, E.H. Tunable wetting mechanism of polypyrrole surfaces and low-voltage droplet manipulation via redox. *Langmuir* **2011**, *27*, 4249–4256. [CrossRef] [PubMed]

61. Zhao, F.; Liu, L.; Yang, F.; Ren, N. E-Fenton degradation of MB during filtration with Gr/PPy modified membrane cathode. *Chem. Eng. J.* **2013**, *230*, 491–498. [CrossRef]

62. Wang, C.; Song, H.; Zhang, Q.; Wang, B.; Li, A. Parameter optimization based on capacitive deionization for highly efficient desalination of domestic wastewater biotreated efficient and the fouled electrode regeneration. *Desalination* **2015**, *365*, 407–415. [CrossRef]

63. Wang, S.; Liang, S.; Liang, P.; Zhang, X.; Sun, J.; Wu, S.; Huang, X. In-situ combined dual-layer CNT/PVDF membrane for electrically enhanced fouling resistance. *J. Membr. Sci.* **2015**, *491*, 37–44. [CrossRef]

64. Xu, L.; Zhang, G.-Q.; Yuan, G.-E.; Liu, H.-Y.; Liu, J.-D.; Yang, F.-L. Anti-fouling performance and mechanism of anthraquinone/polypyrrole composite modified membrane cathode in a novel MFC–aerobic MBR coupled system. *RSC Adv.* **2015**, *5*, 22533–22543. [CrossRef]

65. Jeon, G.; Yang, S.Y.; Byun, J.; Kim, J.K. Electrically actuatable smart nanoporous membrane for pulsatile drug release. *Nano Lett.* **2011**, *11*, 1284–1288. [CrossRef] [PubMed]

66. Curcio, M.; Spizzirri, U.G.; Cirillo, G.; Vittorio, O.; Picci, N.; Nicoletta, F.P.; Iemma, F.; Hampel, S. On demand delivery of ionic drugs from electro-responsive CNT hybrid films. *RSC Adv.* **2015**, *5*, 44902–44911. [CrossRef]

67. Formoso, P.; Muzzalupo, R.; Tavano, L.; De Filpo, G.; Nicoletta, F.P. Nanotechnology for the environment and medicine. *Mini-Rev. Med. Chem.* **2016**, *16*, 668–675. [CrossRef] [PubMed]

68. De Filpo, G.; Siprova, S.; Chidichimo, G.; Mashin, A.; Nicoletta, F.P.; Cupelli, D. Alignment of single-walled carbon nanotubes in polymer dispersed liquid crystals. *Liq. Cryst.* **2012**, *39*, 359–364. [CrossRef]

69. De Filpo, G.; Mormile, S.; Nicoletta, F.P.; Chidichimo, G. Fast, self-supplied, all-solid photoelectrochromic film. *J. Power Sources* **2010**, *195*, 4365–4369. [CrossRef]

70. Kudryashov, M.A.; Mashin, A.I.; Tyurin, A.S.; Chidichimo, G.; De Filpo, G. Metal-polymer composite films based on polyacrylonitrile and silver nanoparticles. Preparation and properties. *J. Surf. Investig X-ray Synchrotron Neutron Tech.* **2010**, *4*, 437–441. [CrossRef]

71. Kudryashov, M.A.; Mashin, A.I.; Logunov, A.A.; Chidichimo, G.; De Filpo, G. Filpo Frequency dependence of the electrical conductivity in Ag/PAN nanocomposites. *Tech. Phys.* **2012**, *57*, 965–970. [CrossRef]

72. Tyurin, A.; De Filpo, G.; Cupelli, D.; Nicoletta, F.P.; Mashin, A.I.; Chidichimo, G. Particle size tuning in silver-polyacrylonitrile nanocomposites. *eXPRESS Polym. Lett.* **2010**, *4*, 71–78. [CrossRef]

73. Shawky, H.; Chae, S.-R.; Lin, S.; Wiesner, M.R. Synthesis and characterization of a carbon nanotube/polymer nanocomposite membrane for water treatment. *Desalination* **2011**, *72*, 46–50. [CrossRef]

74. De Lannoy, C.F.; Jassby, D.; Gloe, K.; Gordon, A.D.; Wiesner, M.R. Aquatic biofouling prevention by electrically charged nanocomposite polymer thin film membranes. *Environ. Sci. Technol.* **2013**, *47*, 2760–2768. [CrossRef] [PubMed]

75. Reddy, A.V.R.; Trivedi, J.J.; Devmurari, C.V.; Mohan, D.J.; Singh, P.; Rao, A.P.; Joshi, S.V.; Ghosh, P.K. Fouling resistant membranes in desalination and water recovery. *Desalination* **2005**, *183*, 301–306. [CrossRef]

76. Mansouri, J.; Harrisson, S.; Chen, V. Strategies for controlling biofouling in membrane filtration systems: Challenges and opportunities. *J. Mater. Chem.* **2010**, *22*, 4567–4586. [CrossRef]
77. Gullinkala, J.; Escobar, I. A green membrane functionalization method to decrease natural organic matter fouling. *J. Membr. Sci.* **2010**, *360*, 155–164. [CrossRef]
78. Cai, G.; Gorey, C.; Zaky, A.; Escobar, I.; Grunden, C. Thermally responsive membrane-based microbiological sensing component for early detection of membrane biofouling. *Desalination* **2011**, *270*, 116–123. [CrossRef]
79. Chen, Y.; Bose, A.; Bothun, G. Controlled release from bilayer-decorated magneto-liposomes via electromagnetic heating. *ACS Nano* **2010**, *4*, 3215–3221. [CrossRef] [PubMed]
80. Frimpong, R.; Fraser, S.; Hilt, J. Synthesis and temperature response analysis of magnetic-hydrogel nanocomposites. *J. Biomed. Mater. Res. Part A* **2007**, *80*, 1–6. [CrossRef] [PubMed]
81. Preiss, M.; Bothun, G. Stimuli-responsive liposome-nanoparticle assemblies. *Expert Opin. Drug Deliv.* **2011**, *8*, 1025–1040. [CrossRef] [PubMed]
82. Stuart, M.A.; Huck, W.T.; Genzer, J.; Muller, M.; Ober, C.; Stamm, M.; Sukhorukov, G.B.; Szleifer, I.; Tsukruk, V.V.; Urban, M.; et al. Emerging applications of stimuli-responsive polymer materials. *Nat. Mater.* **2010**, *9*, 101–113. [CrossRef] [PubMed]
83. Liu, Z.; Wang, W.; Xie, R.; Ju, X.J.; Chu, L.Y. Stimuli responsive smart gating membranes. *Chem. Soc. Rev.* **2016**, *45*, 460–475. [CrossRef] [PubMed]
84. Jiang, Y.; Lee, A.; Chen, J.; Cadene, M.; Chait, B.T.; MacKinnon, R. The open pore conformation of potassium channels. *Nature* **2002**, *417*, 523–526. [CrossRef] [PubMed]
85. Liu, H.; Liu, X.; Meng, J.; Zhang, P.; Yang, G.; Su, B.; Sun, K.; Chen, L.; Han, D.; Wang, S.; et al. Hydrophobic interaction-mediated capture and release of cancer cells on thermoresponsive nanostructured surfaces. *Adv. Mater.* **2013**, *25*, 922–927. [CrossRef] [PubMed]
86. Shannon, M.A.; Bohn, P.W.; Elimelech, M.; Georgiadis, J.G.; Marinas, B.J.; Mayes, A.M. Science and technology for water purification in the coming decades. *Nature* **2008**, *452*, 301–310. [CrossRef] [PubMed]
87. Lalia, B.S.; Ahmed, F.E.; Shah, T.; Hilal, N.; Hashaikeh, R. Electrically conductive membranes based on carbon nanostructures for self-cleaning or biofouling. *Desalination* **2015**, *360*, 8–12. [CrossRef]
88. Wang, J.; Zhu, M.; Holloway, B.C.; Outlaw, R.A.; Manos, D.M.; Zhao, X. Carbon Nanostructures and Methods of Making and Using the Same. Patent WO2005084172 A3, 19 October 2006.
89. Duan, W.; Chen, G.; Chen, C.; Sanghvi, R.; Iddya, A.; Walker, S.; Liu, H.; Ronen, A.; Jassby, D. Electrochemical removal of hexavalent chromium using electrically conducting carbon nanotube/polymer composite ultrafiltration membranes. *J. Membr. Sci.* **2017**, *531*, 160–171. [CrossRef]
90. Lu, X.; Zhao, C. Highly efficient and robust oxygen evolution catalysts achieved by anchoring nanocrystalline cobalt oxides onto mildly oxidized multiwalled carbon nanotubes. *J. Mater. Chem. A* **2013**, *1*, 12053–12059. [CrossRef]
91. Liu, H.; Vajpayee, A.; Vecitis, C.D. Bismuth-doped tin oxide-coated carbon nanotube network: Improved anode stability and efficiency for flow-through organic electrooxidation. *ACS Appl. Mater. Interfaces* **2013**, *5*, 10054–10066. [CrossRef] [PubMed]
92. Meng, F.; Chae, S.R.; Drews, A.; Kraume, M.; Shin, H.S.; Yang, F. Recent advances in membrane bioreactors (MBRs): Membrane fouling and membrane material. *Water Res.* **2009**, *43*, 1489–1512. [CrossRef] [PubMed]
93. Singh, R.; Hankins, N.P. Introduction to membrane processes for water treatment. In *Emerging Membrane Technology for Sustainable Water Treatment*; Hankins, N.P., Singh, R., Eds.; Elsevier: Amsterdam, The Netherlands, 2016; pp. 15–52. ISBN 978-0-444-63312-5. [CrossRef]
94. Gorey, C.; Hausman, R.; Escobar, I.C. Functionalization of polymeric membranes and feed spacers for fouling control in drinking water treatment applications. In *Responsive Membranes and Materials*; Bhattacharyya, D., Schäfer, T., Wickramasinghe, S.R., Daunert, S., Eds.; John Wiley: New York, NY, USA, 2014; pp. 163–186. ISBN 9780470974308. [CrossRef]
95. Ulbricht, M.; Richau, K.; Kamusewitz, H. Chemically and morphologically defined ultrafiltration membrane surfaces prepared by heterogeneous photo-initiated graft polymerization. *Colloid Surf. A* **1998**, *138*, 353–366. [CrossRef]
96. Sun, M.P.; Su, Y.L.; Mu, C.X.; Jiang, Z.Y. Improved antifouling property of PES ultrafiltration membranes using additive of silica-PVP nanocomposite. *Ind. Eng. Chem. Res.* **2010**, *49*, 790–796. [CrossRef]
97. Nils, J.A.; Digiano, F.A. Influence of NOM composition on NF. *J. Am. Water Works Assoc.* **1996**, *88*, 53–66. Available online: http://www.jstor.org/stable/41295534 (accessed on 16 June 2017).

98. Pabby, A.K.; Rizvi, S.S.H.; Sastre, A.M. Membranes in chemical and pharmaceutical industries and in conservation of natural resources: An introduction. In *Handbook on Membrane Separations Chemical Pharmaceutical Food and Biotechnological Applications*; Pabby, A.K., Rizvi, S.S.H., Sastre, A.M., Eds.; CRC Press: Boca Raton, FL, USA, 2008; ISBN 9780849395499. Available online: https://www.crcpress.com/Handbook-of-Membrane-Separations-Chemical-Pharmaceutical-Food-and-Biotechnological/Pabby-Rizvi-Requena/p/book/9781466555563 (accessed on 16 June 2017).

99. Akhtar, S.; Hawes, C.; Dudley, L.; Reed, I.; Stratford, P. Coatings reduce the fouling of microfiltration membranes. *J. Membr. Sci.* **1995**, *107*, 209–218. [CrossRef]

100. Chen, J.; Kim, S.L.; Ting, Y.P. Optimization of membrane physical and chemical cleaning by a statistically designed approach. *J. Membr. Sci.* **2003**, *219*, 27–45. [CrossRef]

101. Simon, A.; Price, W.E.; Nghiem, L.D. Influence of formulated chemical cleaning reagents on the surface properties and separation efficiency of nanofiltration membranes. *J. Membr. Sci.* **2013**, *432*, 73–82. [CrossRef]

102. Childress, A.E.; Elimelech, M. Effect of solution chemistry on the surface charge of polymeric RO and NF membranes. *J. Membr. Sci.* **1996**, *119*, 253–268. [CrossRef]

103. Rautenbach, R.; Linn, T.; Al-Gobaisi, D.M.K. Present and future pretreatment concepts and strategies for reliable and low-maintenance RO seawater desalination. *Desalination* **1997**, *110*, 97–106. [CrossRef]

104. Chellam, S.; Wiesner, M.R. Particle back-transport and permeate flux behaviour in cross-flow membrane filters. *Environ. Sci. Technol.* **1997**, *31*, 819–824. [CrossRef]

105. Elimelech, M.; Philip, W.A. The future of seawater desalination: Energy, technology, and the environment. *Science* **2011**, *333*, 712–717. [CrossRef] [PubMed]

106. Liu, F.; Hashim, N.A.; Liu, Y.; Abed, M.R.M.; Li, K. Progress in the production and modification of PVDF membranes. *J. Membr. Sci.* **2011**, *375*, 1–27. [CrossRef]

107. Nady, N.; Franssen, M.C.R.; Zuilhof, H.; Eldin, M.S.M.; Boom, R.; Schroën, K. Modification methods for poly(arylsulfone)membranes: A mini-review focusing on surface modification. *Desalination* **2011**, *275*, 1–9. [CrossRef]

108. Rana, D.; Matsuura, T. Surface modification for antifouling membranes. *Chem. Rev.* **2010**, *110*, 2448–2471. [CrossRef] [PubMed]

109. Du, J.R.; Peldszus, S.; Huck, P.M.; Feng, X. Modification of membrane surface via microswelling for fouling control in drinking water treatment. *J. Membr. Sci.* **2014**, *475*, 488–495. [CrossRef]

110. Ronen, A.; Walker, S.L.; Jassby, D. Electroconductive and electroresponsive membranes for water treatment. *Rev. Chem. Eng.* **2016**, *32*, 533–550. [CrossRef]

111. Wu, Z.; Chen, H.; Dong, Y.; Mao, H.; Sun, J.; Chen, S.; Craig, V.S.J.; Hu, J. Cleaning using nanobubbles: Defouling by electrochemical generation of bubbles. *J. Colloid Interface Sci.* **2008**, *328*, 10–14. [CrossRef] [PubMed]

112. Agarwal, A.; Ng, W.J.; Liu, Y. Principle and applications of microbubble and nanobubble technology for water treatment. *Chemosphere* **2011**, *84*, 1175–1180. [CrossRef] [PubMed]

113. Wang, Z.; Ci, L.; Chen, L.; Nayak, S.; Ajayan, P.M.; Koratkar, N. Polarity-dependent electrochemically controlled transport of water through carbon nanotube membranes. *Nano Lett.* **2007**, *7*, 697–702. [CrossRef] [PubMed]

114. Wang, S.; Guillen, G.; Hoek, E.M.V. Direct observation of microbial adhesion to membranes. *Environ. Sci. Technol.* **2005**, *39*, 6461–6469. [CrossRef] [PubMed]

115. Zaky, A.M.; Chaplin, B.P. Porous substoichiometric TiO_2 anodes as reactive electrochemical membranes for water treatment. *Environ. Sci. Technol.* **2013**, *47*, 6554–6563. [CrossRef] [PubMed]

116. Huang, J.; Wang, Z.; Zhang, J.; Zhang, X.; Ma, J.; Wu, Z. A novel composite conductive microfiltration membrane and its antifouling performance with an external electric field in membrane bioreactors. *Sci. Rep.* **2015**, *5*, 9268. [CrossRef] [PubMed]

117. Gugliuzza, A. Smart membrane surfaces: Wettability amplification and self-healing. In *Smart Membranes and Sensors Synthesis Characterization and Applications*; Gugliuzza, A., Ed.; Wiley: New York, NY, USA, 2014; pp. 161–184. ISBN 978-1-118-42379-0. [CrossRef]

118. Liu, L.; Liu, J.; Gao, B.; Yang, F.; Chellam, S. Fouling reductions in a membrane bioreactor using an intermittent electric field and cathodic membrane modified by vapor phase polymerized pyrrole. *J. Membr. Sci.* **2012**, *394–395*, 202–208. [CrossRef]

119. Gao, G.; Zhang, Q.; Vecitis, C.D. CNT-PVDF composite flow-through electrode for single-pass sequential reduction-oxidation. *J. Mater. Chem. A* **2014**, *2*, 6185–6190. [CrossRef]
120. Ronen, A.; Duan, W.; Wheeldon, I.; Walker, S.L.; Jassby, D. Microbial attachment inhibition through low voltage electrochemical reactions on electrically conducting membranes. *Environ. Sci. Technol.* **2015**, *49*, 12741–12750. [CrossRef] [PubMed]
121. Duan, W.; Dudchenko, A.; Mende, E.; Flyer, C.; Zhu, X.; Jassby, D. Electrochemical mineral scale prevention and removal on electrically conducting carbon nanotube–polyamide reverse osmosis membranes. *Environ. Sci. Processes Impacts* **2014**, *16*, 1300–1308. [CrossRef] [PubMed]
122. Duan, W.; Ronen, A.; Walker, S.L.; Jassby, D. Polyaniline-coated carbon nanotubes ultra-filtration membranes: Enhanced anodic stability for in situ cleaning and electro-oxidation processes. *ACS Appl. Mater. Interfaces* **2016**, *8*, 22574–22584. [CrossRef] [PubMed]
123. Vecitis, C.D.; Schnoor, M.H.; Rahaman, S.; Schiffman, J.D.; Elimelech, M. Electrochemical multiwalled carbon nanotube filter for viral and bacterial removal and inactivation. *Environ. Sci. Technol.* **2011**, *45*, 3672–3679. [CrossRef] [PubMed]
124. Dudchenko, A.V.; Rolf, J.; Russell, K.; Duan, W.; Jassby, D. Organic fouling inhibition on electrically conducting carbon nanotube–polyvinyl alcohol composite ultrafiltration membranes. *J. Membr. Sci.* **2014**, *468*, 1–10. [CrossRef]
125. Hong, S.; Bae, T.; Tak, T.; Hong, S.; Randall, A. Fouling control in activated sludge submerged hollow fiber membrane bioreactors. *Desalination* **2002**, *143*, 219–228. [CrossRef]
126. Sofia, A.; Ng, W.J.; Ong, S. Engineering design approaches for minimum fouling in submerged MBR. *Desalination* **2004**, *160*, 67–74. [CrossRef]
127. Duan, W.; Ronen, A.; Valle de Leon, J.; Dudchenko, A.; Yao, S.; Corbala-Delgado, J.; Yan, A.; Matsumoto, M.; Jassby, D. Treating anaerobic sequencing batch reactor effluent with electrically conducting ultrafiltration and nanofiltration membranes for fouling control. *J. Membr. Sci.* **2016**, *504*, 104–112. [CrossRef]
128. Sun, X.; Wu, J.; Chen, Z.; Su, X.; Hinds, B.J. Fouling characteristics and electrochemical recovery of carbon nanotube membranes. *Adv. Funct. Mater.* **2013**, *23*, 1500–1506. [CrossRef]
129. Camalet, J.L.; Lacroix, J.C.; Aeiyach, S.; Chane-Ching, K.; Lacaze, P.C. Electrosynthesis of adherent polyaniline films on iron and mild steel in aqueous oxalic acid medium. *Synth. Met.* **1998**, *93*, 133–142. [CrossRef]
130. Brett, C.M.A.; Oliveira Brett, A.-M.C.F.; Pereira, J.L.C.; Rebelo, C. Properties of polyaniline formed at tin dioxide electrodes in weak acid solution: Effect of the counterion. *J. Appl. Electrochem.* **1993**, *23*, 332–338. [CrossRef]
131. Murugesan, R.; Subramanian, E. Effect of organic dopants on electrodeposition and characteristics of polyaniline under the varying influence of H_2SO_4 and $HClO_4$ electrolyte media. *Mater. Chem. Phys.* **2003**, *80*, 731–739. [CrossRef]
132. Gloukhovski, R.; Oren, Y.; Linder, C.; Freger, V. Thin-film composite nanofiltration membranes prepared by electropolymerization. *J. Appl. Electrochem.* **2008**, *38*, 759–766. [CrossRef]
133. Novoselov, K.S.; Geim, A.K.; Morozov, S.V.; Jiang, D.; Zhang, Y.; Dubonos, S.V.; Grigorieva, I.V.; Firsov, A.A. Electric field effect in atomically thin carbon films. *Science* **2004**, *306*, 666–669. [CrossRef] [PubMed]
134. Lachheb, H.; Puzenat, E.; Houas, A.; Ksibi, M.; Elaloui, E.; Guillard, C.; Herrmann, J.-M. Photocatalytic Degradation of Various Types of Dyes (Alizarin S, Crocein Orange G, Methyl Red, Congo Red, Methylene Blue) in Water by UV-Irradiated Titania. *Appl. Catal. B* **2002**, *39*, 75–90. [CrossRef]
135. Jiang, Y.; Biswas, P.; Fortner, J.D. A review of recent developments in graphene-enabled membranes for water treatment. *Environ. Sci. Water Res. Technol.* **2016**, *2*, 915–922. [CrossRef]
136. De Lannoy, C.-F.; Soyer, E.; Wiesner, M.R. Optimizing carbon nanotube-reinforced polysulfone ultrafiltration membranes through carboxylic acid functionalization. *J. Membr. Sci.* **2013**, *447*, 395–402. [CrossRef]
137. De Lannoy, C.F.; Jassby, D.; Davis, D.D.; Wiesner, M.R. A highly electrically conductive polymer–multiwalled carbon nanotube nanocomposite membrane. *J. Membr. Sci.* **2012**, *415–416*, 718–724. [CrossRef]
138. Liu, Y.; Lee, J.H.D.; Xia, Q.; Ma, Y.; Yu, Y.; Yung, L.Y.L.; Xie, J.; Ong, C.N.; Vecitis, C.D.; Zhou, Z. A graphene-based electrochemical filter for water purification. *J. Mater. Chem. A* **2014**, *2*, 16554–16562. [CrossRef]
139. Liu, Y.; Rosenfield, E.; Hu, M.; Mi, B. Direct observation of bacterial deposition on and detachment from nanocomposite membranes embedded with silver nanoparticles. *Water Res.* **2013**, *47*, 2949–2958. [CrossRef] [PubMed]

140. Chou, W.L.; Yu, D.G.; Yang, M.C. The preparation and characterization of silver-loading cellulose acetate hollow fiber membrane for water treatment. *Polym. Adv. Technol.* **2005**, *16*, 600–607. [CrossRef]

141. Tiraferri, A.; Vecitis, C.D.; Elimelech, M. Covalent binding of single-walled carbon nanotubes to polyamide membranes for antimicrobial surface properties. *ACS Appl. Mater. Interfaces* **2011**, *3*, 2869–2877. [CrossRef] [PubMed]

142. Balta, S.; Sotto, A.; Luis, P.; Benea, L.; Van der Bruggen, B.; Kim, J. A new outlook on membrane enhancement with nanoparticles: The alternative of ZnO. *J. Membr. Sci.* **2012**, *389*, 155–161. [CrossRef]

143. Tang, L.; Gu, W.; Yi, P.; Bitter, J.L.; Hong, J.Y.; Fairbrother, D.H.; Chen, K.L. Bacterial anti-adhesive properties of polysulfone membranes modified with polyelectrolyte multilayers. *J. Membr. Sci.* **2013**, *446*, 201–211. [CrossRef]

144. Diagne, F.; Malaisamy, R.; Boddie, V.; Holbrook, R.D.; Eribo, B.; Jones, K.L.; States, U. Polyelectrolyte and silver nanoparticle modification of microfiltration membranes to mitigate organic and bacterial fouling. *Environ. Sci. Technol.* **2012**, *46*, 4025–4033. [CrossRef] [PubMed]

145. Gall, I.; Herzberg, M.; Oren, Y. The effect of electric fields on bacterial attachment to conductive surfaces. *Soft Matter* **2013**, *9*, 2443. [CrossRef]

146. Miller, D.J.; Araújo, P.A.; Correia, P.B.; Ramsey, M.M.; Kruithof, J.C.; van Loosdrecht, M.C.M.; Freeman, B.D.; Paul, D.R.; Whiteley, M.; Vrouwenvelder, J.S. Short-term adhesion and long-term biofouling testing of polydopamine and poly(ethylene glycol) surface modifications of membranes and feed spacers for biofouling control. *Water Res.* **2012**, *46*, 3737–3753. [CrossRef] [PubMed]

147. Li, Z.; Lee, D.; Sheng, X.; Cohen, R.E.; Rubner, M.F. Two-level antibacterial coating with both release-killing and contact-killing capabilities. *Langmuir* **2006**, *22*, 9820–9823. [CrossRef] [PubMed]

148. Li, D.; Lyon, D.; Li, Q.; Alvarez, P.J.J. Effect of soil sorption and aquatic natural organic matter on the antibacterial activity of a fullerene water suspension. *Environ. Toxicol. Chem.* **2008**, *27*, 1888–1894. [CrossRef] [PubMed]

149. Baek, Y.; Yoon, H.; Shim, S.; Choi, J.; Yoon, J. Electroconductive feed spacer as a tool for biofouling control in a membrane system for water treatment. *Environ. Sci. Technol. Lett.* **2014**, *1*, 179–184. [CrossRef]

MDPI

St. Alban-Anlage 66

4052 Basel

Switzerland

Tel. +41 61 683 77 34

Fax +41 61 302 89 18

www.mdpi.com

Membranes Editorial Office

E-mail: membranes@mdpi.com

www.mdpi.com/journal/membranes

www.ingramcontent.com/pod-product-compliance
Lightning Source LLC
Chambersburg PA
CBHW051845210326
41597CB00033B/5784